Current Trends in ODONATOLOGY

Dedicated to
My Goddess Mother
Shrimati Devi Sumitra

to provide sufficient depth of the subjects to satisfy the needs at a level which will be comprehensive and interesting. The book shall be useful to the students, teachers, scientists and researchers in the field of Odonatology and other people with similar interest.

My special thanks and appreciation go to the scientists whose contributions have enriched this volume. I wish to express my sincere gratitude to Professor (Dr.) Indu Dhan, Hon'ble Vice-Chancellor, S.K. University, Dumka who has been a source of constant inspiration. I am especially thankful to Professor Misir Oraon, Hon'ble Pro-Vice Chancellor, S.K. University, Dumka for his blessings and also for writing a Foreword to it. I am indebted to my respected Professor B. Kiauta Executive Editor of *Odonatogica* (Bilthoven. The Netherlands). Professor D.B. Tembhare Chief Editor of *Fraseria* (Nagpur University), Professor S.P. Roy and Shri Tribhuvan Poddar of Department of Zoology, Bhagalpur University, Bhagalpur who have been a source of continuous encouragement and inspiration to me. I also acknowledge the incentives provided by one of my research scholars Dr. Chandan Bohra, Head of Zoology, B.S.K. College, Barharwa for helping me actively in bringing out this book.

I also express my deep sense of gratitude to my mother, Smt. Sumitra Devi, whose blessings have always prompted me to pursue academic activities deeply. I am also thankful to my wife, Professor Kumari Bimla and my two lovely sons, Kumar Pallav Shivshankaran and Kumar Prasun Ramakrishnan whose natural smiles extended to me relief all through this tiresome endeavour.

Last but not the least, I am also thankful to Mr. Anil Mittal, Proprietor of Daya Publishing House, Delhi for publishing this book. Finally, I will always remain a debtor to all my well-wishers for their blessings. But for him, this book would not have come into existence.

Arvind Kumar

PREFACE

Odonates are usually the most fascinating and elegant group of insects essentially found near any inland wetlands. They lead a bimodal life : adults are aerial whereas larvae are purely aquatic. Larval odonates form a high proportion of biomass in freshwaters, and thus, occupy an important position in the energy flow pathway of the freshwater ecosystems. As dominant members of the benthic and littoral fauna, they are considered as promising organisms for biomonitoring the organic pollution. Larval odonates devour mosquito larvae as well as other harmful organisms and prove themselves a friend of mankind. Since odonates are at the top of predatory food chain. Their mere presence in any habitat shows the conditions are varied, congenial and healthy, supporting diversified faunal components, which in turn are supported by complex flora.

Although odonates have attracted attention of odonatologists over the years because of their peculiar habit and structure, however, the major study in this field has been faunistic or taxonomic. In addition to these, the odonata population is steadily on decline due to rapid industrialization, urbanization as well as the related disappearance of natural habitats.

So, it is high time to take effective measures to stop the disappearance of odonata habitat, to increase their population and conservation of these elegant flies. Needless to say, the dragonflies and damselflies (Odonata) have their own role in ecobalance.

Hence, the main purpose of this publication is to highlight some of the lesser-known facts of this group of insects with the hope that odonatologists may undertake further investigations on their undeciphered facts.

In all 16 chapters are written by authors who are acknowledged experts in their respective fields, with the intention the insect environment for its proper management. The common goal is to gain a universally better quality of life.

This book is very much a culmination of concepts and research works of scientists from various fields of Odonatology throughout the country and abroad as well and of those who have been associated with the field of Entomology.

The book hopefully will be found stimulating the interests of young scientists and kindle in them a desire to work on Odonatology with keen interest. I heartily congratulate Professor Arvind Kumar for his academic initiatives and feel proud and happy at the outcome of my colleague efforts in putting together a coherent total picture of Odonatology through this publication.

Professor Misir Oraon
Pro-Vice Chancellor
S.K. University, Dumka (Jharkhand)

FOREWORD

Odonata represent the ancient and primitive insects as the first odonate fossil was discovered from the carboniferous rocks. They are carnivorous in habit and help in controlling noxious insects of aquatic and aerial habitat. Adults are terrestrial while larvae are aquatic. Fishes feed on larvae while insectivorous birds, reptiles and amphibia feed on both larvae and adults. The odonata, thus, form the integral part of aquatic as well as terrestrial ecosystems and are also the natural biological agents particularly the larvae, which are biological indicators of aquatic pollution. They are actively used in controlling causative agents of malaria and filaria throughout the watery planet.

The present work is an attempt to make the people acquainted thoroughly with the knowledge of odonata and the factors concerned with the deterioration of the environment and of its valued resources with the view to plan and manage the conservation programme in such a way that the exploitation of natural resources may not upset the equilibrium of nature.

The book entitled *Current Trends in Odonatology* is provided by what one would like to think of as basic knowledge of different aspects of odonata. The first chapter deals with the potential of dragonflies as biocontrol agents of insect pests. Second chapter provides an excellent check-list of odonata of south-west province of Cameroon (U.K.). Third chapter very well illustrates the endocrine glands of odonata. The fourth chapter depicts the ecology of odonata larvae in aquatic environment. Fifth chapter narrates the biology of odonata of Indian sub-continent while sixth chapter throws light on conservation ecology and provides strategic guidance for the management of the threatened odonata and so on. It is hoped that the book will definitely serve developing a proper concern to the readers and researchers and also encourage them to become actively involved in matters related to

First Published, 2002

Reprinted, 2024

ISBN: 978-81-70352-74-7 (HB)

Published by : **Daya Publishing House®**
A Division of
Astral International Pvt. Ltd.
– ISO 9001:2015 Certified Company –
4736/23, Ansari Road, Darya Ganj
New Delhi-110 002
Ph. 011-43549197, 23278134
E-mail: info@astralint.com
Website: www.astralint.com

Printed at : **Replika Press Pvt. Ltd.**

Current Trends in ODONATOLOGY

Arvind Kumar

2024

Daya Publishing House®

A Division of

Astral International Pvt. Ltd.

New Delhi – 110 002

Contents

Foreword (vii)

Preface (ix)

1. Potential of Dragonflies as Bio-Control Agents of Insect Pests of Rice – ***Prof. Abdul Khaliq*** 1
2. A Community Structure of Odonata of the South-West Province of Cameroon with the Description of *Phyllogomphus corbetae* Spec. Nov. (Anisoptera : Gomphidae) – ***Prof. G.K. Vick*** 27
3. Endocrinology of Odonata – ***Prof. D.B. Tembhare*** 83
4. Ecology of Larval Odonates in Lentic Freshwater Ecosystems – ***Prof. Arvind Kumar*** 115
5. Biology of Odonata of Indian Sub-continent : A Review – ***Prof. B. Suri Babu and Prof. A. Kumar*** 151
6. Conservation Assessment and Management Plan (CAMP) Process: A Tool to Provide Strategic Guidance for the Management of Threatened Odonata – ***Dr. B.A. Daniel, Sanjay Molur and Sally Walker*** 169
7. Functional Morphology of Rectal Gills and Oxygen Consumption by Dragonfly Nymphs in Aquatic Ecosystem – ***Prof. Arvind Kumar*** 189
8. Analytical Studies on the Food and Feeding Habits of Tropically Distinct Aquatic Odonate Larvae of Udhuwa Lake in Santal Pargana, Jharkhand, India – ***Dr. Chandan Bohra*** 207
9. Odonata Diversity in Western Himalaya, India – ***Dr. M. Prasad*** 221

10. Description of Territoriality and Reproduction of *Agriocnemis pygmaea* (Rambur, 1842) (Zygoptera : Coenagrionidae) – ***Dr. B. Suri Babu*** 255

11. Comparative Studies on Eco-energetics of Nymphs of Anisoptera and Zygoptera (Odonata) in the River Mayurakshi in Santal Pargana, India – ***Prof. Arvind Kumar*** 273

12. Synopsis of Progress in Taxonomical Studies on Odonata in India – ***Dr. A.R. Lahiri*** 280

13. Odonate Diversity in Jharkhand State with Special Reference to Niche Specialization in Their Larval Forms – ***Prof. Arvind Kumar*** 297

14. Trophic Biology and Energy Contents of Larval Odonates with Special Reference to Their Role in the Management of Aquatic Ecosystem – ***Prof. S.P. Roy*** 315

15. Species Composition and Seasonal Fluctuations in Biomass of Zygopteran Nymphs in a Wetland of Santal Pargana, India – ***Dr. P. Kumari*** 329

16. On the Structure and Life History of Three New Species of Septate Gregarines (Apicomplexa : Conoidasida) from Odonates of West Bengal – ***D.P. Haldar and S. Biswas*** 357

List of Contributors

A.R. Lahiri, Zoological Survey of India, Calcutta-700053.

Abdul Khaliq, Department of Entomology, University of Arid Agriculture, Murree Road, Rawalpindi, Pakistan.

Arvind Kumar, Environmental Biology Research Unit, Post Graduate Department of Zoology, S.K. University, Dumka-814 101 (Jharkhand), India.

B. Suri Babu, Forensic Science Laboratory, Police Control Room, E-4 Saket Colony, Jagdalpur-494001 (Chhatishgarh), India.

B.A. Daniel, Zoo Outreach Organisation, P.O. Box #1683, Peelamedu, Coimbatore-641004, India.

Chandan Bohra, Pollution Research Laboratory, Department of Zoology, B.S.K. College, Barharwa-816101 (Jharkhand), India.

D.B. Tembhare, Post-Graduate Department of Zoology, Nagpur University, University Campus, Nagpur-440 010, India.

D.P. Haldar, Protozoology Laboratory, Department of Zoology, University of Kalyani, Kalyani-741235, West Bengal, India.

G.S. Vick, Crossfields, Little London, Tadley, Hants, United Kingdom RG26 5ET.

M. Prasad, Zoological Survey of India, M-Block, New Alipore, Calcutta.

Poonam Kumari, Department of Zoology, S.G. Women's College, Sheikpura-811105, Bihar, India.

S. Biswas, Protozoology Laboratory, Department of Zoology, University of Kalyani, Kalyani-741235, West Bengal, India.

Sally Walker, Zoo Outreach Organisation, P.O. Box #1683, Peelamedu, Coimbatore-641004, India.

Sanjay Molur, Zoo Outreach Organisation, P.O. Box #1683, Peelamedu, Coimbatore-641004, India.

Surendra Prasad Roy, University Department of Zoology, T.M. Bhagalpur University, Bhagalpur-812007, Bihar, India.

1
Potential of Dragonflies as Bio-Control Agents of Insect Pests of Rice

• *Prof. Abdul Khaliq*

Abstract

The feeding capacity of 11 dragonfly species on some insect pests of rice in a day (10 hours) was determined by forced feeding. The adults of *Orthetrum sabina* consumed, on an average, 9.05, 8.50, 7.15, 7.35, 71.20, 12.40 113.15, 4.25, 6.90 and 4.75 while those of *O. pruinosum neglectum* 3.95, 4.50, 8.00, 9.00, 64.60, 3.65, 32.20, 3.80, 4.85 and 7.00, *O. glaucum* 3.20, 3.75, 5.50, 4.50, 59.95, 2.80, 30.65, 3.30, 3.50 and 5.60, *O. triangulare triangulare* 3.00, 3.70, 5.60, 4.05, 61.50, 3.00, 29.80, 7.00, 3.95 and 4.50, *Crocothemis servilia* 4.00, 4.55, 7.50 8.10, 43.30, 2.80, 21.00, 2.40, 3.50 and 2.90, *C. erythraea* 3.45, 3.35, 6.00, 5.60, 23.85, 2.95, 16.25, 2.25, 2.50 and 4.60, *Pantala flavescens* 2.40, 2.70, 2.20, 7.60, 16.25, 2.20, 9.55, 2.05, 3.00 and 6.65, *Sympetrum commixtum* 2.35, 2.25, 8.80, 4.70, 9.35, 2.75, 6.70, 3.65, 7.40 and 3.40, *Palpopleura sexmaculata sexmaculata* 2.65, 2.00, 8.35, 3.80, 8.65, 1.80, 6.10, 3.10, 1.30 and 3.10, *Acisoma panorpoides panorpoides* 2.10, 1.75, 3.40, 5.55, 7.95, 1.65, 5.90, 3.05, 3.70 and 1.65 and *Ttithemis festiva* 1.45, 1.35, 2.25, 3.30, 7.00, 1.25, 4.40, 3.10 and 2.60 yellow stem borers, white stem borers, hairy caterpillars, rice skippers, white-backed planthoppers, white leafhoppers, green leafhoppers, rice bugs, cicadellid leafhoppers and grasshoppers (*Oxya* spp.) respectively in 10 hours. The females of all the species proved to be the most voracious feeders as compared with their males, *Orthetrum sabina* was most voracious feeder followed by *O. pruinosum neglectum*, *O. triangulare triangulare*, *O. glaucum*, *Crocothemis serevilia*, *O. erythraea*, *Pantala flavescens*, *Palpopleura sexmaculata sexmaculata*, *Sympetrum commixium*, *Acisoma panorpoides panorpoides* and *Trithemis festiva*.

The population density of dragonflies was maximum from middle of August to the end of September. However, it was at peak from last week of

August to 3rd week of September. The population then declined as the crop matured and the pest population become low. The studies on the seasonal abundance of insect pests of rice, namely rice skippers, grasshoppers, green leafhoppers, white leaf hoppers, hairy caterpillars, rice bugs, rice stem borers and *Poophilus* species were conducted at four localities, i.e. Hajeera, Mandhole, Harighale, and Banipasari during the crop season 1996-97. The population of rice skippers and grasshoppers was somewhat higher than other insect pests. However, the population of different insect pests was higher from the 2nd week of August to the 4th week of September. The population then decreased afterwards as the crop matured. The efforts were made to evaluate the feeding potential of dragonflies in the field cages. They did not show any activity and interest in their prey due to their captivity.

In view of this dragonfly attitude, the relationship between the population of dragonflies and insect pests of rice was observed by the multiple linear regression equation. The statistical analysis revealed that the dragonflies had negative regression coefficient with the population density of all the insect pests. It indicated that the pest population was decreased with the corresponding increase in dragonflies population. However, their feeding behaviour was not similar at all the localities but differed from locality to locality.

Key words : Potential, dragonflies, bio-control agents, insect pests, rice.

Introduction

Rice, *Oryza sativa* L. is an important cereal crop and its grain is an excellent poultry feed and can be used as a source of vitamin B. Its husk is fed to poultry and straw is used as packaging material and fodder for animals. The crop is attacked by 128 insect pests in Pakistan (Ahmad, 1983) and of these stem borers, rice skippers, white-backed plant hopper, green leafhopper, rice bug and hairy caterpillar are the most serious. Insecticides are mostly applied for the control of these insect pests which are not only expensive but also hazardous for environment. Therefore, the focus should be shifted from chemical control to the pest management strategy. It needs the evaluation of the natural control agents already operating in the crop fields.

The odonates have been observed feeding on some insect pests of rice. Feijen (1977) reported the odonates on the rice stem borer, *Diopsis thoracica*. Agassiz (1978) stated that *Sympetrum flavcolum* feeds on pyralid moths in an aquatic nursery. Dean (1978) listed some odonate species as potential predatros of the rice insect pests. Lippold (1979) observed some dragonflies feeding on rice leafhopper, gall midges, stem borers and defoliators. Ali (1983) studied the feeding of some dragonflies on insect pests of cotton. Van Vreden and Ahmadzabidi (1986) listed 11 odonate species as predators of rice pests. Xia and Wang (1988) mentioned 19 odonate species as predators of rice pests.

Honghai *et al.* (1995) stated that the statistical analysis of the data on rice stem borer from 1980-93 indicated a significant correlation between overwintering mortality of the pest and average temperature in December and rainfall in April. Therefore, the population density of the pest at second cropping rice can be used in forecasting. Khaliq and Siddique (1995) collected six dragonflies from rice fields of Poonch and Bagh districts of Pak occupied Kashmir (POK). Suglura (1995) discussed the forecasting activities of rice pests at Nara, Japan. Matsumura (1996) described the effects of growth stage of rice on the population dynamics of *Sogatella furcifera*. The population was the highest on rice plant 17-30 days after transplanting which suggested that it preferred the rice plants at the tillering stage. Spurgeon *et al.* (1996) used synthetic pheromones to observe the nightly activity patterns of *Eoreuma loftini* and indicated that optimal sampling times of this pest differed according to sex. Yousuf *et al.* (1998) conducted studies on population dynamics and feeding capacity of dragonflies on insect pests of rice and concluded females of all the species consumed higher number of pests as compared with their males.

Keeping this in view, the studies were conducted (i) to observe the population dynamics of major insect pests of rice and dragonflies during the crop season 1996-97 and (ii) to evaluate the potential of dragonflies as bio-control agents.

Materials and Methods

A. Feeding capacity of different dragonfly species on rice pests in captivity

The following procedure was followed to evaluate the feeding capacity of dragonflies on the insect pests of rice in captivity. The fields were visited before at day-break, when dragonflies were not active and collected easily with the help of an aerial net. At least

20 specimens of each species (10 males and 10 females) were captured alive and kept in the glass bottles separately. The adults of the rice stem borers, rice skippers, white-backed planthopper, green leafhopper, white leafhopper, hairy caterpillar, rice bug and small grasshoppers (*Oxya* spp.) were collected with the help of an aeiral net and kept alive in collection tubes. These were then picked up one by one in left hand and were offered to dragnflies with their wings help up vertically on their back in the right hand. The dragonflies started to eat the insect pests when they were brought near their mouth. Some water was also offered to dragonflies after forced feeding. The number of insect pests consumed by individuals of each species were counted. Feeding procedure was repeated thrice a day, i.e., at 8.00 a.m., 1 p.m. and 6 p.m. The captive dragonflies were again released in the rice fields after completing the feeding procedure.

B. *Population dynamics of dragonflies and insect pests in rice fields during the crop season 1996-97 at different localities*

The studies were carried out in the rice growing areas of Poonch and Bagh districts of AJK. The localities selected were Harighale, Banipasai, Hajeera and Mandhole. The procedure followed to study the population of insect pests of rice and dragonflies was as follows : The population of dragonflies and rice skippers were counted visually from 5 rows (each of 33 m length and selected after leaving 12 rows) in each plot at weekly intervals throughout the crop season. These rows were at an equal distance from each other and thus covered the plots uniformly. The population of insect pests of rice (except rice skippers) was recorded by selecting 20 plants from each field randomly and population per plant was then calculated. This procedure was followed at weekly intervals during the crop season.

C. *Evaluation of dragonflies as bio-control agents of rice insect pests in the field*

The procedure adopted to evaluate the role of dragonflies as bio-control agents against insect pests of rice was as under :

(i) Determination of the feeding potential of dragonflies in the field cages

The known number of insect pests, namely, rice skipper, grasshoppers, green leafhopper, white leafhopper, hairy caterpillar, rice bug, rice stem borers and *Poophilus* species were released in the two field cages (2×2×2m). The cages with four pots of rice plants were placed near the rice fields. The identified specimens of dragonflies, male and female, separately and in pairs, were then released in these cages at 8.00 a.m. The number of different insect pests consumed were counted at 6.00 p.m. The dragonflies were captured from the field day before at 5.00 p.m. and were kept in the laboratory for releasing in the cages next day.

(ii) Comparison of the seasonal abundance of dragonflies and insect pests of rice

The seasonal abundance of dragonflies and insect pests of rice recorded in experiment B was compared to have an idea about the potential of dragonflies as bio-control agents of insect pests of rice.

(iii) Application of multiple linear regression equation to determine the role of dragonflies against inset pests of rice

The seasonal abundance of dragonflies and insect pests of rice was analysed statistically to evaluate the potential of dragonflies as bio-control agents. The multiple linear regression equation used for this purpose was

$$Y = a + b_1x_1 = b_2x_2 + b_3x_3 + b_4x_4 + b_5x_5 + b_6x_6 + b_7x_7 + b_8x_8$$

Y = Dragonflies

a = constant $b_1 \ldots b_8$ = Regression coefficients

x_1 = Rice skipper x_2 = Grasshopper

x_3 = Green leafhopper x_4 = White leafhopper

x_5 = Hairy caterpillar x_6 = Rice bug

x_7 = Rice stem borers x_8 = *Poophilus* species

Results

A. *Feeding capacity of dragonflies on rice pests in captivity*

The number of rice insect pests consumed by different dragonfly species is presented in Figs. 1.1-1.11. It can be seen (Fig. 1.1) that females of *Orthetrum sabina* Drury consumed, on an average, 9.40, 9.50, 7.70, 7.50, 73.90, 13.30, 126.30, 4.50, 7.60, and 5.90 while the males 8.70, 7.50, 6.60, 7.20, 68.50, 11.50, 100.00, 4.00, 6.20, and 3.60 with an overall average of 9.05, 8.50, 7.15, 7.35, 71.20, 12.40, 113.15, 4.25, 6.90 and 4.75 yellow stem borers, white stem borers, hairy caterpillars, rice skippers, white-backed planthoppers, white leafhoppers, green leafhoppers, rice bugs, cicadellid leafhoppers and grasshoppers (*Oxya* spp.) respectively in a period of 10 hours. It revealed that the female eat higher number of insect pests as compared with their males.

The number of insect pests of rice consumed by *O. pruinosum neglectum* (Rambur) are given in Fig. 1.2. It shows that its females, on an average, consumed 4.50, 5.00,8.00, 8.20, 74.80, 3.90, 35.00, 4.00, 4.70, and 7.00 yellow stem borers, white stem borers, hairy caterpillar, rice skippers, white-backed planthoppers, white leafhoppers, green leafhoppers, rice bugs, cicadellid leafhoppers and grasshoppers (*Oxya* spp.) in a period of 10 hours. It indicated that the females take higher number of insect pests as compared with their males.

Fig. 1.3 shows that the females of *O. glaucum* (Brauer) consumed on an average 3.50, 4.20, 6.00, 5.00, 69.20, 3.00, 32.50, 3.50, 4.00 and 5.30 white the males 2.90, 3.30, 5.00, 4.00, 50.70, 2.60, 27.60, 3.10, 3.00 and 5.90 with an overall average of 3.20, 3.75, 5.50, 4.50, 59.95, 2.80, 30.05, 3.30, 3.50 and 5.60 yellow stem borers, white stem borers, hairy caterpillar, rice skippers, white-backed planthoppers, white leafhoppers, green leafhoppers, rice bugs, cicadellid leafhopper and grasshoppers (*Oxya* spp.) respectively in a period of 10 hours. It is also clear that females are more voracious predators as compared with their males.

The females of *O. triangulare triangulare* (Selys) (Fig. 1.4) took, on an average, 3.30, 4.30, 6.00, 4.50, 70.00, 3.30, 32.60, 7.00, 4.00 and 5.00 while the males 2.70, 3.10, 5.20, 3.60, 53.00, 2.70, 27.00,

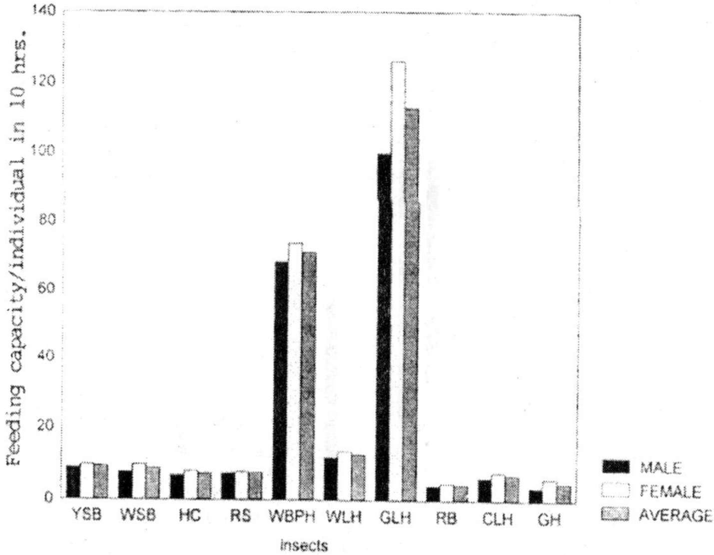

Fig. 1.1 : Number of insect pests of rice consumed by *Orthetrum sabina* under forced feeding.

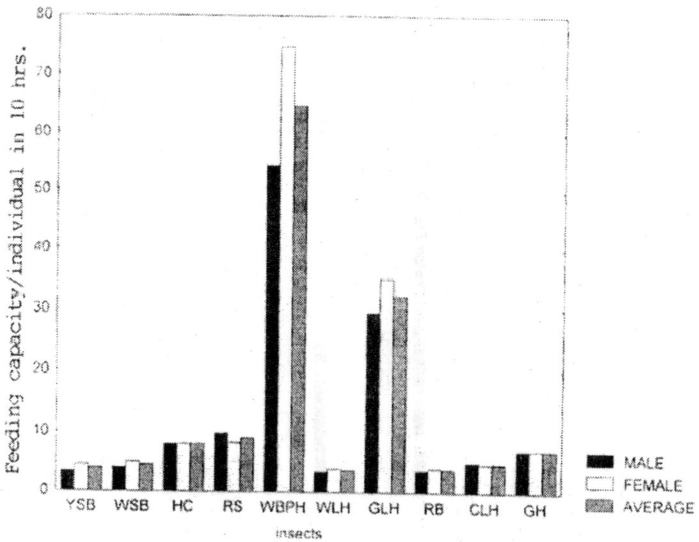

Fig. 1.2 : Number of insect pests of rice consumed by *O. pruinosum neglectum* under forced feeding.

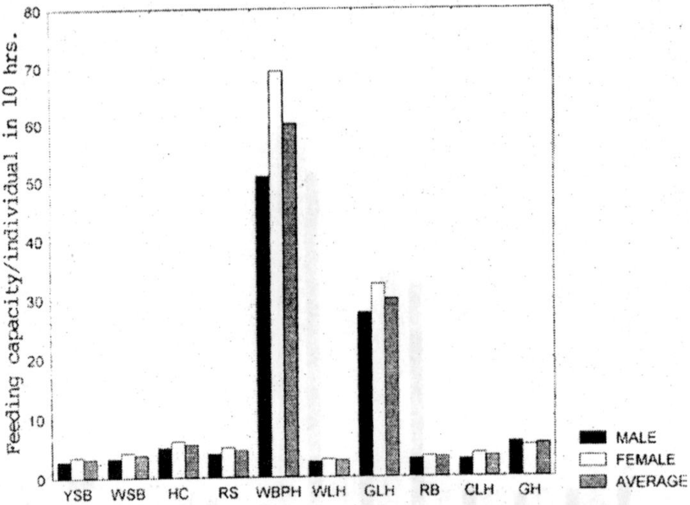

Fig. 1.3 : Number of insect pests of rice consumed by *O. glaucum* under forced feeding.

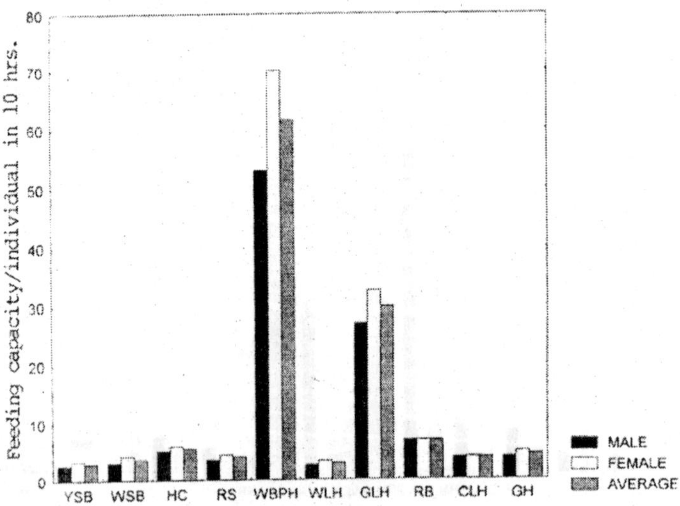

Fig. 1.4 : Number of insect pests of rice consumed by *O. triangulare triangulare* under forced feeding.

7.00, 3.90, and 4.00 with an overall average of 3.00, 3.70, 5.60, 4.05, 61.50, 3.00, 29.80, 7.00, 3.95 and 4.50 yellow stem borers, white stem borers, hairy caterpillar, rice skippers, white-backed planthoppers, white leafhoppers, green leafhoppers, rice bugs, cicadellid leafhopers and grasshoppers (*Oxya* spp.) respectively in a period of 10 hours. It also showed that females are more voracious feeders as compared with their males.

The females of *Crocothemis servilia* (Drury) (Fig. 1.5) consumed on an average 4.20, 4.80, 8.00, 9.00, 46.60, 3.00, 23.00, 2.50, 3.90 and 3.30 while the males 3.80, 4.30, 7.00, 7.20, 40.00, 2.60, 19.00, 2.30, 3.10 and 2.50 with an overall average of 4.00, 4.55, 7.50, 8.10, 43.30, 2.80, 21.00, 2.40, 3.50 and 2.90 yellow stem borers, white stem borers hairy caterpillar, rice skippers, white-backed planthoppers, white leafhopers, green leafhoppers, rice bugs, cicadellid leafhoppers and grasshoppers (*Oxya* spp.) respectively in a period of 10 hours. It revealed that females are more voracious feeders are compared with their males.

The females of *C. erythraea* (Brulle) (Fig. 1.6) consumed on an average 3.70, 3.60, 7.00, 6.30, 26.50, 3.10, 17.40, 2.40, 3.00 and 5.10 while the males 3.20, 3.10, 5.00, 4.90, 21.20, 2.80, 15.10, 2.10, 2.00 and 4.10 with an overall average of 3.45, 3.35, 6.00, 5.60, 23.85, 2.95, 16.25, 2.25, 2.25, 2.50 and 4.60 yellow stem borers, white stem borers, hairy caterpillars, rice skippers, white-backed planthoppers, white leafhoppers, green leafhoppers, rice bugs, cicadellid leafhoppers and grasshoppers (*Oxya* spp.) respectively in a period of 10 hours. It is obvious that the females took higher number of insect pests as compared with their males.

The females of *Pantala flavescens* (F.) (Fig. 1.7) consumed on a average 2.80, 3.00, 2.40, 8.10, 17.90, 2.30, 10.70, 2.20, 3.40 and 7.20 while the males 2.00, 2.40, 2.00, 7.10, 14.60, 2.10, 8.40, 1.90, 2.60 and 6.10 with an overall average of 2.40, 2.70, 2.20, 7.60, 16.25, 2.20, 9.55, 2.05, 3.00 and 6.65 yellow stem borers, white stem borers, hairy caterpillars, rice skippers, white-backed planthoppers, white leafhoppers, green leafhoppers, rice bugs, cicadellid leafhoppers and grasshoppers (*Oxya* spp.) respectively in a period of 10 hours. It showed that females are more voracious feeders as compared with their males.

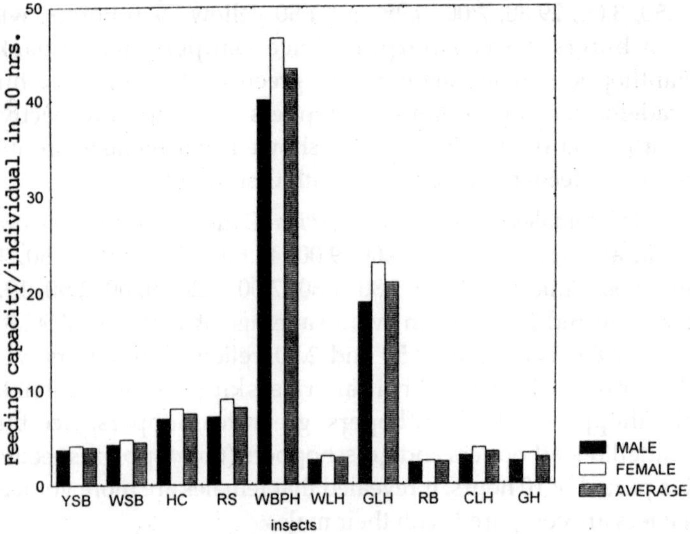

Fig. 1.5 : Number of insect pests of rice consumed by *Crocothemis servilia* under forced feeding.

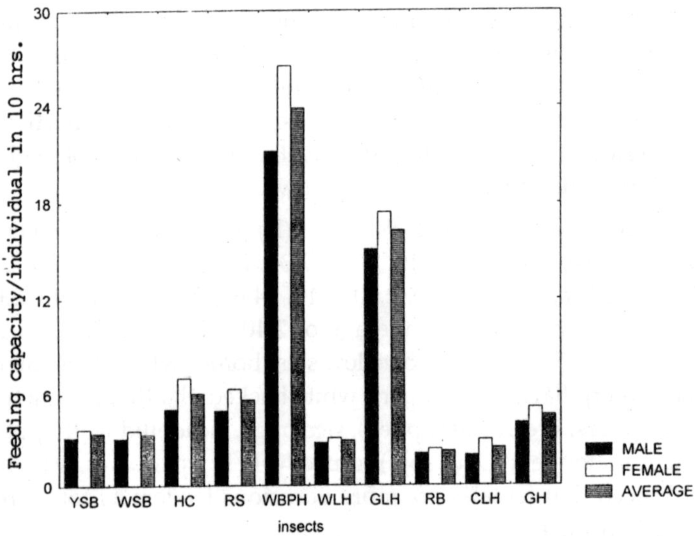

Fig. 1.6 : Number of insect pests of rice consumed by *C. erythraea* under forced feeding.

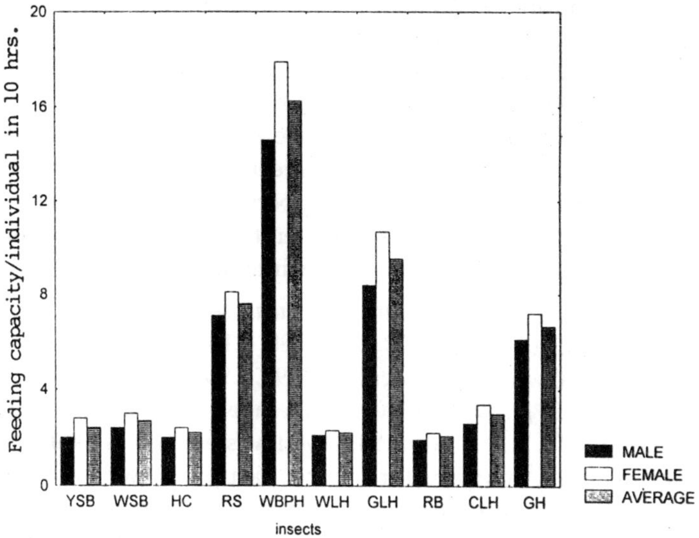

Fig. 1.7 : Number of insect pests of rice consumed by *Pantala flavecens* under forced feeding.

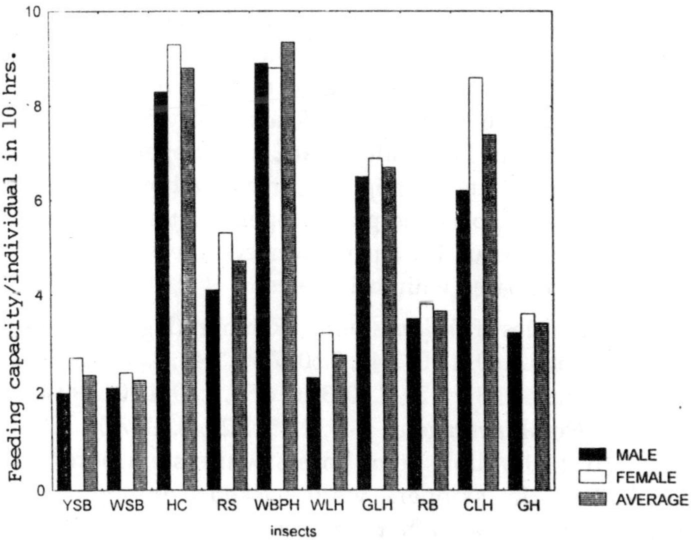

Fig. 1.8 : Number of insect pests of rice consumed by *Sympetrum commixtum* under forced feeding.

The females of *Sympetrum commixtum* (Selys) (Fig. 1.8) consumed on an average 2.70, 2.40, 9.30, 3.30, 9.80, 3.20, 6.90, 3.80, 8.60 and 3.60 whereas the males 2.00, 2.10, 8.30, 4.10, 8.90, 2.30, 6.50, 3.50, 6.20 and 3.20 with an overal average of 2.35, 2.25, 8.80, 4.70, 9.35, 2.75, 6.70, 3.65, 7.40 and 3.40 yellow stem borers, white stem borers hairy caterpillar, rice skippers, white-backed planthoppers, white leafhoppers, green leafhoppers, rice bugs, cicadellid leafhoppers and grasshoppers (*Oxya* spp.) respectively in a period of 10 hours. It can be seen that the females are more strong feeders as compared with their males.

The females of *Palpopleura sexmaculata sexmaculata* (F.) (Fig. 1.9) consumed on a average 3.20, 2.10, 8.80, 4.10, 9.20, 1.90, 6.30, 3.20, 1.40, and 3.10 while the males 2.10, 1.90, 7.90, 3.50, 8.10, 1.70, 5.90, 3.00, 1.20 and 3.10 with an overall average of 2.65, 2.00, 8.35, 3.80, 8.65, 1.80, 6.10, 3.10, 1.30 and 3.10 yellow stem borers, white stem borers, hairy caterpillars, rice skippers, white-backed planthoppers, white leafhoppers, green leafhopers, rice bugs, cicadellid leafhoppers and grasshoppers (*Oxy* spp.) respectively in a period of 10 hours. It showed that the females haves better predaceous ability as compared with their males.

The females of *Acisoma panorpoides panorpoides* Rambur (Fig. 1.10) consumed on an average 2.20, 1.90, 3.70, 6.10, 8.50, 1.80, 6.60, 3.30, 4.30 and 1.70 while the males 2.00, 1.60, 3.10, 5.00, 7.40, 1.50, 5.20, 2.80, 3.10 and 1.60 with an overall average of 2.10, 1.75, 3.40, 5.55, 7.95, 1.65, 5.90, 3.05, 3.70 and 1.65 yellow stem borers, white stem borers, hairy caterpillars, rice skippers, white-backed planthoppers, white leafhoppers, green leafhoppers, rice bugs, cicadellid leafhoppers and grasshoppers (*Oxy* spp.) respectively in a period of 10 hour. The females seemed to be the more voracious feeders as compared with their males.

The females of *Trithemis festiva* (Rambur) (Fig. 1.11) consumed on an average 1.60, 1.50, 2.40, 3.50, 7.50, 1.40, 4.70, 3.30 and 2.80 while the males 1.30, 1.20, 2.10, 3.10, 6.50, 1.10, 4.10, 2.90 and 2.40 with an overall average of 1.45, 1.35, 2.25, 3.30, 7.00, 1.25, 4.40, 3.10 and 2.60 yellow stem borers white stem borers, hairy caterpillars, rice skippers, white-backed planthoppers, white leafhoppers, green leafhoppers, rice bugs, and cicadellid leaf hoppers respectively in a period of 10 hours. *T. festiva* refused to

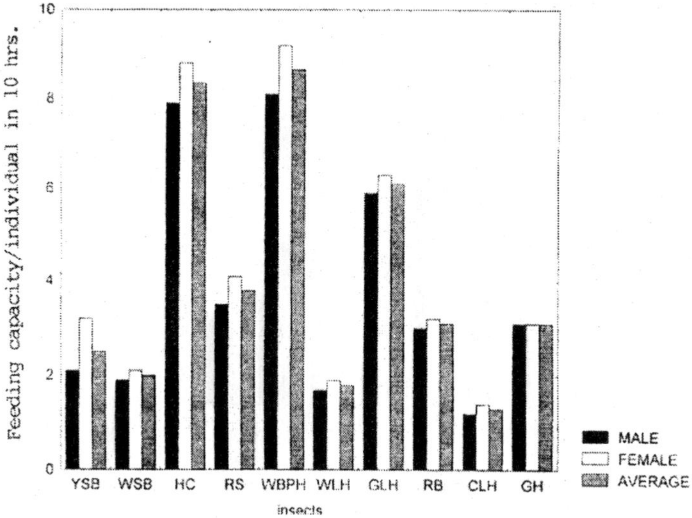

Fig. 1.9 : Number of insect pests of rice consumed by *Palpopleura sexmaculata sexmaculata* under forced feeding.

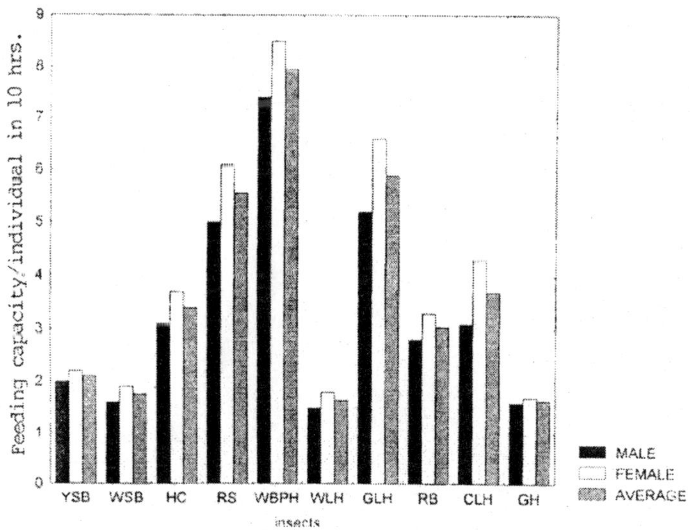

Fig. 1.10 : Number of insect pests of rice consumed by *Acisoma panorpoides* under forced feeding.

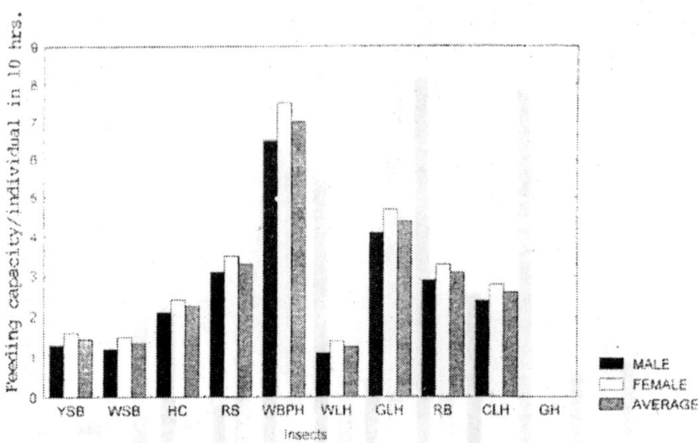

Fig. 1.11 : Number of insect pests of rice consumed by *Trithemis festiva* under forced feeding.

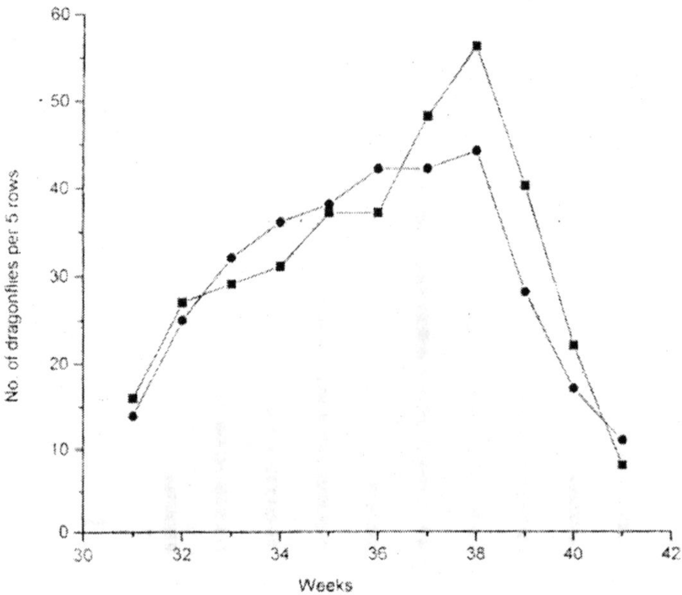

Fig. 1.12 : Population density of dragonflies in rice fields in two localities of Poonch district.

take grasshoppers when offered probably due to its small size. The females appeared to be the better feeders as compared with their males.

B. Population dynamics of dragonflies and rice insect pests in rice fields throughout the crop season at different localities

(a) Population dynamics of dragonflies in rice fields throughout the crop season

The population density of dragonflies in the rice crop at four different localities, i.e. Hageera, Mandhole, Banipasari and Harighale is given in Fig. 1.12 and Fig. 1.13. Poulation was recorded from 18-1996 at Hajeera and Mandhole while from 25-7-1996 at Banipasari and Harighale. The population was low in the beginning in all the localities. Then the population started increasing from 2nd week of August. The peak seasonal population of dragonflies recorded on 6-9-1996 at Banipasari and Harighale was 93 and 31 respectively. Likewise, the peak seasonal population of dragonflies recorded at other two localities, viz. Hajeera and Harighale on 21-9-1996 was 56 and 44 respectively. The overall trend showed that the maximum population was recorded from 2nd week of August to the end of September. The crop matured afterwards and the population of dragonflies decreased due to the absence of the insect pests on rice.

(b) Population dynamics of insect pests of rice throughout the crop season

The studies on population density of insect pests of rice were started on 1-8-1996 at both the localities of Poonch district, i.e. Hajeera and Mandhole and are presented in Tables 1.1 and 1.2 respectively. The population of rice skipper was maximum in the third week of August at Hajeera and in the second week of this month at Mandhole. The grasshoppers were first seen in nymphal stage, they matured afterwards and their population was at peak in second week of August at Hajeera and then decreased with the passage of time and was again at peak in the last week of September. After this, their population decreased in the rice fields

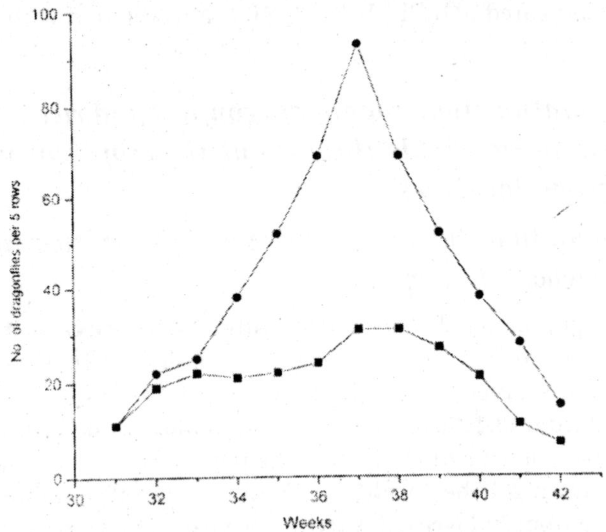

Fig. 1.13 : Population density of dragonflies in rice fields in two localities of Bagh district.

probably due to the maturity of crop. The population build up at Mandhole was somewhat different and was maximum during the last three weeks of September. The green leafhopper appeared in the second week of August and its population was maximum in the second and third week of September and then the population was declined. Almost similar trend was observed in case of white leafhopper.

Table 1.1 : Average seasonal population of some insect pests of rice at Hajeera, Poonch district

Date	Rice skipper	Grass-hopper	Green leaf-hopper	White leaf-hopper	Hairy cater-pillar	Rice bug	Rice stem borers	*Poophilus* spp.
1-8-96	1.40	0.52	0.00	0.08	0.00	0.04	0.00	0.00
9-8-96	2.60	0.84	0.16	0.12	0.00	0.00	0.00	0.08
16-8-96	2.40	0.52	0.12	0.08	0.08	0.08	0.20	0.00
23-8-96	2.60	0.52	0.12	0.12	0.08	0.08	0.20	0.00
31-8-96	1.40	0.52	0.12	0.12	0.08	0.12	0.16	0.04
7-9-96	1.00	0.32	0.12	0.00	0.08	0.00	0.16	0.00
14-9-96	1.00	0.32	0.32	0.08	0.04	0.12	0.08	0.04
21-9-96	1.20	0.60	0.44	0.20	0.02	0.08	0.08	0.04
28-9-96	1.20	0.24	0.04	0.00	0.04	0.16	0.04	0.00
5-10-96	1.00	0.52	0.00	0.08	0.04	0.00	0.00	0.08

Table 1.2 : Average seasonal population of some insect pests of rice at Mandhole, Poonch district

Date	Rice skipper	Grass-hopper	Green leaf-hopper	White leaf-hopper	Hairy cater-pillar	Rice bug	Rice stem borers	*Poophilus* spp.
1-8-96	1.00	0.24	0.00	0.00	0.04	0.04	0.00	0.00
9-8-96	1.60	0.24	0.08	0.04	0.08	0.04	0.00	0.04
16-8-96	4.00	0.32	0.08	0.04	0.08	0.08	0.00	0.08
23-8-96	2.40	0.40	0.20	0.04	0.16	0.08	0.00	0.12
31-8-96	2.80	0.44	0.28	0.12	0.24	0.08	0.24	0.12
7-9-96	2.80	0.32	0.28	0.20	0.32	0.16	0.16	0.00
14-9-96	2.80	0.52	0.40	0.40	0.44	0.16	0.12	0.08
21-9-96	1.00	0.52	0.60	0.44	0.32	0.16	0.12	0.00
28-9-96	1.20	0.60	0.24	0.08	0.24	0.08	0.24	0.08
5-10-96	1.00	0.44	0.16	0.08	0.04	0.12	0.36	0.00

Hairy caterpillar appeared in the second week of August at Hajeera, although, their population was very low. It was maximum up to first week of September and was decreased afterwards. It was observed at Mandhole in the first week of August, increased gradually and was at pea in the middle of September. Its population then decreased, however, it was somewhat higher at Mandhole than at Hajeera. Rice stem borers appeared in the middle of August at Hajeera and their population was maximum up to first week of September. It was recorded at the end of August at Mandhole and was maximum in the first week of October. The rice bug started its activity in the middle of August and remained up to the maturity of the crop. There were some variations in its population build up at both the localities. The *Poophilus* spp. were also observed in the rice fields, their population fluctuated during the season, however, it was ignorable due to its low population.

The results of the studies conducted to observe the seasonal abundance of some insect pests of rice during the crop season in Bagh district are given in Tables 1.3 and 1.4. The research was started in the last week of July after transplantation of the rice in the field. Some insect pests appeared in the field in the first week of August but other started their activities at the end of this month. the rice skipper started its activities in the first week of August and was at peak at the end of this month. A decreasing trend was then seen with slight variations. The grasshoppers were in the nymphal stage at this time and their mature stage appeared after

Table 1.3 : Average seasonal population of some insect pests of rice at Harighale, Bagh district

Date	Rice skipper	Grass-hopper	Green leaf-hopper	White leaf-hopper	Hairy cater-pillar	Rice bug	Rice stem borers	*Poophilus* spp.
25-7-96	0.00	0.00	0.00	0.00	0.00	0.00	0.00	0.00
3-8-96	0.67	0.20	0.20	0.00	0.00	0.00	0.00	0.00
10-8-96	2.67	0.70	0.13	0.00	0.00	0.00	0.00	0.00
17-8-96	3.33	0.60	0.13	0.07	0.07	0.00	0.20	0.00
24-8-96	4.00	0.60	0.13	0.07	0.07	0.00	0.07	0.00
31-8-96	4.67	0.46	0.07	0.13	0.07	0.00	0.13	0.07
7-9-96	3.67	0.40	0.13	0.13	0.00	0.13	0.20	0.00
14-9-96	2.33	0.60	0.13	0.07	0.00	0.13	0.20	0.00
21-9-96	2.67	0.46	0.33	0.07	0.00	0.20	0.13	0.07
28-9-96	0.33	0.66	0.20	0.00	0.07	0.07	0.13	0.07
5-10-96	0.67	0.33	0.13	0.00	0.00	0.07	0.00	0.00

Table 1.4 : Average seasonal population of some insect pests of rice at Banipasari, Bagh district

Date	Rice skipper	Grass-hopper	Green leaf-hopper	White leaf-hopper	Hairy cater-pillar	Rice bug	Rice stem borers	*Poophilus* spp.
25-7-96	0.00	0.00	0.00	0.00	0.00	0.00	0.00	0.00
3-8-96	0.50	0.20	0.10	0.00	0.00	0.00	0.00	0.00
10-8-96	1.00	0.80	0.40	0.00	0.00	0.00	0.00	0.20
17-8-96	2.00	0.70	0.10	0.00	0.00	0.00	0.20	0.00
24-8-96	2.50	0.90	0.10	0.10	0.00	0.00	0.00	0.00
31-8-96	3.00	1.30	0.20	0.00	0.10	0.00	0.10	0.00
7-9-96	2.00	1.20	0.20	0.10	0.10	0.10	0.10	0.10
14-9-96	2.50	1.20	0.20	0.10	0.10	0.00	0.10	0.00
21-9-96	2.50	0.80	0.60	0.10	0.10	0.10	0.10	0.00
28-9-96	0.50	0.90	0.70	0.00	0.00	0.10	0.00	0.00
5-10-96	0.00	0.60	0.70	0.00	0.00	0.00	0.00	0.10

three weeks. Their population was at peak from second week of August to the end of September. The population decreased after this as the crop matured. The population of grasshoppers at Banipasari was somewhat higher than at Harighale. The green leafhopper appeared in the first week of August. Their population fluctuated during the crop season and was at peak in the last week of September and then declined afterwards. The population of white leafhopper was first observed in the third week of August. The population was very low, however, it was maximum almost during the first two weeks of September.

The population of the hairy caterpillar was also very low at both the localities of the district Bagh. Rice bug appeared in the first week of September and its population was higher at Harighale than at Banipasari. The rice stem borers were first recorded in the second week of August. Their population was low, fluctuated during the season and was somewhat higher at Harighale than at Banipasari. The population of *Poophilus* spp. was very low and rarely observed during the crop season.

C. Evaluation of dragonflies as bio-control agents of rice insect pests in the field

(i) Determination of the feeding potential of dragonflies in the field cages

Different species of dragonflies, males and females separately or in pairs, were released in the field cages to evaluate their feeding potential in the field. The dragonflies refused to feed on the insect pests released on the rice plants in the field cages. This practice was repeated again and again for several days but in vain. It was due to the fact that dragonflies have strong senses of captivity and they do not feed under captive conditions.

(ii) Comparison of the seasonal abundance of dragonflies with inset pests of rice

The average seasonal population of dragonflies and some insect pests of rice at Hajeera and Mandhole is given is Fig. 1.12, and Tables 1.1 and 1.2. It can be seen that the population of rice skippers was maximum (2.60) on 23-8-96 at Hajeera and (4.00) on 16-8-96 at Mandhole. It was then decreased afterwards with corresponding increase in dragonflies population at both the localities. Similar trends were also observed in case of grasshoppers, rice stem borers and *Poophilus* species. Then green leafhopper, white leafhopper, hairy caterpillar and rice bug showed no correlation with dragonflies population at Mandhole. Likewise, same trend was observed at Hajeera. However, the seasonal abundance of white leafhopper, hairy caterpillar and rice bug fluctuated with corresponding increase or decrease in dragonflies population.

The seasonal abundance of dragonflies and insect pests of rice at Harighale and Banipasari is presented in Fig. 1.13 and Tables 1.3 and 1.4 respectively. The population density of rice skipper at both the localities revealed that it was decreased with increasing population of dragonflies. Similar trends were seen for grasshopper, green leafhopper and hairy caterpillar at Harighale and rice stem borers and *Poophilus* species at Banipasari. However, white leafhopper, rice bug, rice stem borers and *Poophilus* species at Harighale while grasshopper, green leafhopper, white leafhopper, hairy caterpillar and rice bug at Banipasari seemed to be not much affected by the seasonal abundance of dragonflies.

(iii) Application of multiple linear regression equation to determine the role of dragonflies against insect pests of rice in the field.

The multiple linear regression equation was applied to determine the relationship between the population of dragonflies and insect pests of rice at four localities viz. Hajeera, Mandhole, Harighale and Banipasari and the results are given in Tables 1.5-1.8 respectively. Table 1.5 shows that the population density of rice skipper, grasshopper, white leafhopper and rice stem borers had negative regression coefficient with the population of dragonflies. It indicated that the population of these pests decreased with corresponding increase in the population of dragonflies. Therefore, it can be concluded that the dragonflies predated on these insects. The regression coefficient of dragonflies with other insect pests was positive which revealed an increasing trend in their population with the increasing of dragonflies population.

Table 1.6 reveals that the population density of grasshopper, white leafhopper and rice stem borershad negative regression coefficient with the poulation of dragonflies which showed a decreasing trend in the population of these insect pests with corresponding increase in the population of dragonflies. It can be concluded from the result that dragonflies played an important role to check the population of these insect pests. The regression coefficient of other insect pests with dragonflies was positive which indicated that the population of these pests was also increased when the population of dragonflies was increased.

Table 1.5 : Statistical analysis of multiple linear regression equation for seasonal abundance of dragonflies and insect pests of rice at Hajeera

Predictor	Coef.	stdev	t-ratio	p
Constant	28.00	16.07	1.74	0.332
Rice skipper	-0.910	9.83	-0.09	0.941
Grasshopper	-20.60	68.21	-0.30	0.813
Green leafhopper	78.30	25.14	3.11	0.198
White leafhopper	-10.00	195.40	-0.05	0.968
Hairy caterpillar	197.10	683.50	0.29	0.821
Rice bug	40.00	117.30	0.34	0.791
Rice stem borers	-62.20	313.60	-0.20	0.875
Poophilus spp.	91.70	134.70	0.68	0.619

Analysis of variance

Source	DF	SS	MS	F	p
Regression	8	1237.19	154.65	3.30	0.403
Error	1	46.91	46.91		
Total	9	1284.10			

Table 1.6 : Statistical analysis of multiple linear regression equation for seasonal abundance of dragonflies and insect pests of rice at Mandhole

Predictor	Coef.	stdev	t-ratio	p
Constant	14.105	4.632	3.05	0.202
Rice skipper	3.906	1.615	2.42	0.250
Grasshopper	-19.80	12.86	-1.54	0.367
Green leafhopper	70.78	15.22	4.65	0.135
White leafhopper	-48.97	19.95	-2.45	0.246
Hairy caterpillar	36.18	14.40	2.51	0.241
Rice bug	21.06	51.42	0.41	0.753
Rice stem borers	-10.97	10.55	-1.04	0.488
Poophilus spp.	5.55	28.76	0.19	0.879

Analysis of variance

Source	DF	SS	MS	F	p
Regression	8	1003.49	125.44	20.53	0.169
Error	1	6.11	6.11		
Total	9	1.009.60			

Table 1.7 shows that the population density of rice skipper, hairy caterpillar and rice bug had negative regression coefficient with the population dragonflies. It revealed that there was a predator-prey relationship between these pests and dragonflies. The regression coefficient of other pests with dragonflies

population was positive which showed no predator-prey relationship between these pests and dragonflies.

Table 1.7 : Statistical analysis of multiple linear regression equation for seasonal abundance of dragonflies and insect pests or rice at Harighale

Predictor	Coef.	stdev	t-ratio	p
Constant	9.218	4.280	2.15	0.164
Rice skipper	-3.689	5.622	-0.66	0.579
Grasshopper	26.45	21.58	1.23	0.345
Green leafhopper	31.14	31.72	0.98	0.430
White leafhopper	199.9	211.2	0.95	0.444
Hairy caterpillar	-143.05	93.59	-1.53	0.266
Rice bug	-59.40	62.71	-0.95	0.443
Rice stem borers	12.90	56.74	0.23	0.841
Poophilus spp.	15.32	71.38	0.21	0.850

Analysis of variance

Source	DF	SS	MS	F	p
Regression	8	403.17	50.40	2.49	0.318
Error	240.46	20.23			
Total	10	443.64			

Table 1.8 shows a negative regression coefficient between the population of rice skipper, green leafhopper and *Poophilus* species. It showed that there was a predator-prey relationship between their populations. The regression coefficient of other pests with the population of dragonflies was positive which showed no relationship between the population of these pests and the dragonflies.

Discussion

The feeding behaviour of different dragonfly species during the act of feeding was observed in detail. The individuals of almost all the species, on offering the prey insects, start eating the abdomen first and finished it in 3-4 bites and then thorax in 1-2 bites rejecting the wings, legs and head of the insect pests. When heads of the insect pests were offered, they chewed and vomit out and totally refused it when vomited head was again offered. However, *Orthetrum sabina* consumed the hairy caterpillars without leaving any part. Similar behaviour was observed when hairy caterpillars and grasshoppers (*Oxya* spp.) were offered to *O. pruinisum neglectum*. Some water was offered and drunk by

Table 1.8 : Statistical analysis of multiple linear regression equation for seasonal abundance of dragonflies and insect pests of rice at Banipasari

Predictor	Coef.	stdev	t-ratio	p	
Constant	14.168	3.266	4.34	0.049	
Rice skipper	-6.539	2.520	-2.60	0.122	
Grasshopper	49.193	6.336	7.76	0.016	
Green leafhopper	-34.880	6.792	-5.14	0.036	
White leafhopper	141.36	38.96	3.63	0.068	
Hairy caterpillar	152.35	44.42	3.43	0.076	
Rice bug	75.88	39.34	1.93	0.194	
Rice stem borers	29.82	27.33	1.09	0.389	
Poophilus spp.	-8.0	21.82	-0.37	0.749	
Analysis of variance					
Source	DF	SS	MS	F	p.
Regression	8	5989.52	748.69	41.47	0.024
Error	2	36.11	18.06		
Total	10	6025.64			

dragonflies after forced feeding. The females seemed to be the most voracious feeders as compared with the males. However, in some cases the males consumed greater number of insect pests. When the voraciousness of different dragonfly species was compared among themselves. *Orthetrum sabina* was found to be the most voracious feeder followed by *O. pruinosum neglectum, O. triangulare triangulare, O. glaucum, Crocothemis servilia, C. erythraea, Pantala flavescens, Palpopleura sexmaculata semaculata, Sypetrum commixtum, Acisoma panorpoides panorpoides* and *Trithemis festiva.*

The population density of dragonflies was maximum from middle of August to the end of September. However, it was at peak from the last week of August to the third week of September. It declined afterwards as the crop matured and population of the pests become lower. It shows that the dragonflies are attracted to the rice fields with increase in the pest population. The population of dragonflies was higher at Banipasari than all other localities. The results regarding the activity of dragonflies differ slightly from Yousuf *et al.* (1998) who observed the most favourable period for the activity of dragonflies from September 15 to October 1 at Sheikhupura and Gujaranwala, Punjab (Pakistan). It may be due to the difference in climatic conditions of this region with the rice growing areas of Punjab.

The studies on seasonal abundance of insect pests of rice revealed that the population density of rice skippers and grasshopper was higher as compared with the other pests. The population of grasshoppers was higher in the rice fields, therefore condition of the crop was thoroughly examined throughout the crop season to observe the status of grasshoppers as rice pests. It was noted that the damage was considerably very low rather ignorable. It may be due to the reason that the rice fields in Pak Occupied Kashmir (POK) are surrounded by the grassy areas because of heavy rains. Although the grasshoppers visited rice fields but they feed on the grasses and damage caused to the crop is unnoticeable.

The feeding potential of dragonflies was determined under the field conditions and many alternatives were tried for this purpose. The dragonflies were released in the field cages and insect pests of rice were provided as food on potted rice plants. A single individual or a pair of the same species were released separately in the field cage. Sometimes, water continuously to familiarized them with the captive conditions but they did not show any activity or response to their prey. It seems that the dragonflies have a strong sense of their captivity, for they again started their feeding activities when released in the field. The results are in agreement with those of Yousuf and Ali (1986) who failed to evaluate the feeding potential of the dragonflies in the laboratory and field cages.

In view of this dragonflies attitude, the most appropriate method to evaluate the predation of dragonflies against the insect pests of rice was to use some statistical method. Therefore, the multiple linear regression equation was applied and seasonal abundance of dragonflies and insect pests of rice was analysed statistically to determine the predation-prey relationship. It revealed that dragonflies had negative regression coefficient with all the insect pests of rice. It showed strong predator-prey relationship, however, feeding behaviour of dragonflies at various localities was not similar and varied slightly from locality to locality.

On the basis of the results of the present studies it can be concluded that the natural control agents have the ability to control the population of insect pests if not disturbed by the human practices. The farmers of this area (AJK) have small holdings and are not applying insecticides. Even then the population of the pests is not so high to cause economic loss to the rice crop. It seems to be controlled by the naturally occurring biological control agents including dragonflies. Thus, the area is ideal for biological control studies as the chemicals are not applied and the balance between the pests and bio-control agents is naturally maintained.

Acknowledgements

The author is thankful to Pakistan Science Foundation, Islamabad for providing financial assistance to conduct this research under the Project No. AJK-UCR/BIO (218).

List of Insect pests of rice consumed by dragonfies during the crop season 1995-97

S.No.	Common Name	Scientific Name	Family	Order
1.	Yellow stem borer (YSB)	*Scirpophaga incertulas* (Wlk.)	Pyralidae	Lepidoptera
2.	White stem borer (WSB)	*S. innotata* (Wlk.)	Pyralidae	Lepidoptera
3.	Hairy caterpillar (HC)	*Epizygaena cashmirensis* Koll.	Zygaenidae	Lepidoptera
4.	Rice skipper (RS)	*Parnara guttata* B. & G.	Hesperiidae	Lepidoptera
5.	White-backed planthopper (WBPH)	*Sogatella furcifera* (Horv.)	Delphacidae	Homoptera
6.	White leafhopper (WLH)	*Cofana spectra* (Dist.)	Cicadellidae	Homoptera
7.	Green leafhopper (GLH)	*Nephotettix* spp.	Cicadellidae	Hemoptera
8.	Rice bug (RB)	*Leptocorisa acuta* (Thun.)	Coreidae	Homoptera
9.	Cicadellid leafhopper (CLH)	—	Cicadellidae	Hemoptera
10.	Grasshopper (GH)	*Oxya* spp.	Acrididae	Orthoptera

References

Agassiz, D. 1978. Five introduced species, including one new to science, of China mark moths (Lepidoptera : Pyralidae) new to Britain, *Entomol. Gaz.*, 29(3): 117-27 (*Odonatologica*, 12(1) : Abst. No., 95, 1983).

Ahmad, I. 1983. Studies on the rice pests of Pakistan with reference to their systematics and pheromone glands. Fin. Rept., P.L. 480., Deptt. Zoo., Univ. Karachi, Pakistan.

Ali, M. A. 1983. Studies on population and feeding habits of dragonflies on insect pests of cotton. M.Sc. Thesis, Deptt. Agric. Ento., Univ. Agric., Faisalabad, Pakistan.

Dean, G.J.W. 1978. Insect pests of rice in Laos. *Pans* 24(3) : 280-89.

Feijen, H. R. 1977. Research on the ecology and natural enemies of the rice stemborers, *Diopsis thoracica* in Malawi. *Int. Rice Res. Newls.,* 2(5) : 17-18.

Honghai, Z., Yuhuo, H., QiBiao, B., Chun Yu, C., Can, C. and Zheng, H. H. 1995. Improvement of forecasting method for occurrence amount of first generation of rice stem borer. *Pla. Prot.,* 21(5); 20-21.

Khaliq, A., and Siddique, M. 1995. Some rice field Odonata in the districts of Poonch and Bagh, Azad Kashmir, Pakistan. Notul. *Odonatol.,* 4(6) : 106.

Lippold, P. C. 1979. Guidelines for integrated control of rice insect pests. FAO Plant Production and Protection Pap. 14, vii+115 pp., FAO, Rome, Italy (*Odonatologica,* 20(3) : Abst. No. 7733, 1991).

Matsumura, M. 1996. Population dynamics of the white-baked planthopper, *Sogatella furcifera* (Hemiptera : Delphacidae) with special reference to the relationships between its population growth and the growth stage of rice plants. *Res. Popul. Ecol.,* 38(1) : 19-25.

Spurgeon, D.W., Shaver, T.N., Raulston, J.R. and Lingren, P.D. 1996. Activity patterns of adult Mexican rice borers (Lepidoptera : Pyralidae) in sugarcane. *Southwest Entomol.,* 21(1) : 23-32.

Suglura, T. 1995. Forecasting of pest occurrence by the plant protection Centre-activities in Nara Prefecture. *Agrochem. Japan,* No. 66; 2-5.

Van Vreden, G., and Ahmadzabidi, A.L. 1986. Pests of rice and their natural enemies in Peninsular Malaysia. Pudoc, Wageningen, vii+230 pp. ISBN 90-220-0890-8 (*Odonatologica,* 20(4) : Abst. No. 7891, 1991).

Xia, S., Wu, U. and Wang, Z. 1988. Primary coloured iconograph of entomophagous insects in paddy fields. Hunan Press Sci. and Technol. Hunan iv+96 pp., 205 col. Phot. Incl.ISBN 7-5357-02538 (*Odonatologica,* 21(3) : Abst. No.8320, 1992).

Yousuf, M. and Ali, M.A. 1986. A new and easy method of feeding dragonflies in captivity. *Notul. Odanatol.,* 2(7) : 117-18.

Yousuf, M.; Khaliq, A. and Najam, M.A., 1998. Population and feeding capacity of dragonflies on insect pests of rice in Pakistan (Anisoptera : Libellulidae). *Notul. Odonatol.,* 5(2) : 17-19.

2

A Community Structure of Odonata of the South-West Province of Cameroon, with the Description of *Phyllogomphus corbetae* Spec. Nov. (Anisoptera: Gomphidae)

• *G.K. Vick*

Abstract

A check-list of the dragonflies of the South-West Province of Cameroon, based upon field work undertaken between 1995 and 1998 and a survey of historical records, is given. Notes on seasonal occurrennce, habitat requirements and taxonomy are provided. As new is described: *P. corbetae* sp. n. (holotype male: Kumba, outlet stream from Borombi Mbo, 20-IX-1997; allotype female: Limbe, Bimbia, Elephant River, 4-VII-1996).

Introduction

Political : Cameroon occupies an area of about 475,000 km^2 and is therefore approximately the same size as France of Spain. It covers latitudes between 2° and 13° N and longitudes of 8° and 16° E. The South-West Province occupies about 5% of the national territory and lies adjacent to the Nigerian border and the Gulf of Biafra. Its areas is approximately equal to that of Belize, or half that of Costa Rica or Switzerland; this is roughly equivalent to about six English countries. Before independence and reunification in 1960-1961, it was part of the British Cameroons, and together with the North-West Province, it forms the anglophone part of the country. The population is 0.82 million, giving an average density of 33 people/km^2 (Ministry of Planning and Regional Development, 1989).

For the purpose of a dragonfly survey, it forms a very workable homogeneous recording unit over which the climatic regime is relatively constant, apart from the natural local variations due to orographic uplift associated with mountains and topographic diversity.

Physical-Geze (1943) discusses the physical geography of the region; NGWA (1967) and NEBA (1987) provided excellent summaries. The basement rock of the region is part of the great pre-Camabrian African shield which consists of granite and metamorphic rock (mica schists, quartzite, gneiss and migmatites). A large tectonic tough which runs across the Province from SW to NE is marked out by isolated volcanos and prequaternary volcanic deposits stretching from Mount Cameroon (active six times this century) to Mount Kupe (2070 m) and the Rumpi Hills (1760m) to Mounth Manenguba (2250m) and the Mamfe region in the West. These form the southern part of the Cameroon Highlands. Recent basalts are present.

Crater lakes are present, including Debundsha, Barombi Mbo, Edib and the Twin Lakes of Manengouba. The province is drained by the Manyu-Cross River system in the Mamfe Depression in the north, and by the River Ndian, Mungo and Wouri and their tributaries in the south. All flow into the Atlantic. There are extensive sedimentary deposits in these valleys.

On the whole, it appears that climate is more significant than geology in determining dragonfly distributions in the tropics and our preliminary work would support this. We have not yet found any correlations between dragonfly distributions and geology or soil type.

Climate (See Table 2.1 and Table 2.2) : The climate in this part of Cameroon is usually described as 'monsoon-equatorial': rain falls in virtually every month but there is a single rainfall maximum which occurs in the northern hemisphere summer. This is different from the climate in the rest of southern Cameroon which is closer to the equatorial pattern with two double rainfall peaks. The climate

Table 2.1 : Rainfall figures in mm: various dates and durations of records

Station	J	F	M	A	M	J	J	A	S	O	N	D	Total	Source
Mamfe	33	79	130	206	325	437	513	465	564	452	152	38	3424	Hawkins & Brunt, 1965
Nyasoso	21	107	230	182	298	402	822	720	645	453	77	45	4002	Ejedepang Koge, 1986.
Tombel	53	121	248	240	267	507	489	480	554	474	160	65	3658	Ejedepangkoge, 1986
Ndian	40	127	258	300	395	715	887	952	752	544	319	93	5382	Ndian Estate, 1987
Kumba	17	70	197	206	194	225	276	335	390	264	78	46	2298	FAO-UDNP Onarest 1977B.
Debundsha	242	314	500	520	743	1285	1424	1371	1619	1158	610	358	10144	NGWA, 1967

Table 2.2 : Sunshine figures in hours/month

Station	J	F	M	A	M	J	J	A	S	O	N	D	Total	Source
Mamfe	169	184	171	148	153	108	85	65	81	139	175	180	1658	Hawkins & Brunt 1965
Ndian	135	159	150	157	151	87	11	1	44	109	124	128	1256	Ndian Estate, 1987

of northern Cameroon is different again in that there is a strongly monsoonal, or tropical, pattern with a prolonged severe dry season.

In South-West Cameroon, the prevailing moisture-laden south-west monsoons deposit heavy rainfall on the windward side of the mountain massifs. The effect of the dry harmattan which blows from the north is minimal. In the lowlands, Tombel and Ndian usually have no month with less than 40 mm rain, although recently there appears to have been a tendency for longer periods without rain.

Debundsha, on the windward side of Mount Cameroon, receives about 1000mm rain (approximately 400") and is one of the wettest place in the world, usually regarded as second only to Cherrapunji in Assam. The mean annual temperature in the lowlands is about 25°-26°C and there is little fluctuation during the year. Even Kumba, which lie in the rain-shadow of Mount Cameroon receives nearly 2300mm and usually only has one month with below 40mm.

The mountainous areas, lying at altitudes over 700m, have a modified version of this climate with lower average temperatures, but a similar rainfall pattern (see fugures for Nyasoso), with a general trend for higher percipation.

The relative humidity at all station is above 80% even in the driest months and reaches 88% in the wettest month; in cloud forest above 1000m it approaches 100%. Almost all stations have at least 40mm of rain per month and as the relative humidity is so high, the term 'dry season' is only a relative term: in most areas trees are evergreen, and rivers and the larger strems are perennial, although very small streams may dry up. Watercourses reach their highest levels in September and October and their lowest levels in February and March.

Table 2.3 shows in simplified from the climatic characteristics of each month, and a very brad breakdown into the seasons: dry (occasional rainfall in most months); transition (short, heavy storms in most weeks); wet (prolonged rainfall and overcast conditions).

Table 2.3 : Summary of seasonal pattern

Season	Month	Character of climate	
dry	November	fairy wet	sunny
	December	dry	sunny
	January	dry	sunny
	February	dry	sunny
transition	March	wet	sunny
	April	wet	cloudy
	May	wet	cloudy
wet	June	very wet	very cloudy
	July	very wet	very cloudy
	August	very wet	very cloudy
	September	very wet	very cloudy
	October	very wet	cloudy

Natural Vegetation: Letouzey (1968a, 1968b) discusses the vegetation of province. Apart from the coastal mangroves, dense evergreen equatorial `Biafran' forest is the natural vegetation at low and medium altitudes (0-1200 m), dominated by the Caeslpiniodae. These forests are part of a large association which extends from south-eastern Nigeria and east towards the R. Congo and the Sangha river (Ministry of Planning and Regional Development, 1989). The province lies at the core of this region: gradients of increasing species-poverty extend from the core in all directions; this has been observed for plants, birds, mammals, reptiles, amphibians and butterflies. From our preliminary observations it also appears to be true for Odonata.

The rainforest was drastically reduced during the Pleistocence. During glacial periods in the North Africa cooled and became drier. A few areas persisted as lowland rainforest refuges. There are generally coastal and are presumed to be in regions which have the heaviest rainfall now: Mount Cameroon, Korup, Mount Kupe and the Bakossi Mountains. The fauna and flora have radiated out again from these foci as the climate became wetter and warmer (Hamilton, 1976; Livingstone, 1982; Grubb, 1982).

It is believed that the evergreen forest at Korup could have persisted even when temperatures were reduced by an average of 5°C and the rainfall reduced by as much as 70%. It is richer botanically than any other African forest for which data are available (Ministry of Planning and Regional Development, 1989). However, recent work by the Royal Botanical Garden (Kew) also suggests that Mount Kupe is comparatively rich. Probably the entire province forms a fairly dense centre of endemism for most moisture-adapted organisms.

Above 1600m, a dense humid submontane forest occurs with a very different plant ensemble. Mount Kupe (2050m) provides an excellent example of this association. On Mount Cameroon, there is a full altitude zonation which is summarised by Letouzey (1968a, 1968b) as : 0-1200m dense evergreen equatorial forest; 1200-1800m submontane forest; 1800-2600m montane scrub and svannah; 2600m and above subalpine and alpine.

As far as odonata fauna of the province is concerned we have observed a lowland fauna and a distinct 'montane' fauna, which is found in cool streams at altitudes of above 700-800m. The cool streams investigated (on Mounts Cameroon and Kupe) arise at much higher altitude than this and I have referred to these species as montane even though this appears to be contradicting the botanical evidence where the lower limit would be 1800m. They do however show a lower altitudinal limit and as they are associated with mountainous areas, the term 'montane' is justified; a similar situation has been observed with certain amphibians, according to recent work out by C. Wild of WWF (Cameroon). We have not explored at altitudes higher than 2000m - an endemic dragonfly fauna at this altitude is unlikely. In general, tropical Africa does not appear to possess a genuinely montane species assemblage of odonata.

There has been extensive clearing for agriculture throughout the region, especially for the cultivation of oil-palm (native to the region), cocoa, rubber, banana/plantain and coffee. However, compared with many parts of the equatorial regions of the world, there are still extensive areas of pristine forest present.

Water Temperatures: In an equatorial climate in which there is so little change in the average daily temperature it is useful to measure stream temperatures. The daily fluctuations in air temperature are greater than the monthly hanges in average air temperature and as the heat capacity of water is high, buffering diurnal changes, it is not crucial to measure water temperature at exactly the same time of day in order to gain a useful measure of the character of the stream.

It is fairly easy to divide the habitats into lowlands and highland, with the faunistic border for Odonata being at about 700m in altitude. But water temperature is the crucial factor, lowlands streams can have montane species if cool. This usually occurs when a montane stream enters the lowlands the maintains its character, especially when flowing through primary forest.

The water temperature of montane streams is dense primary forest is within a range of 19°-23°C (altitude range 700-1200m). In the lowlands, the figure is about 25°-26° in heavy forest, rising to 27°-28° in secondary forest with farm bush. The large crater lake near Kumba, Barombi Mbo (altitude 380m) had water at 29°C; the crater lake at Debundsha at about 50m altitude was 30.5°C. In open habitats very much higher temperatures are recorded, e.g. 35°C at the lowland coastal marshes at Debundsha. All of the above temperatures were measured at about noon.

Local Names for Dragonfly

In the areas of survey, we asked the local people what was their word for dragonfly. In the village of Nyasoso, centre of the territory of the Bakossi tribe, they used the word n'sisag for the adults and n'gon for the larval stages. In fact, one of the older men in the village described how, when he was a boy, they used to collect dragonfly larvae for food from the streams in Mount Kupe. We assume that they were eating the larvae of *Phyllomacromia* spp. which are the only large species which are at all common; certainly aeshnid spp. are only rarely obtained in these streams.

In the region around Debundsha, on the south-west side of Mount Cameroon, the local tribe is the Bakweri and local men identified dragonfly adults as *ecuve ecuve* (presumably a plural). The isolated Barombi tribe, who fish the lake of that name near Kumba, used the term *mbang mbang*. All of these tribes speak true Bantu languages.

Table 2.4 : Comparison of *Phyllogomphus corbetae* spec. nov. with *P. montanus*

Character (male/female)	Corbetae	*Montanus*
antheumeral stripe (m/f)	confluent with prothoracic collar, forming pair of 7s	separated from prothoracic collar
separated upper humeral spot	yes	no
labrum (m/f)	black with one extensive yellow spot occupying 50% of area	black with two separated yellow spots occupying 30% area
genital hamule (m)	inner branch>outer branch	outer branch>inner branch
superior appendage (m) lateral aspect	basal swelling with weak tooth on ventral surface	strong basal tooth on ventral surface
dorsal aspect	narrower in apical half, weakly recurved at apex	broader in apical half, strongly recurved at apex
segment 10 (m)	weakly formed dorsal ridge scarcely rising any higher than that on segment 9	strongly formed dorsal rige rising at least twice as high the ridge on segment 9
(f)	dorsal ridge vestigial for whole length of segment	dorsal ridge present but weak, extending three quarters of length of segment
ratio length of seg 10 to seg 9		
(m)	1.0	1.3
(f)	1.2	1.7
valvar scale (f)	full length of segment 9 just overlapping onto base of segment 10	full length of segment 9 and three quarters of segment 10
occipital ridge (f)	2 central spines, sparse hairs	no spines, even distribution of long black hairs.
Illustrations in literature		
(m)	Corbet (1977) as *P. montanus*	Buchholz (1958) as *P. hartwigi spec. nov.*
(f)	not previously figured	Fraser (1957)- holotype
larva	Corbet (1977) as *p. montanus*	Fraser (1957)-poor figure

History of Recording in the South-West Province

When the kings of the coastal Duala tribes agreed to give sovereignty to Germany in 1884 this only meant that a small coastal strip became the German colony of Kamerun (Mbuagbaw, Brain & Palmer, 1987). After 1885, and the 'award' to Germany of a larger area at the Berlin Conference, there was an intense period of exploration, searching for the best routes from the coast into the 'useful' interior (e.g. the grassfields of the Bamenda Uplands) through the dense forest of what is now the South-West Province. The Germans planned a line of 'stations' which ran from Mount Cameroon to the interior. A 'factory' was established at Mundame (Mungo River) and a 'station' was set up at Barombi. The latter became the administrative centre of Johann Albrechtshohe. There was severe resistance from the Bakweri tribe at Buea on Mount Cameroon between 1891 and 1894, but a 'station' was eventually built there and it became the capital of Kamerun from 1901-1909.

During the German period of colonialisation, much odonate materials was sent back to Europe. Selys (1879, 1886) described material from Cameroon, including *Umma mesostigma**, *Chlorocypha glauca**, *C. cancellata** and *Chlorocnemis nigripes** from Mount Cameroon (Locally called Mongo-Ma Loboh). Also *Acanthagyna sextans** was described from the mountain by McLachlan (1896). At this time, species were sometimes described with data 'Kamerun' and, if they are not further localised, it is not possible to include them in the records of the province.

McLachlan's material, representing many species, in the NHM collection was collected in this period (about 1880-1890), but unfortunately the data labelling is perfunctory: Camaroons River (now called the lower Wouri River which flows into the Atlantic near Douala and thus just outside the province); merely 'Camaroons' whch generally referred to 'Old Camaroon' which was the settled area which developed between Victoria (now usually called Limbe), Tiko and Douala and probably mostly in our area and Mongo-Ma-Loboh (entirely within the South-West Province). I have included all of these records which are represented by specimens in the NHM in Table 2.5.

---

* Described as new to science from material taken in the region.

Table 2.5 : Summary of records by species, with summary of altitude, forest cover and habitat

Family/species	Historical records (prior to 1980)	D'andrea & Carfi 1997	Records by CDP (1995-1998)	Altitude	Forest	Habitat
PROTONEURIDAE						
Chlorocnemis contraria Schmidt 1951	77, 78, 96, 104, 106, 107	91	1, 5, 6, 11, 17, 25, 31	M	F	I
C. eisentrauti Pinhey, 1974	88					
C. flavipennis Selys, 1863	78, 105	91	1, 3, 5, 8, 12, 25, 30, 35	M/L	F/S	I
C. nigripes Selys, 1886	60, 65, 73, 78, 85, 104	91	8, 11, 12, 17, 23	L	F/S	2
Elattoneura balli Kimmins, 1938	93, 104		8	L	S	2
E. nigra Kimmins, 1938			1, 5, 6, 8, 10, 11, 14, 16, 17, 19, 23, 25, 26, 31, 35	M/L	F/S	1,2
E. pruinosa (Selys, 1886)	65, 68, 69, 70, 71, 72, 73, 79, 82, 93, 96, 104					
Isomecocnemis cyanura (Forster, 1909)	85, 104					
I. subnodalis (Selys, 1886)	83, 106					
Prodasineura vittata (Selys, 1886)	62, 63, 79, 93, 94, 96, 104		35			
COENAGRIONIDAE						
Agriocnemis maclaclani (Selys 1877)			23, 24, 25, 26, 28	L	F/S	2,4,5
Ceriagrion glabrum (Burmeister, 1839)		89	6, 13, 21, 25, 26	L	S/O	2
C. rubellocerinum, Fraser,1947			11	L	S	2
Enallagma buchholzi Pinhey, 1971	61, 87		27	L	F	6
E. camerunense Karsch, 1899	72					
E. vansomereni Pinhey, 1955						

Contd...

Table 2.5 – *contd...*

Family/species	Historical records (prior to 1980)	D'andrea & Carfi 1997	Records by CDP (1995-1998)	Altitude	Forest	Habitat
Ischnura senegalensis (Rambur, 1842)			23, 24, 26	L	O	4, 5
Pseudarion angelicum Fraser, 1947		89, 92	23, 24, 26, 28	L	S	4
P. epiphonematicum Karsch, 1891	72, 73, 78, 79, 82, 88, 96 104		4, 5, 6, 8, 10, 11, 12, 19	I	S/O	2
P. flavipes Sjostedt, 1900	62, 63, 65, 68, 72		16, 17, 35	L	F	2
P. kersteni (Gerstacker, 1869)		91	8, 12, 17	L	S/O	2
P. melanicterum Selys, 1876	72, 73. 79, 82, 87, 88, 102		4, 8, 9, 10, 11, 12, 15, 17, 23, 24, 33, 35	L	S/O	2, 4
P. nubicum Selys, 1876	63					
P. risi Schmidt, 1936	104		11, 14	L	S	2
P. serrrulatum Karsch, 1893	60, 82, 96		31	L	F	2
P. S. Sjoestedti forster, 1906	60, 79, 80, 81, 82, 104		11, 16	L	S	2
P. sublacteum doualae Pinhey, 1961			36			
PLATYCNEMIDIDAE						
Mesocnemis singularis Karsch, 1891	67, 73	89, 92	7, 8, 16, 17, 18, 21, 30	L	S/O	2, 6
Platycnemis congolensis Martin, 1908			19, 32	L	F	2
Stenocnemis pachystigma (Selys, 1886)	71, 72, 78, 79		1, 5, 6, 10, 11, 23	M	F	1, 3
PERILESTIDAE						
Nubiolestes diotima (Scdmidt, 1943)	93, 106		1, 3, 6	M	F	1

contd....

Table 2.5 – *contd...*

Family/species	Historical records (prior to 1980)	D'andrea & Carfi 1997	Records by CDP (1995-1998)	Altitude	Forest	Habitat
MEGAPODAGRIODNIDAE						
Neurolestes trinervis Selys, 1885	78, 86		1, 11, 31, 32, 34	M	F	1
AMPHIPTERY GIDAE						
Pentaphlebia stahli Forster, 1909	60, 78, 79, 86, 88, 106		1, 3, 5, 6, 11, 19, 20, 34, 42	M/L	F	1, 2
CHLOROCYPHIDAE						
Africocypha lacuselephantum (Karsch, 1899)	72, 78, 79, 83, 106		1, 4, 5, 19, 20	M/L	F	1, 2
Chlorocypha concellata (Selys, 1879)	60, 76, 79, 82, 83, 96, 104, 105	91	4, 8, 10, 11, 12, 19, 23, 24, 25, 30, 34, 38, 42	L	F/S/O	2, 4, 5
C. centripunctata Gambles, 1975		1, 3	M	F	1	
C. curta (Hagen, 1853)	62, 72		8, 11, 14, 16, 18, 30, 36	L	F/S	2
C. cyanifrons (Selys, 1873)	62, 68, 72, 102					
C. dispar (p. de Beauvois, 1805)	73					
C. glauca (Selys, 1879)	60, 72, 79, 96, 104		11, 16, 17, 18, 23, 34, 35, 36	L	F/S	2, 6
C. gracilis (karsch, 1899)	67, 72, 79, 83, 99, 104, 106		4, 6, 42	M	F/S	1
C. hintzi (Grunberg, 1914)	82					
C. neptunus (Sjostedt, 1900)	63, 103	89, 92				
C. rubida (Hagen, 1853)	79, 93	89				
Chlorocypha sp. 2(nr. *rubida*)			30	L	F	2
C. selysi Karsch, 1899	60, 63, 72, 78, 79, 93, 96, 106	91	3, 4, 8, 10, 11, 12, 17, 19, 42	L	F/S	2
C. victoriae Forster, 1914		92				

contd....

Table 2.5 – *contd...*

Family/species	Historical records (prior to 1980)	D'andrea & Carfi 1997	Records by CDP (1995-1998)	Altitude	Forest	Habitat
Platycypha lacustris (Forster, 1914)		4		M	S	2
P. rufitibia (Pinhey, 1961)	80, 83. 104, 106		15, 17, 34, 36	L	F	2
CALOPTERYGIDAE						
Phaon camerunensis Sjostedt, 1900	63, 79		7, 8, 11, 15, 17, 22, 32, 34, 38	L	F	2
P. iridipennis (Burmeister, 1839)	70, 73, 75	90, 92	8, 9, 11, 12, 21, 23, 25, 28, 30, 31, 32	L	S/O	2
Sapho bicolor Selys, 1853			3, 4, 35, 38, 42			
S. orichalcea McLachlan, 1869	60, 62, 63, 65, 67, 72, 73, 76, 78, 82, 83, 86, 88, 93, 97, 99, 101 106		1, 8, 11, 12, 13, 16, 22, 38, 42	M/L	F	1
S. orichalcea (small)			15, 17	L	F	2
Umma longistigma (Selys, 1869)	62, 63, 65, 68, 72, 73, 93, 95, 102		8, 11, 15, 17, 19, 22, 41	L	F	2
U. longistigma (narrow inf.)	83 (?), 104(?)		30, 31, 32, 34, 35, 36, 38	L	F	2
U. mesostigma (Selys, 1879)	60. 72, 78, 88, 93, 95, 96, 104, 105, 106		1, 3, 5, 6, 10, 11, 12, 13, 15, 19, 20, 22, 34, 41, 42	M/L	F/S	1, 2
U. mesumberi vic, 1996			1, 3, 42			
U. puella Sjostedt, 1917	91		33, 35, 38			
U. purpurea Pinhey, 1961	79, 83, 104		32, 34, 35, 36			
U. saphirina Forster, 1916	83, 102					
COMPHIDAE						
Crenigomphus sp.	79					

contd....

Table 2.5 – *contd...*

Family/species	Historical records (prior to 1980)	D'andrea & Carfi 1997	Records by CDP (1995-1998)	Altitude	Forest	Habitat
Diastatomma multilineata Fraser, 1949			19	L	F	2
D. tricolor (P.de Beauvois, 1805)	63, 72, 104		15, 17, 35	L	S	2
Gomphidia gamblest Gauthier, 1987	75, 76		8, 11, 15, 16, 17, 38	L	F/S	2
Ictinogomphus fraseri Kimmins, 1958	72, 76, 77		18	L	F	6
Lestinogomphus angustus Mertin, 1911			36			
Microgomphus camerunensis Longfield, 1951	72, 73		11	L	S	2
Neurogomphus fuscifrons Karsch, 1890	72, 73					
Notogomphus spinosus Karsch, 1890	73		8	L	S	2
Notogomphus sp. 1			1	M	F	1
Notogomphus sp. 2			42			
Notogomphus sp. 3			1	M	F	1
Omchogomphus sp. 1			1	M	F	
Omchogomphus sp. 2						
Paragomphus abnormis Karsch, 1890	72, 73, 83, 104		1, 4, 7, 12, 13, 22 37	M/L	F/S	1, 2
P. genei Selys, 1841			11, 15, 17	L	F/S	2
P. nigroviridis Cammaerts, 1969			1	M	F	1
Paragomphus sp. 1			1, 5	L	S	2
Paragomphus sp. 2						

contd....

Table 2.5 – *contd...*

Family/species	Historical records (prior to 1980)	D'andrea & Carfi 1997	Records by CDP (1995-1998)	Altitude	Forest	Habitat
Paragomphus sp. 3			4			
Phyllogomphus montanus						
Fraser, 1957	95		8, 11, 12, 19, 34, 40	L	F/S	2
Tragogomphus aurivillii	72, 76		16, 21, 25	L	F	2
Sjostedt, 1900	63					
T. mamfei Pinhey, 1961	78, 83					
AESHNIDAE						
Aeshna scotias Pinhey, 1952			1	M	F	1
Anax chloromelas Ris, 1911			1, 26, 40	M/L	F	1, 5
A. congoliath Fraser, 1953			9, 12, 32	L	S	2
A. imperator Leach, 1815			18, 23, 24, 26, 27	L	S/O	2, 4, 5, 6
A. tristis Hagen, 1867			1, 2, 10, 11, 19, 26, 32	M/L	S	2, 5
Acanthagyna africana						
(P. de Beauvois, 1805)	98					
A. bullata (Karsch, 1891)	62, 63, 96		6, 17	L	F/S	2
A. cylindrata (Karsch, 1891)	79, 93					
A. nigeriensis Gambles, 1956			23	L	F/S	2
A. sextans (McLachlan, 1896)	60, 96, 99, 100					
Heliaeschna cynthiae	88	90				
Fraser, 1939						
H. fuliginosa Karsch, 1893	60, 63, 68, 93		17	L	F	4
H. lanceolata LeRoi, 1915			38			
H. longfieldae Gambles, 1967	60					
H. trinervulata Fraser, 1955			1			
Hemianax ephippiger						
(Burmeister, 1839)			26	L	S/O	5

contd....

Table 2.5 – *contd...*

Family/species	Historical records (prior to 1980)	D'andrea & Carfi 1997	Records by CDP (1995-1998)	Altitude	Forest	Habitat
CORDULIDAE						
Idomacromia proavita Karsch, 1896	65, 72, 79		1, 15, 17, 19	M/L	F	2, 3
Phyllomacromia aeneotharax (Nunney, 1895)	79, 86, 88		1	M	F	1
P. bicrustulata (Legrand, 1975)			9, 15, 32, 35	L	F	2
P. caneri (Gauthier, 1987)	72, 73		1, 9, 10, 11, 12, 15, 17, 19, 34, 35	M/L	F/S	1, 2
P. funicularia (Martin, 1907)	83, 104		8, 9, 11, 12, 15, 17, 19, 35	L	F/S	2
P. insignis (Kirby, 1889)	93					
P. melania (Selys, 1871)	62, 73		18	L	F	6
P. paula (Karsch, 1892)	60					
LIBELLULIDAE						
Acisoma panorpoides Rambur, 1842		89	12, 24, 25, 26	L	S/O	2, 5
Acisoma trifidum Kirby, 1889	79	89, 90				
Allorhizucha klingi Karsch, 1890	69, 70, 72, 73, 75, 78, 79, 96		4, 5, 8, 10, 11, 15, 16, 19, 22, 23, 24, 25, 33, 37	L	S	2
A. preussi Karsch, 1891	62, 63, 68, 94					
Allorhizucha sp. nr. campioni Ris, 1915			11	L	S	2
Atoconeura b. pseudeudoxia Karsch, 1899	78, 79		7, 8, 19	L	F/S	2
Bradinopyga strachani (Kirby, 1900)	94		42			

contd....

Table 2.5 – *contd...*

Family/species	Historical records (prior to 1980)	D'andrea & Carfi 1997	Records by CDP (1995-1998)	Altitude	Forest	Habitat
Chalcostephia flavifrons Kirby, 1889		89, 90	8, 26, 31, 34	L	F/S	2, 5
Crocothemis divisa Karsch, 1898	79					
C. erythraea (Brulle, 1832)	87		8	L	O	2
C. sanguinolenta (Burmeister, 1839)	79		17	L	S/O	2
Cyanothemis simpsoni Ris, 1915	99, 104		15, 16, 17	L	F	2
Diplacodes lefebvrii (Rambur, 1842)			24, 26, 28	L	S/O	2, 4, 5
Eleuthemis buettikoferi Ris, 1910			8, 16, 21, 32	L	F/S	2
Hadrothemis camarense (Kirby, 1889)	60, 65, 72, 73, 94, 96		15, 37	L	F	2(?)
H. coacta (Karsch, 1891)	63, 65, 72, 73, 74		7, 13, 17, 19, 25	L	F/S	4(?)
H. defecta (Karsch, 1891)			16	L	S	2(?)
H. infesta (Karsch, 1891)	73, 94					
H. versuta (Karsch, 1891)	62, 65, 74					
Hemistigma albipuncta (Rambur, 1842)			26, 31, 34, 35, 36	L	S	5
Lokia incongruens (Karsch, 1893)	62, 94					
Malgassophlebia nigeriae Pinhey, 1961			11, 19, 34	L	F	2
Micromacromia camerunica Karsch, 1890	62, 63, 64, 72, 73, 88		8, 11, 12, 15, 16, 17, 22, 23, 24, 42	L	F/S	2

contd....

Table 2.5 – *contd...*

Family/species	Historical records (prior to 1980)	D'andrea & Carfi 1997	Records by CDP (1995-1998)	Altitude	Forest	Habitat
Neodythemis africana Fraser, 1954	88		39			
N. gorillae Pinhey, 1961	78, 79			M	F	1
Notiothemis robertsi Fraser, 1944			19	L	F	2
Olpogastra lugubris (Karsch, 1895)	79, 86, 94		8, 16, 28	L	S/O	2
Orthetrum africanum (Selys, 1887)	93		8, 25, 30	L	S	2
O. angustriventree (Rambur, 1842)		78				
O. austeni (Kerby, 1900)	77		17, 19, 26	L	F/S	2, 5
O. brachiale (P.de Beauvois, 1805)	62, 63, 68, 86		1, 12	L	S	2
O. camerunense Gamble, 1959	87		3, 14, 37, 42	M	S	4(?)
O. chrysostigma (Burmeister, 1839)	68, 73	91	31, 35			
O. guineense Ris, 1910	86	89	5, 8, 10, 11, 12, 16, 17, 30	L	S	2
O. julia Kirby, 1900	88	91, 92	1, 3, 8, 10, 11, 12, 17, 19, 23, 32, 34, 35, 36, 37, 42	M/L	F/S	1, 2, 5
O. kalai Longfield, 1936	79, 83	90, 91, 92	8, 11, 12, 15, 17, 23, 26, 30	L	F/S	2, 5
O. machadoi Longfield, 1955	79	90, 91	12, 17, 25, 37	L	F/S	2
Oxythemis phoenicosceles Ris, 1910	77	89	17, 18, 30	L	F/S	2, 4, 6
Palpopleura deceptor (Calvert, 1899)	79		11	L	S	2
P. lucia (Drury, 1773)	62, 63, 65, 68, 73, 87, 88, 89	90, 91	3, 8, 10, 11, 12, 21, 23, 25, 26, 30, 33, 34, 35	L	S/O	2, 5

contd....

Table 2.5 – *contd...*

Family/species	Historical records (prior to 1980)	D'andrea & Carfi 1997	Records by CDP (1995-1998)	Altitude	Forest	Habitat
Pantalal flavescens (Fabricius, 1798)	68, 73		6, 8, 11, 12, 17, 23, 24, 26, 28	L	O	2, 5
Porpax asperipes Karsch, 1896	78, 79					
P. bipunctus Pinhey, 1966			11	L	S	2
Rhyothemis notata (Fabricius, 1781)	65, 66, 73					
R. semihyalina (Desjardins, 1832)		26		L	S	5
Sympetrum fonscolombii (Selys, 1840)			6, 8	L	O	2
Tertrathemis bifida Fraser, 1941	89, 90, 91		17, 19, 24	L	F	2, 4
T. camerunensis Sjostedt, 1900	62, 99					
Tetrathemis sp. nr. godiardi Lacroix, 1921			17	L	F	4
Thermochoria equivocata Kirby, 1889	63, 68, 71, 94, 96					
Tholymis tillarga (Frabicius, 1798)	63, 64, 68, 100		23, 26	L	S/O	2, 5
Trapezostigma basilaris (p. de Beauvois, 1805)	86		12, 23, 26	L	S/O	2, 5
Trithemis aconita Lieftinck, 1969			10, 11, 12, 15, 23, 30, 31	L	S	2

contd....

Table 2.5 – *concld.*

Family/species	Historical records (prior to 1980)	D'andrea & Carfi 1997	Records by CDP (1995-1998)	Altitude	Forest	Habitat
T. anmulata (P.de Beauvois, 1805)	64, 68		24	L	S	4
T. arteriosa (Burmeister, 1839)	72, 73, 77, 79, 87	89, 90, 91	4, 11, 12, 17, 18, 24, 34, 35, 36	L	O	2, 4
T. basitincta Ris, 1912	79					
T. dichroa Karsch, 1893	63, 83	91	8, 9, 12, 15, 17, 21, 35	L	S	2
T. furva Karsch, 1899	87					
T. grouti Pinhey, 1961	79	90				
T. hartwigi Pinhey, 1970			26	L	S	2
T. nuptialis Karsch, 1894			8, 12, 15, 31	L	S	2
T. pruinata Karsch, 1899			1, 31, 32, 36	M	F	1
Urothemis assignata (Selys, 1872)		90	23, 24, 26, 28	L	S/O	2, 4, 5
U. edwardsi (Selys, 1849)			26	L	S/O	5
Zygonyx fallax (Schouteden, 1934)		90				
Z. flavicosta (Sjostedt, 1900)	78, 82, 83		4, 7, 8, 10, 11, 12, 17, 21, 35, 42	L	S	1, 2
Z. regisalberti (Schouteden, 1934)	88					
Z. speciosa Karsch, 1891	73, 83, 94		4, 7, 8, 11, 12, 16, 17, 19, 21, 22, 38, 42	L	S	1, 2
Z. speciosa Karsch, 1891	73, 83, 94		4, 7, 8, 11, 12, 16, 17, 19, 21, 22, 38, 42	L	S	1, 2
Z. torrida (Kirby, 1889)		91	21	L	S	1, 2
Zyxomma atlanticum Selys, 1889			16, 24, 30	L	S	2, 5

Karsch in a series of papers (1890, 1891a, 1891b, 1892, 1893, 1894, 1896, 1899) laid the foundations for the study of the dragonflies of Cameroon and many of his records are from areas which are now within the province. He reported on the findings of Dr. Paul Preuss (Karsch, 1890, 1891b) at Barombi Station and Kribi (now in Province). From Barombi Station he records 18 species, including *Mesocnemis singularis**, *Pseudagrion epiphonematicum**, *Paragomphus abnormis**, *Neurogomphus fuscifrons** (type female), *Notogomphus spinosus**, *Phyllomacromia caneri (as sophia)*, *Zygonyx speciosa**, and Z. *pretiosa** (female sex described as different species, now synonymised with *speciosa*), *Hadrothemis coacta**, *H. camarensis*, and *Allorrhizucha klingi**.

He described a female taken by Dr. Preuss at Buea on Mount Cameroon as *Phyllomacromia paula** (Karsch, 1892) and added *Sapho orichalcea*, *Heliaeschna fuliginosa** (Karsch, 1893) from Preuss' specimens taken near Victoria (now Limbe).

The next contribution (Karsch, 1894) dealt with material from Yaunde (Yaounde), now in Central Province, Karsch (1899) dealt with Leopold Conradt's material from Johann-Albrechthohe (now called Kumba). He recorded 26 species taken at Elephantsee (now called Barombi Mbo) and includes: *Chlorocypha gracilis**, *C. lacus-elephantum** (described from a female), *stenocnemis pachystigma*, *Pseudagrion flavipes* (as *hemicolon**), *Enallagma camerunense**, *Neurogomphus fuscifrons* (first male), *Diastatomma tricolor*, *Phyllomacromia caneri* (as *sophia*) and *Idomacromia proavita*.

Ingve Sjostedt was resident in Cameroon from 1890-1892 and he collected many groups of insects. In 1899, he published a useful list of 41 species found in the coastal area between Mount Cameroon and the Nigerian border, and northwards to Itoki (perhaps the village of that name situated at 5°0 1'N, 9° 15'E). He refers to Kitta (probably Kita) and Bonge (probably the German 'station' of Mbonge), both in or near the lower reaches of the R. Meme; Kottasee=Richardsee (now Lake Kotto) SW of Kumba; N'dian (the village Ndian in Korup on the Ndian River which becomes the Rio del Rey) and Ekundu. I cannot place his site Wewoka. Most notable of the records are *Thermochoria picta** (spp. of *equivocata*), *Zygonyx flavicoasta**, *Allorrhizucha preussi*, *Tetrathemis camerunensis**, *Idomacromia proavita*, *Diastatomma tricolor*,

* Described as new to science from material taken in the region.

Acanthagyna bullata, Heliaeschna fuliginosa, Phyllomacromia melania (as *Hylaeschna paludis*), *Tragogomphys aurivillii**, *Phaon Camerunensis* Chlorocypha cyanifrons* (as *glaucifrons**), *C. neptunus**, *C. selysi* (as *camerunensis**), *C. agracilis as grandis**), *Prodasineura vittata, Pseudagrion flavipes**, *Pseudagrion nubicum* (as *Coenagrion glaucum**).

By 1900, the groundwork had been done and a total of 51 presently, recognised species had been recorded for our area, if one includes unpublished material in the NHM Collection. Latter, Forster (1906) described, *Pseudagrion sjoestedti** from Bipindi (Mount Cameroon) and the amphipterygid *Pentaphlebia stahli** from the same mountain (Forster, 1909), Grunberg (1914) described *Chlorocypha hintzi* from Ekona-Bavinga. Leroi (1915) reported on the German Central Africa Expedition. Forster (1916) described *Umma saphirina* and Sjostedt (1917) described *U. puella*, from Kamerun but neither gave precise locations.

After the First World War, Cameroon was split into French and British sectors and governed by mandate. The Province fromed the southern part of British South Cameroon until independence and reunification in 1960-61. During this period, the amount of dragonfly work seemed to slacken. In 1926 Miss M. E. Fountaine collected *Chlorocypha selysi* at Ekona (Gambles, 1975) and in 1931, Miss Steele captured a female *Heliaeschna* in a cave on the frist plateau of Mount Cameroon and this was described but not named by Longfield (1936). It was later named by Gambles (1967) as *H. longfieldae**.

Schmidt (1943) described the perilestid *Nubiolestes diotima** (from Essossung in the Bakossi Mountains, on the north flank of Kupe), as *Eolestes diotima*. In 1951, he added *Chlorocnemis contraria** from Ekona-Bavinga on Mount Cameroon, although it had been taken in 1910. Longfield (1951) described *Microgomphus camerunensis** from a female taken near Kumba.

Fraser (1956) reported on the Danish expedition to Cameroon.

In 1958, just prior to independence of these regions, the leading expert on African Odonata, Elliot Pinhey, and T. Coffin-Grey made a Land Rover journey from Bulawayo, via the Belgian Congo and the French Equatorial territories and Cameroon to Ikom

* Described as new to science from material taken in the region.

in Nigeria. Odonata were collected at many sites on the route and these papers, although they contain errors, must rank as some of the most exciting in the literature (Pinhey, 1961a, 1961b). In the Province, the region around Mamfe was surveyed, and the Gorilla Mountain (Nta Ali) was visited. Widdicombe is just on the border between the South-West and North-West Province; the village is actually in the latter but I have included the records in Table 2.5.

He records *Chlorocypha greyi** and its synonym *C. ntaali** (borth later synonymised with *A. lacuselephntum*), *C. rufitibia** *Pentaphlebia stahli, Phaon camerunensis, Umma purpurea, Tragogomphus mamfei*, Idomacromia proavita, Phyllomacromia aeneothorax, Phyllomacromia funicularia, Neodythemis gorillae*, Orthetrum angustriventre, Trithemis grouti*, Stenocnemis pachystigma, Neurolestes trinervis, Pseudagrion superbum* (syn. of *serrulatum*).

Pinhey (1962a) lists *Sapho bicolor* (as superba) and *Acanthagyna cylindrata* from Mamfe. In Pinhey (1971), he mostly details records from Fernando Po but adds records of five species from Malende, Koto Barombi See, Mueli, and Diebo-Efote in the Province, including the elusive *Thermochoria equivocata picta* from Malende.

Odonates collected by W. Hartwig are recorded by Pinhey (1974) in a report on the Eisentraut Expedition to Cameroon. Sites visited were: Nyasoso (Lager 1), Manenguba Lake, Dikume-Balue in the Kumpi Hills and from Victoria (now Limbe). A total of 27 species was listed. Most notable were *Neurolestes trinervis* (Nyasoso). *Pentaphlebia stahli* (Nyasoso, Dikume-Balue), *Onychogomphus* sp., (?*styx* or? *supinus*) (Nyasoso), *Phyllomacromia aeneothorax* (as*camerunica* n. sp.) (Nyasoso, Dikume-Balue), *Chloronemis eisentrauti** (Dikume-Balue), *Enllagma buchholzi* (Manenguba), *E. vansomereni* (Manenguba), *Umma saphirina* (Dikume-Balue), *Heliaeschna cynthiae* (Dikume-Balue), *Neodythemis africana* (Dikume-Balue), *Orthetrum camerunense* (Manenguba), *Zygonyx regisalberti* (Dikume-Balue) and *Trithemis furva* (Manenguba).

Robert Gambles was resident at Von, Nigeria in the 1950s and 1960s. He spent a considerable time working on the fauna of the region and he was preparing a text on the Dragonflies of Nigeria

* Described as new to science from material taken in the region.

to include the anglophone regions of Cameroon. Sadly, this was never completed, but he gave me a draft copy in 1980 and I have found it to be invaluable. His many contributions to the taxonomy of the region have been seminal and he made two trips into the province during his stay in Nigeria : the frist in December and January 1957/58; the second in October 1962. In 1959, he described *Orthetrum camerunense* from the grassfields of Bemenda, now North-West Province and 20 km from the border of our region, but later found in the Province (see above). On his 1962 visit to the province he collected *Chlorocypha glauca, Cyanothemis simpsoni* and *Diastatomma tricolor* from the Mamfe area, and *Pseudagrion sjoestedti* (Mamfe and sites 30 mi E and W) (Gambles, 1975). He took *Chlorocnemis flavipennis* (=nubilipennis) on Gorilla Mountain (18 October, 1962). In 1967, he studied the type specimens of *C. nigripes* and *C. contraria,* both having their type localities on Mount Cameroon (=Mongo-maLobah), and clarified their status. He recorded *nigripes* from Dechang, near Mamfe. He revised *Phyllomacromia* of the *picta* and *sophia* groups (Gambles, 1979) and added more sites, but not new species to the list, for the province. He clarified the status of *P. aeneothorax* as the correct for *P. selysi* (Kirby, 1900), *P. lieftincki* (Fraser, 1954) and *P. camerunica* (Pinhey, 1974); and *P. melania* as the correct name for *P. contumax* (Selys, 1879), *P. biflava* (Martin, 1906), and *Hylaeschna paludis* Sjostedt, 1899. The Gambles' Collection in NHM contains many unpublished records and I have extracted these data from the specimens and incorporated them into Table 2.5.

Interesting work was done on the crater lakes of Barombi Mbo, Mboandong, Lake Kotto, Lake Soden and Lake Debundsha in the 1970s by a group whose focus was primarily ecological and ichthyological, but dragonflies were recorded. Green (1972) provides an introduction to the ecology of the first four lakes.

Dragonfly records are given for Lake Kotto and Mboandong (Corbet *et al.*, 1973) and for the crater lake at the Cape Debundsha (*Green et al.*, 1974). Corbet (1977) describes a collection of gomphids from Lake Kotto and Barombi Mbo. There is material of *Acanthagyna sextans* from Barombi Kotto in the NHM collection taken by Sarah Corbet.

Vick (1996) reported on studies made in the areas of Mount Cameron and Mount Kupe in 1996 and gives a description of the habitat of Mount Kupe as well as a checklist of 72 species recorded in the 1995 expedition. *Umma mesumbei** was described from Kupe. As this visit was in effect the first CDP survey, I have included the records again in this paper (i.e. not as 'historical' records).

The larvae of *Nubiolests diotima* and *Stenocnemis pachystigma* (assumption) were described in Vick (1998), notes were also provided on the larva of *Pentaphlebia stahli* in the same article.

D'Andrea and Carfi (1997) records 35 species from four sites in the province in addition to making a number of valuable records and observations. These are shown in a separate column in Table 2.5. The sites of importance here are Barombi Mbo and three sites near Mount Cameroon : Itaka (400m), Malende (300m) and Munyenge (400m). The most significant records are : *Umma puella* (Munyenge), *Chlorocypha neptunus* (Barombi Mbo, Ikata), *C. rubida* (Ikata), *C. victoriae* (Barombi Mbo), *Ellatoneura balli* (Munyenge), *Mesocnemis singularis* (Ikata, Barombi Mbo), *Pseudagrion angelicum* (Ikata, Barombi Mbo), *Heliaeschna cynthiae* (Malende) *Orthetrum austeni* (Malende), *O. kalai* (as brachiale) (Malende, Barombi Mbo), *Oxythemis phoenicosceles* (Ikata), *Tetrathemis bifida*(Ikata, Malende, Munyenge), *T. grouti* (Malende), *Zygonyx fallax* (Malende), *Z. torrida* (Munyenge).

Methods

The first visit was made in March and April of 1995. The staff at the Birdlife Internation (ICBP) project at Mount Kupe were keen to instigate some odonate survey work and my wife and I decided to concentrate on this area. We stayed in the village of Nyasoso with a local family. Two local guides, Otto Mesumbe and Elvis Njume, were made available and we spent much of the time teaching them about odonata : basic biology and life histories, fieldcraft and specimen processing.

The scientific coordinator, O'kah Ebwekoh also helped with habitat information and Ebong Harrison provided us with a detailed map of the stream system on the Nyasoso-side of the mountain.

---

* Described as new to science from material taken in the region.

It is extremely difficult to record adequately the dragonfly fauna of a tropical forest. Many species are easily recorded and a strong sunny spell often brings adults of these species to the stream. However, it is interesting to note how infrequently some species have been seen over the three years of intensive recording at Mount Kupe. For example, two females of *Aeshna scotias* were taken on Shrike Trial by M.C. Vick in 1995 while flying in heavy shade, looking for oviposition sites; the species has not been recorded since then at the time of writing (August 1998), although larvae were found in 1997. *Anax chloromelas* has been found only once in the forest. One the seepage at Max's Trail, a female of *Idomacromia proavita* was caught in March, 1997; again this is the only adult that has been seen, although larvae have been found. A single male of the montane forest species *Trithemis pruinata* has been taken in three years. We suspect that many of these species spend most of their time in the tree canopy and the females only come to water to oviposit. Males may wait for females at a height too great for the human observer to see them. Also, the lack of seasonality undoubtedly makes the numbers of individuals per species at any given time fairly low. How many species have not been seen at all?

In order to overcome some of these problems we investigated two additional modes of recording: these are not unusual in temperate regions, but appear to have only rarely been carried out in a tropical country owing to the difficulty of continuous manning. Fortunately, the ability and enthusiasm of Otto Mesumbe made him an ideal person to do this. Larvae have been collected using metal sieves and reared in buckets by Otto and his helpers in the village. The local farm stores have provided us with many suitable sieves, called 'shifters' in Pidgin English. This work has enabled us to make association between larvae and the adults which have emerged. Both adult and exuviae have been preserved together in 70% ethanol. We are now able to record the presence of certain species form larval collection. Another innovation was the use of the emergence trap on the seepage at Max's Trail. We used a large rectangle of cheap curtain netting (approximately 5m × 5m) which was suspended over the seepage. A supporting framework was constructed using available branches and the netting was held down at ground level using large rocks. This has been so successful that other traps are now planned in the lowlands, where it will be possible to target small streams in dense forest.

In 1995, we decided that this would be a long-term project and we set up the Cameroon Dragonfly Project; this was formalised in 1996 with Professor Philip Corbet as president, Otto Mesumbe as Cameroon coordinator, David Chelmick as treasurer, and myself as secretary. Our aim is to further knowledge about dragonflies in Cameroon in order to their conservation. We have decided to concentrate our efforts in the South-West Province initially. Our objectives are to produced a species list for the region : to publish a key to the adults; to identify areas of greatest conservation importance : to describe as many larvae as possible and to write larval keys, at least to restricted areas where we know the fauna well. Perhaps most important of all is to gain the interest of local people. This contribution is the first of these objectives; the second is now in draft awaiting illustrations; other objectives are in progress.

A Magellan 2000 global positioning system (GPS) device was used in the 1997 survey: latitude and longitude is therefore quoted for many sites visited in 1997 and this is given to the nearest second (accurate to about 1/60 of a nautical mile of about 30 m in each direction). The data from the Takamanda Survey conducted by C. Wild of World Wildlife Fund were also obtained using a GPS device and these positions ar quoted to the nearest minute (accurate to about I nautical mile in each direction).

Description of Sites Surveyed 1995-1998

Sites (8)-(9) are on the eastern flanks of Mount Kupe and located in Littoral Province, although on the border with South-West Province; the rest are all in South-West Province. The numbering is somewhat irregular owing to the need to add sites to the paper while it was being prepared.

Mount Kupe and adjacent premontane lowland (1995-1998)

1. Nyasoso, Mount Kupe
(a) Water catchment and Shrike Trail 920m, GPS (04°49'18" N, 09°41'29" E)

(b) Max's Trail and seepage 1000 m

(c) Nature Trail 980m

Primary forest with montane streams, fast flow, rocky bed, locally sandy with coarse silt, heavy shade. Seepage on Max's Trail with slow flow, pebbly bed with fine silt, heavy shade.

2. Nyasoso village, GPS 04°49′45″ N, 09°41′13″E

5. Nsuke - R. Muawane, R. Kengele 450m. farmbush and secondary forest.

6. Ndom - Ebul Kack 850m, open farmbush.

8. Lala - R. Ndibe 400m, lowlands stream, nutrient rich, sand and silty bed, in open farmbush GPS 04°46′48″, 09°45′33″E.

9. Nlohe - Nyelle Stream 400m, open farmbush.

10. Tombel-R. Peng 400m, open farmbush.

12. Ngusi - R. Tubere 300m.

Secondary forest with farmbush, stream with moderate, flow, sandy bed, locally with sand and mud, moderately nutrient rich, GPS 04°51′16″N, 09°3857″E.

19. Atop - Taw Stream 250m, farmbush and secondary forest.

20. Ndom Essosong - Nweh Stream 850m, farmbush and young secondary forest.

22. Mbabe 300m, rocky and sandy stream, farmbush/ secondary forest.

Bakossi Mountains (1995-1998)

3. Kodmin (=Kumin) village,1450m, muddy pool in otherwise fast clear rocky stream, open to sun on village edge in farmbush, grassland, forest mosaic, GPS 04°59′N, 09°42E.

4. Ngombo (=Mgombo-Aku), R. Nkincho, R. Mbombe 800m

Fast, rocky stream with gravel and sandy substrate, some mud locally, GPS 04°55′30″N, 09°42′42″E, farmbush and secondary forest.

7. Mbomekoged - Ngwese Stream 200m and sandy stream.
13. Edib village - Bakossi Mountains, 1100m, large fast rocky stream in primary forest.
14. Bangem - R. Mbwe, R. Mukwene, 1150m, small rivers in primary and secondary forest.
39. Baseng to Edib road.
40. Messaka
41. Molongo
42. Muambong (1k, west o), 1100m, Jide River, large fst rocky river and its tributaries, primary forest, GPS04°58′N, 09°43E.

Bakossi Lowlands and Upper Mungo River (1995-1997)

11. Ebonji - Sandwater Falls, Blackbush Water, Ekom River 300m.

    Secondary forest, stream moderate to fast flow, sandy bed, some parts rocky with waterfalls
15. Etam : Camp Water
16. Etm - First stream W of Mungo River, GPS 04°43′48″N, 09°33′45″E
17. Etam - second stream W of Mungo River, GPS 04°42′43″N, 09°32′36″E

    Sites (15)-(17)-primary and secondary forest with farmbush in parts : stream with slow flow and sandy beds, with silt and mud locally, all about 300m

Kumba-Mungo River (1998)

18. Kumba - Barombi Mbo 380m

    Crater lake, deep, surrounded by dense primary forest GPS 04°38′57″N, 09°24′43″E
21. Kumba-Barombi Mbo outlet stream 300m

Mount Cameroon Premontane Coastal Lowlands (1995-1997)

23. Botake stream descending from Little Mount Cameroon 50m, GPS 04°01′47″N, 09°05′49″E

24. Bimbia - lowland forest reserve near Limbe 50m
26. Debundsha - marsh at Njonji 2m

    Lowland marsh, almost dry in dry season, no shade, surrounded by farm bush and oil-palm but primary forest in vicinity on slopes of Mount Cameroon

    GPS, 04°05′03″N, 08°59′45″E
27. Debundsha Crater Lake - 50m surrounded by primary forest, GPS 04°05′48″N, 08°58′43″E
28. Limbe Botanical Gardens 10m, open habitat, fast stream
29. Mutengene - Catholic Mission

Korup National Park (29-31 January, 1997)

30. Mundemba - stream near Park HQ and stream at Iriba Inene Camp

Takamenda Forest Reserve (720km^2) - north of Mamfe (all 11 December, 1997 to 22 February, 1998)

31. Assam - 140m GPS 06°01′N, 09°18′E
32. Takpe - 500m, GPS 06°02′N, 09°21′E
33. Mfakwe - 200m, GPS 06°04′N, 09°26′E
34. Kekpane - stream Manita, GPS 06°06′N, 09°24′E
35. Obony III - stream Meyingepung, 115m, GPS 06°08′N, 09°18′E
36. Takamanda - village 568m, GPS 06°02′N,09°17′E
37. Bahe - 200m, GPS 05°57′N. 09°18′E (just south of Reserve border)
38. Matene - 'montane', GPS 06° 17′N, 09°22′E

Published Historical Records (prior to 1980)

Mount Cameroon Premontane Coastal Lowlands

60. Mount Cameroon=Mongo Ma Loboh (Selys, 1879, 1886; Karsch, 1892, 1893; McLachlan, 1896; Forster, 1906, 1909; Longfield, 1936; Schmisdt,1951)
61. Lake-Debundsha (Green *et al.*, 1974)

62. Kitta (now Kita?) (Sjostedt, 1899)
63. Bonge (now Mbonge) (Sjostedt, 1899)
64. Ekundu (Sjostedt, 1899)
65. N'dian (now in Korup NP, on N'dian River (Sjostedt, 1899)
67. Wewoka (Cannot place) (Sjostedt, 1899)
68. Itoki (possibly=Itoki at 5°01'N, 09°15'E) (Sjostedt, 1899)
69. Mueli 550m (Pinhey, 1971)
70. Diebo-Efoye 130m (Pinhey, 1971)
71. Malende 125m (Pinhey, 1971)

Kumba and District

72. Kumba=Johann-Albrechthohe: Barombi Mbo= Elephantsee 400m (Karsch, 1899; Corbet, 1977; Gambles, 1979; D'Andrea and Carfi, 1997)
73. Kumba : Barombi Station (Karsch, 1891b; Longfield, 1951)
74. Mukonje Farm (between Kumba and Nguti)(RIS, 1909-1919)
75. Lake Kotto-Richardsee 110m (SW of Kumba) (Sjostedt, 1899; Pinhey, 1971)
76. Lake Kotto (Tung Nusia, Tung Nsuria inlet streams) 110m (Corbet *et al.*, 1973; Corbet, 1977)
77. Mboandong 130m (Corbet *et al.*, 1973)

Bakossi Mountains including Mount Kupe and Adjacent Premontane Lowland

77a. Bakossi Mountains - Essossong 1060 or 1600m Carl Rathke leg, Specimen in Hamburg Museum (Schmidt, 1943)
86. Nyasoso (Lager 1) 900m (Pinhey, 1974)
87. Manenguba Lake 1900m, open grassland and scrub (Pinhey, 1974)
88. Dikume-Balue (Rumpi Hills) 110m, secondary forest (Pinhey, 1974 (86-88 are Eisentraut Expedition records)

Mamfe and District

78. Nta Ali=Gorilla Mountain, summit=1266m, primary forest (Pinhey, 1961a,b; 1962a)
79. Mamfe (Pinhey, 1961a, 1961b; 1962a)
80. Mamfe (30mi E of) (Pinhey,1961a, 1961b; 1962a)
81. Mamfe (30 mi Wof) (Gamble, 1975)
82. Mamfe (40-50 mi E of (Pinhey, 1961a, 1961b; 1962a)
83. Widdicombe (Mamfe Road)-on border of SWP and NWP-possibly in latter (Pinhey, 1961a, 1961b; 1962a)
84. Ajassor, (Pinhey, 1961a, 1961b, 1962a)
85. Dechag, 5 mi from Mamfe (Gambles, 1967)

Article Published Since 1980

Vick (1996): all records included under heading of Sites Surveyed 1995-1998 (CDP records) in this article D'Andrea and Cafri (1997) : from vicinity of Mount Cameroon, records of collections made in 1979 :

89. Ikata 400m
90. Malende 300m
91. Munyenge 400m
92. Barombi Mbo 400 m

Unpublished Record

The following sites are represented by material in the collections of the NHM (London)

93. 'Camaroons' (McLachlan Collection)
94. 'Camaroons River (McLachlan Collection)
95. Isubu, Cameroons (paratype of *Phyllogomphus montanus*, collector unknown)
96. Mongo-Ma-Lobah (McLachlan Collection)
97. Port Victoria (=Limbe)
98. Old Tal. 'Camaron (McLachlan collection) (*Acanthagyna africana* - na data)

99. Kumba - without further details

14 October, 1949, H. Oldroyd leg., *Tetrathemis camerunensis, Spaho orichalcea*

23 October, 1962, R.M. Gambles leg., *Cyanothemis simpsoni*

20-24 March, 1970, S.A. Corbet leg., *Chlorochypha gracilis*

10 April, 1970, no collector, *Acanthagyna sextans*

22 December, 1973, G. Popov leg., *Palpopleura lucia*

100. Barombi Kotto

Tholymis tillarga 28 October, 1963, C.A. Wright leg.

Acanthagyna sextans 11 April, 1970, S.A. Corbet leg.

101. Korup National Park-R. Mana and The Rock

2-3 February, 1989, G. Spiers leg.

102. Kunde Bassum -3 May, 1914

103. Bipindi (F.C. Fraser bequest) (*Chlorocypha neptunus* - no data)

104. Mamfe and district (within 40 miles of Mamfe), 24 December, 1957-5 April, 1958; 6-19 October, 1962; R.M. Gambles leg. and det.

105. Nta - Ali=Gorilla Mountain, Bakabe, 18 October, 1962, R.M. Gambles leg. and det.

106. Widdicombe (Mamfe road)-borders of South-West and North-West Province, 31 December, 1957-4 January, 1958; 9-12 October 1962, R.M. Gambles and leg and det.

107. Manyemen (south of Nguti on Kumba road) - no date (1958 or 1962)

(*Chlorocnemis contraria*) R.M. Gambles leg. and det.

Etymology : Named after Dr. Sarah Corbet who took the first male speciemen in 1970 (described as *P. montanus*

Male (Holotype) adult : Colour preservation fairly good, acetone used but drying difficult in high humidity. Difficult to distinguish between yellow and green markings.

Head : Width across eyes 11.0μm. Labium (median and lateral lobes) brown, extensively marked with lack on apical and lateral margins. Mandibles brown, bases greenish yellow, shading to brown basally. Labrum (Fig. 2.4a) black with single median yellow marking occupying about 50% of area. Anteclypeus green. Postclypeus brown with one central and two lateral green spots. Frons brown with greenish yellow strips on ridge, extending over anterior and dorsal surfaces to occupy about 50% of each. Vertex and epicranium black. Antennal bases black and antennae brown tipped with yellow. Occiput black bearing dense black hair fringe, Rear of head black.

Thorax : Prothorax black. Synthorax black marked with well-defined greenish yellow markings on each side as follows. Dorsun (mesepisternum) with antehumeral stripes confluent with separated stripes on mesothoracic collar, forming pair of of inverted 7s. spot on lower part of carina. Distinct dorsal mesepisternal spot. Side of thorax with two broad (about 1.6mm wide) subparallel bands, one of fairly constant width on mesepimerum, the other widening centrally on metepimerum. Dorsal spot (about 0.6mm circular) on metepisternum adjacent to the second lateral suture. spot on metasternum. Ventral surfaces on thorax brown. Antealar triangles black and axilaries of wings yellow.

Legs : Coxae brown. Trochanters brown. Femora, tibiae and tarsi entirely black.

Wings : Forewings clear. Hindwings with tawny suffusion, strongest between base and nodus. Veins black. Nodal index 14.19.19.14/16.14.13.16. First and fifth Ax strengthened in all wings. Prefurcal crossveins between sectors of arculus 2.2/1.1. Discoidal field commenting 3-celled in forewing, 3 or 4-celled in hindwing. Mspl distinct in all wings, arising 2-3 cells proximal to level of nodus. Anal triangle 3-celled. Anal loop 4 or 5-celled and 5 cells between loop and posterior margin. Pt black surmounting 4-5 cells. Membranula full length, very narrow, white.

Abdomen : Tergites 1-2 reddish brown, 3-7 black and 8-10 reddish brown, marked as follows. Terfite 2 with ventral yellow band extending the full length of the segment, a green spot above the auricle and a minute dorsal streak at base of tergite. Tergite 3 with ventral band forming a continuation of that on tergite 2, exending to middle of segment; dorsum with paired yellow longitudinal streaks on dorsum, extending over the basal fifth of

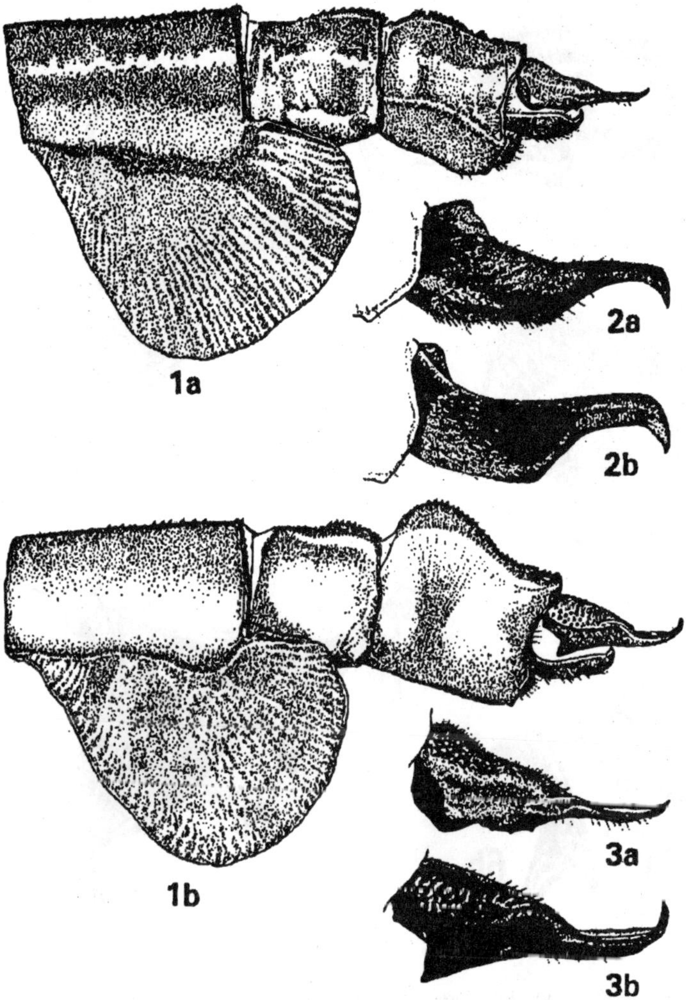

Figs. 2.1-2.3 : *Phyllogomphoides corbetae* 2.1 Abdominal segments
2.2 Superior appendages (dorsal view),
2.3 Superior appendages (lateral view)

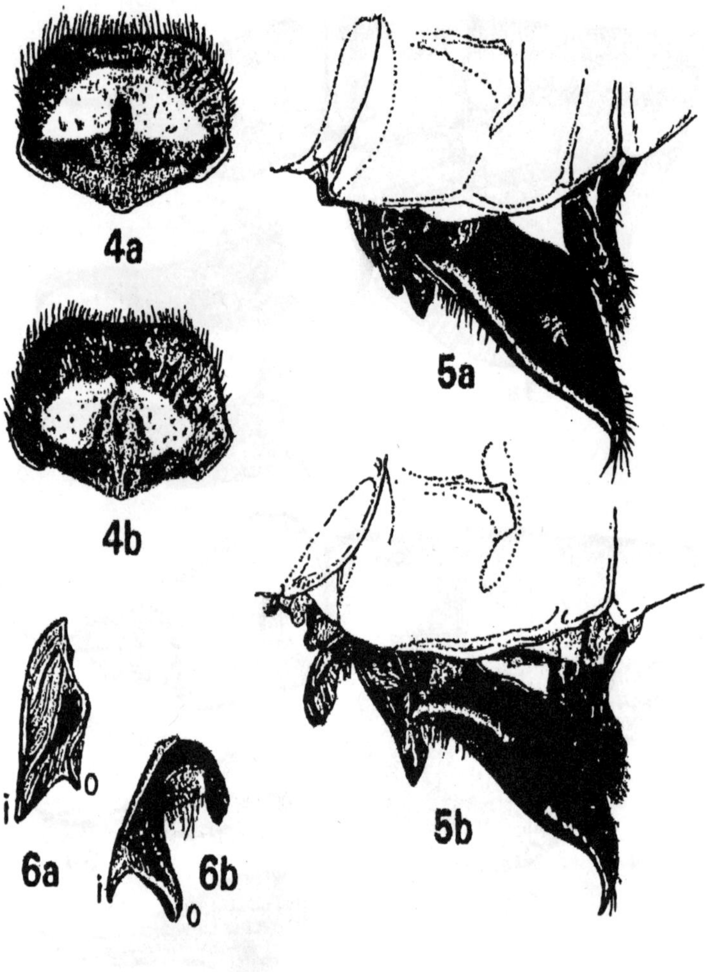

Figs. 2.4-2.6 : (a) *Phyllogomphoides corbetae* and (b) *P. montanous*, 2.4 labrum; 2.5 accesory genitalia, lateral view; 2.6 apex of anterior hamule (i = inner branch, o = outer branch).

the segment. Tergite 7 with large basal yellow spot on dorsum extending over basal quarter of segment, the spot consisting of a larger basal spot joined by a narrow neck to a smaller apical spot; ventrally with narrow longitudinal yellow stripes adjacent to sternite. Tergites 8-10 without markings.

Lateral foliations on segment 8 massive, reaching to end of tergite 9, coloured deep black. Length of tergites 8, 9, 10 as measured on dorsum in ratio 1.6.0 : 1.0. Weak lateral ridge on tergite 8. Dorsal ridges absent on 8; very small, and in apical half of, 9; very small, and mostly in basal half, of 10 (cf. extensive ridges of *montanus*).

Appendages : Superiors brown basally and darker apically. In lateral view (Fig. 2.3a) with basal swelling forming a ventral ridge which bears a weak basal tooth. In dorsal view, apical half very narrow and strongly recurved (Fig. 2.2a). Inferior visible in dorsal view, widely divaricate, about half length of superiors.

Accessory genitalia (See Fig. 2.5a) : Note anterior hamule (Fig. 6a) with wide inner branch whichis much thicker and longer that the outer branch (cf. *montanus* which has outer branch larger).

Measurements (in mm) : Abdomen with appendages 64 mm, hindwing 46 mm.

Female (Allotype), adult : Colour preservation very good, acetone used, dried in hot air stream. Extremely similar in all markings to holotype, except that all pale head marking are clearly seen to be green, and all thoracic and abdominal markings are yellow; these are likely to be true living colours of the male also.

Head : Width (across eyes) 11.5 mm. Occipital ridge at rear of head very characteristic (Fig. 2.9a), bearing two spines and sparse black hairs, especially between the spines. This is very distinct from *montanus* which lacks the spines and has a uniform mantle of long black spines (Fig. 2.9b).

Wings : All with extensive tawny suffusion tawny suffusion. Veins black. Nodal index 15.19.18.16/16.13.14.17. Strengthened A x 1st and 5th (4th in one wing). Prefurcal cross veins between sectors of arculus 2.2/1.1 Discoidal field starting with 3 rows in all wings. Mspl beginning 3-4 cells proximal to the level of the nodus. Anal loop 4-celled, 5-6 cells between loop and posterior margin. Pt black, surmounting 4-5 cells. Membranule very narrow, white, 4 cells long.

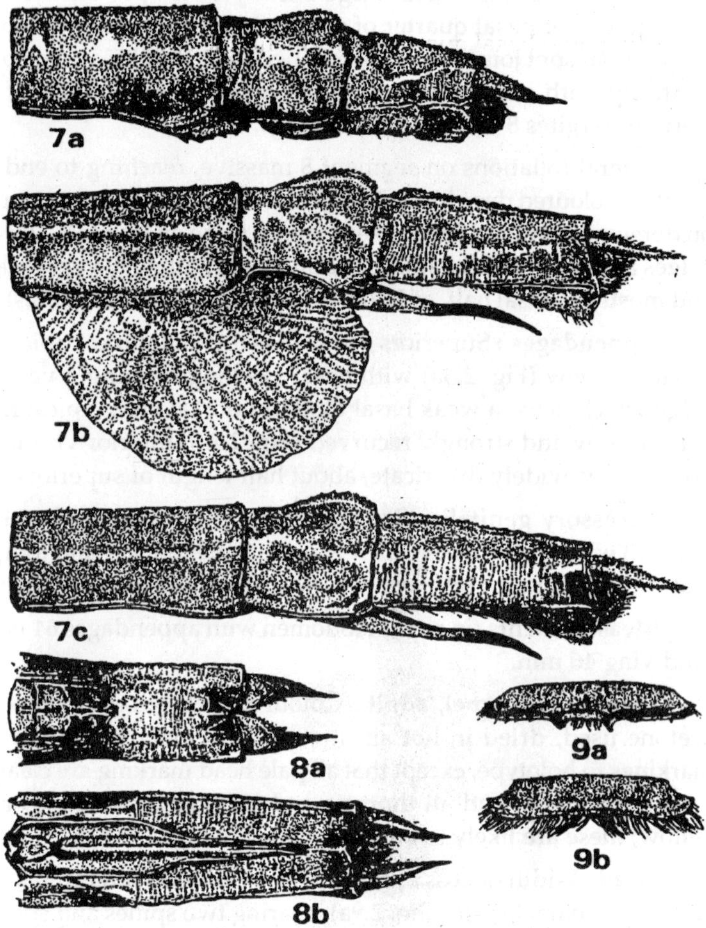

Figs. 2.7-2.9 : (a) *Phyllogomphoides corbetae* (b) *P. montanous*, (with large abdominal foliations) (c) *P. montanus* (with vestigal foliations); 2.7 Abdominal segments 8-10, lateral view; 2.8 Abdominal segemnts 9-10, ventral view 2.9 Occipital ridge, caudial view.

Abdomen : Lenghts of tergites 8, 9, 10 in ratio 1.5:1.0:1.2 Only vestigial foliation on tergite 8, overlapping slightly into 9 (true of paratype also, but presumably females exist with full-size foliations. Dorsal ridges absent on tergite, 8, present in apical half of 9 and scarcely detectable on 10 (Fig. 2.7a).

Appendages : Pointed, conical, pale with black apices, about half the length of segment 10. Vulvar scale short, scarcely overlapping onto tergite 10 (Fig. 2.8a).

Measurements (in mm) : Abdomen with appendages 60.5. hindwing 50.

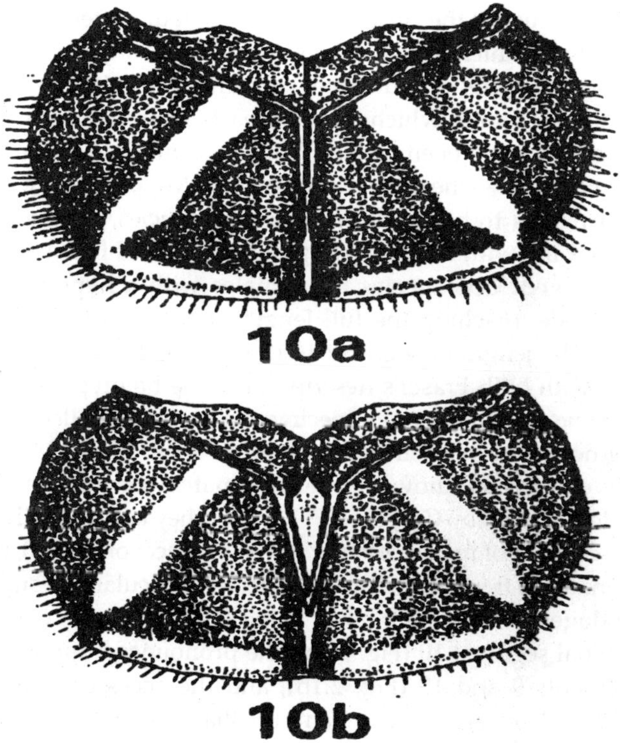

Fig. 2.10 : (a) *Phyllogomphoides corbetae* (with large abdominal foliations) and (b) *P. montanus* (mesepisternum, dorsal view).

Discussion

This is a very difficult genus to study as the species are so elusive, living in the tree tops and rarely visiting water. Material in collections is scarce and often in bad condition. Also several of the taxa are poorly known and in some cases described from females. I have been fortunate to have reasonably good material which has been collected in forest streams of the Province since 1995 and possess five males and eight females. All are large insects with hindwing lengths of at least 44mm, and they all have three lateral thoracic stripes on each side. They clearly belong to two different species. It has been possible to associate males and females because of the close similarity of colour pattern of thorax (especially the antehumeral stripes), head (especailly labrum), and the relative lengths of abdominal segments 9 and 10. Finally one male and one female which resemble each other in the characters stated above have been collected at the same time at the same stretch of stream—the outlet from Barombi Mbo. Deciding on the current names to be used for these two species, however, has presented me with an interesting puzzle which I believe I have now resolved. One species has females with a hypertrophied vulvar scale, reaching the full length of segment 9 and three-quarters the length of segment 10 (Figs. 2.7b, 2.8b). These agree exactly with both Fraser's description of the holotype female of *P. montanus* and the lectotye specimen in the NHM collection and I have no doubt in determining them as such. Males which are clearly conspecific, however, do not match the description of *P. montanus* given by Corbet (1977). In fact they agree exactly with the description of male *P. hartwigi* given by Buccholz (1958), based upon material from Koto Barombi See. In particular, the superior appendages, with strong basal spine (Fig. 2.3b), the extended abdominal segment 10 (Fig. 2.1b), the pronounced dorsal ridges of segments 9 and 10 (Fig. 2.1b), and the accessory genitalia (Fig. 2.5b) all match perfectly. I am sure that this is indeed the male of *montanus* as quoted in Pinhey (1962b) and I agree with his synonomy.

The other species is very different as the female lacks the hypertrophied vulvar scale of *montanus* : it only just reaches the base of segment 10 (Fig. 2.7a, 2.8a). Male material which I believe also belongs to this species is identical to Corbet's description of material from Lake Kotto which she incorrectly attributed to *montanus.* The superior appendages lack the strong basal tooth, only possessing swelling with a weak tooth at tis site (Fig. 2.3a); the dorsum of segments 9 and 10 is only weakly ridged (Fig. 2.1a) and the accessory genitalia (Fig. 2.5a) match her figures. This is a very distinctive species with differs from all others of the genus and which I call. *P. corbetae* in appreciation of the contribution of Sarah Corbet to the odonatology of Cameroon. I list in Table 2.4 the main characters which separate these two taxa.

The new species is clearly distinct from all other described taxa with respect to the morphology of the superior appendages, abodominal segment 9 and 10 or the accessory genitalia.

An interesting observation is that female specimens received of *montanus* are dimorphic: some have the enormous lateral foliations (Fig. 2.7b) on segment 8 which reach the end of segment 9 and closely resemble those in the male; another only has vestigial foliations (Fig. 2.7c). Of my six females, five have large foliations and one has not. Of *corbetae,* I only have two females and both have vestigial foliations. This dimorphism has also been noted in other taxa of the genus; in the NHM there are two females of *P. dundomajoricus* Fraser 1957 (1 female, Belgian Congo, Ituri Forest, 4000 ft, 21-IV-1930, Howards de Walden Expedition; 1 female N. Angola, Lunda Province Dundo February, 1948) which both lack foliations. Also *P. brunneus* Pinhey, 1976 is known to have females without foliations.

Checklist of Dragonflies from South-West Cameroon

The checklist is summarised in Table 2.5. Column 1 shows all taxa which have been recorded from the area of the South-West Province. Column 2 shows historical records for each species. I have attempted to trace the first records of each taxon from the area and provide useful historical site data. I do not claim that these data are complete, although I have tried to be as inclusive as possible. Citations to 'Kamerun' which are not localised to this region cannot be included. Column 3 shows separately the records of D'Andrea and Cafri (1997) because their data all refer to specimens collected in 1979. Column 4 shows all 'recent records', i.e. the records made by the Cameroon Dragonfly Project which started in March, 1995. These include the data in Vick (1996), which although appearing before D'Andrea and Carfi (1997), refers to more recently collected material. Finally the last three columns show very basic ecological information (where known) for most of those species recorded between 1995 and the present. I have attempted to divide the requirements of each species into broad categories: montane or lowland; primay forest, secondary forest and 'bush' and open country; the type of water body. These appears to me to typify the species requirements using information based mostly on adult collecting. I hope they will be useful for further studies.

The list given comprises 179 species, including a few unnameable specimen which nevertheless belong to addition taxa and should be counted into any total. I have included the two different forms of both *Umma longistigma* and *Sapho orchalcea* as additional species.

Notes on Certain Species

Ceriagrion rubellocerinum – The material has been compared with NHM material from Nigeria determined by R.M. Gambles (e.g.a. male from Vom 2-XI-1959 R.M. Gambles leg.) and another male from Ikoyi 8-IX-1955 standing under this taxon.

Pseudagrion F. Flavipes – This distinctly marked taxon was found on two lowland forest streams near Etam in April 1997. The material agrees well with NHM material from Cameroon (a male 39.5 mi E. of Mamfe, 9-X-1962, R. M. Gambles leg./det.).

Pseudagrion S. stoestedti – Material taken agrees with Cameroon material in the NHM (e.g.a male 20 mi E of Mamfe, 19-X-1962, R.M. Gamble leg/det.).

Pseudagrion risi – The specimens from Ebonji and Bangem agree with NHM material from Cameroon collector and determined by R.M. Gambles (a male from Bambui, 27-XII-1957 and a male from Bamberi 27-XII-1956).

Platycnemis congolensis – I have compared my material with NHM material from Uganda (Kamengo Forest - van Someren, June,

1953). The male inferior appendage is longer than the superior, and the latter bears a long ventral tooth.

Sapho. – *S. orichalcea* is the dominant calopterygid on rocky montane streams in the Province, provided there is some surviving riparian forest cover. *S. bicolor* should still occur in the Mamfe region where Pinehy had found it. To the south of the Province, *S. orichalcea* is replaced by *S. gloriosa*, e.g. a specimen in NHM from Batouri District (4°30'N, 14°15'E) taken in 1935 by the 'gorilla collector', F.G. Merfield.

In the lowlands around Etam, we have found very small specimens of *Sapho* which appear to belong to *orichalcea* (hindwing 33 mm compared with an average of 44mm for normal *orichalcea*). I have listed them separately in the checklist as they may be specifically distinct.

Umma – The presence of six species of this genus is impressive and the South-West Province is an important centre of diversity for these forest-adapted calopterygids. The newly-described *U. mesumbei* (Vick, 1996) is a very distinctive species of large size which has only so far been found on Mount Kupe and at Kodmin in the Bakossi Mountains. It is possibly endemic to the region. *U. mesostigma* and *U. longistigma* appear to be occur more generally in Bakossiland, especially in the lowlands. In the Takamanda region, north of Mamfe, we have recorded four species : *longistigma, mesostigma,* and apparently very local *puella* and *purporea*. The sixth species, *saphirina* has only found so far from the area of the Rumpi Hills near the Nigerian border. There appear to be two different forms of *U. longistima*, based on the shape of the inferior appendage: the narrower forms occurring in Korup and north of Mamfe.

Diastatomma multilineata – A male taken at Atop (26-XI-1997, near Etam) belongs to this taxon and it agrees with a Fraser male from Eala (Belgian Congo) in the NHM Collection in markings, superior appendages (with basal section and apical projection making 100°) and shape of Hagen's tubercle. GAM-BLES (1987) published an interesting paper on this genus and its distinction from *Gomphidia*. The species of *Diastatomma* possess a 'Hagen's tubercle, the shape of which is constant and specifically diagnostic.' However, another male which is almost certainly conspecific, having similar marking, accessory genitalia and appendages, from the same site, has a much shorter tubercle. It appears that the shape of the tubercle is not an obsolutely reliable character for identification.

Diastatomma tricolor – This species appears to be fairly common in the forested lowlands around Etam. Specimens concord with material in NHM from Calabar Road, South Cameroons, (R.M. Gambles leg. 17-X-1962).

Gomphidia gamblesi – My material is clearly conspecific with Nigerian specimens in the NHM standing under the manuscript name *G. nigeriensis* Gambles. These specimens include 2 males from Sobo plain, near Sapele (25/30-XI-1957, B.J. MacNulty leg., R.M. Gambles det. 1973) and one male (labelled 'type') from Sapoba Forest Reserve, nr Benin (10/12-III-1972, J.T. Medler leg., R.M. Gambles det.). The names was never published. Gauthier (1987) described the same species from Togo as *G. gamblesi*. The appendages and accessory genitalia are identical to the Nigerian and Cameroonian material, but the Togo material appears to be smaller and there are a few minor wing venational differences. There is actually some variation in anal appendage morphology and specimens from Takamanda differ slightly from more southern material (e.g. Kumba are).

Notogomphus sp. 1, sp. 2, sp. 3. – There are three species of *Notogomphus* which are characterised by the very long hind legs (reaching to middle of second abdominal segment), bearing long spines. Two have been taken at Kupe, and are represented by single females: the larger species (sp.1) in the lowlands at Lala; the smaller species (sp. 2) in the submontane forest above Nyasoso. Species 2 seems to be closest to *dendrohyrax* (Forster, 1906) from the Usambaras of Tanzania: the vulvar scale is close; the markings are close but there are differences in the size of the Pt and the leg spination. Species 3 is a teneral male from Takamanda, softened in KOH, preserved in alcohol but I have not been able to identify it.

Onychogomphus sp. 1 sp. 2 – A female (sp. 1) from Mount Kupe is similar to *O. supinus*, with similar markings, but the abdomen widens slightly; there is a similar very dark Pt; the antehumerals are clearly detached, unlike *O. septemflavum* from Congo to which it is otherwise close. There is also another female of this genus from Kupe which appears to be distinct (sp. 2).

Paragomphus Abnormis – I discussed the difficulties of deciding whether material collected at and near Mount Kupe belonged to taxon or to the later taxon. *P. moka* Longfield, in my earlier paper (Vick, 1996). In the past three years I have received a very large series of specimens (50+) and I find that all of the characters described by Cammaerts (1969) to separate the two taxa intergrade or break down. However, the type of *abnormis,* and the single specimen used by Cammaerts, both possess a basal subcostal nervure, which is not present in *moka,* nor any of my specimans, I do feel this character is sufficient to separate the two taxa and therefore regard my material as *abnormis.*

Paragomphus nigroviridis – I have material from lowland forest streams at Etam which agrees well with Cammaerts' description. It is a dark species, with antehumeral strips joined to the prothoracic collar in a pair 7s. Kimmins has determined a male in the NHM from the Belgian Congo (Bambesa, February 1958) as this species and my male agrees with that one.

Paragomphus sp. 1, sp. 2, sp. 3 – A number of gomphid species have been obtained which cannot be determined as they are represented only by females; in this genus even males can be very difficult. Sp. 1 is a female from the forest on Mount Kupe; sp. 2 is male from Etam; and sp. 3 is a female from Ngombo-Mbong.

Anax chloromelas – This taxon is normally considered to be rare. At the lowland marsh near Debundsha, three females were captured in July 1996 flying rapidly over the water, which was about 60 cm deep at this time of the year. A larva (which later emerged in the UK) was also colleceted. The *chloromelas* were flying with *A. tristis* which was much commoner. On 27 March, 1997, at the start of the rains, we returned to the marsh to find it almost entirely dry and no larvae could be found. Males of *A. chloromelas* were flying rapidly over the few pools of water that had formed. Other less common species which has been found at this interesting site are *Orthetrum austeni* and *Trithemis hartwigi.* The waterbody is temporary and some parts are inundated fields, partly planted with crops. The larva must grow rapidly during the wet season so that it can emerge before the desiccation of the habitat and the dry season must be passed an an adult in the forest. The presence of the lowland rainforest on the lower slopes of Mount Cameroon in the vicinity of the marsh may be an important factor which is responsible for the richness of this habitat. To some extent this is supported by the record of another specimen, a male, which was taken by O.Mesumbe in the forest at Mount Kupe in the dry season on 17 February, 1997. There do not appear to be suitable habitats for the species on the mountain and I assume that it breeds in an undiscovered marsh in the nearby lowlands.

Heliaeschna fuliginosa – A male which was taken above a shaded forest pool at Etam on 5 April, 1997 agrees well with NHM material from Nigeria (det C. Longfield) and Congo, Eta (det. F.C. Fraser). In particular, the pattern of denticles around the genital fossa distinguishes the taxon from the otherwise similar *H. lanceolata.* The anal appendages also match well, but there is a trace of black T-marking of the frons. We also have a female of *lanceolata* with strongly-marked T on the frons from montane forest in Takamanda (27 January, 1998).

Heliaeschna trinervulata – I am unable to place for certain a female which was taken on 29 December, 1995 Mount Kupe. It is medium-sized species but agian there is a trace of a black T-mark on the frons, which *trinervulata* should not have. The size and shape of the pterostigma, the anal loop and other details of venation suggest *trinervulata;* this resemblance is strengthened by the shape and smaller size of the two-pronged process in my specimen which differs from that in *fuliginosa, cynthiae* and *libysns.* Perhaps the trace of the black T on the frons is due to *post-mortem* decomposition?

Idomacromia proavita – The two species of this genus are specialised forest insects and are probably under-recorded rather than being rare. In April 1997, we observed a female of *proavita* ovipositing in a damp seepage in upland rainforest on Kupe at an altitude in axcess of 1200m. This seepage had a gradient of about I in 20 and a width of 2m. The substrate was fine gravel and silt. Water depth varied from 0 to 6mm and the mid-day temperature was 20°C. It was situated in heavy shade of the submontane forest and only shafts of sunlight penetrated through the canopy. Later, a male, flying in a most erratic manner, was seen patrolling a small shaded stream, in a lowland forest near Etam, at an altitude of approximately 300m. The stream varied from 60 to 120cm in width and about 30cm deep; the water temperature was 26°C at midday. The bed was sandy with some gravel and much silt and leaf litter in the region where the male was taken. Both insects were observed in the middle of the day. Other observes have also mentioned the erractic flight of *Idomacromia* males. It appears the *proavita* is not exclusively a seepage-breeder and it appears to be one of the few species which can cope with both montane and lowland conditions.

Phyllomacormia. – I have followed the convincing opinion of May (1997) in regarding the African species of *Macromia* s.l. as belonging to a separate genus, *Phyllomacromia,* which appears to be monophyletic.

Phyllomacromia bicrustulata. – Material which I have placed in this taxon has been found at Etam, near Kumba. It is one of the species with twin tufts on either side of the swelling in segment 10. My male agrees well with a male in NHM (from Ikom, S. Nigeria 1956) determined by Fraser in 1957 as *P. pinheyi* sp. n. This was never described: Gambles later determined the specimen as *bicrustulata.* I have compared my specimen with *P. lamottei* Legrand, 1993 which is very close to *bicrustulata,* and it agrees more precisely with the latter, especially in the shape of the ventral surface of the hamule. However, I am not totally convinced that my determination is correct.

Phyllomacromia melania. – Identified using the key in Gambles (1979).

Cyanothemis simpsoni. – Several males were perching on bushes bordering a sluggish meandering lowland-forest stream at Etam. They were extremely wary and difficult to secure. The male abdomen bears a striking pale blue pigment which disappears on death; this is quite the more permanent pruinose blue of *Orthetrum* and *Hadrothemis.* Material agrees with paratypes in the NHM det. Ris (and det. Kimmins 1967) from Sierra Leone (Mongheri, 15-XI-1912, J.J. Simpson leg.) and with Gambles' material from Calabar Road, South Cameroons (taken 17-X-1962).

Malgassophlebia nigeriae. – I have one male and two females of this taxon. The male from Ebonji (13 September, 1957) agrees well with the taxon *nigeriae* Pinhey, 1960. This is separated from *bispina* Fraser, 1958 as the dorsal spine on the appendages is stituated at three-quarters of the length from the base in *bispina* and only one-half in *nigeriae.* The genus is keyed in Legrand (1986) and my specimen runs through to the '*bispina* group', but this is not keyed further, Pinhey describes *nigeriae* as a subspecies of *bispina* but it seems to me that the difference is sufficient to warrant specific to warrant specific status, and I have recorded it thus in the checklist.

Neodythemis africana. – A female specimen of this rare taxon was found in the emergence trap over the extensive seepage of Max's Trail on 29 April, 1997. This female has appendages, vulvar scale, thoracic markings and specialised shape of segment 10 which agree with a female in Fraser's collection, now in NHM (London), from Moyen Congo (Ketta Forest October, 1959) determined by Fraser as *africana.* This is distinct from *gorillae* (Pinhey) especially with regard to thoracic marking and *gorillae* lacks the specialised shape of segment 10 as is shown by a Gambles specimen from Nigeria (Obudu, 25 February, 1971). The teneral adult was preserved with its exuviae. A male of this genus was later taken at Kodmin on 19 November, 1997 in primary montane rainforest and it agrees fairly closely with material of *africana in* NHM.

Orthetrum camerunense. – Material from the drier parts of South-West Cameroon, in the rain shadow areas, belongs to this taxon which appears to be a good species (Gambles, 1959). It is probably commoner in the grasslands of the Bamenda Highlands of the North-West Province.

Porpax asperipes – A male from Ebonji agrees well with material in NHM collection from the Congo (Bambesa, Uela, XII-1938, J. Vrijdagh leg., F.C. Fraser det.).

Tetrathemis bifid and *T. godiardi.* – The first species was found at two lowlands sites. At Etam, males were seen in bushes above a shady pool in a partially dried up stream bed; it was also caught flying over a shallow shady stream on the edge of rainforest. At

Seme Beach, a male was found in dense vegetation near coastal marshes. The specimens agree well with material in the NHM from Nigeria (R. Idodo, 21-III-1954, R.M. Gambles leg/det.) At Etam, another species of *Tetrathemis* close to *godiardi* was flying with *bifida*. Comparing my single male specimen with a series of *godiardi* from northern Nigeria (Jemaa, 21-II-1960 and 12-III-1961) determined by R.M. Gambles in the NHM, mine appears to have the tips of the superior appendages more strongly divergent, although less so than in Fraser's *denticauda* but as I only have one specimen I cannot be sure that this difference is of importance.

Trithemis hartwigi. – A single male of this taxon was taken at Debundsha marsh of 14 July, 1996. It has previously only been recorded from the type locality of Fernando Po (Bioko). Several features are diagnostic : black wing venation, large black Pt. the shape of the accessory genitalia, black Iabrum, wide body, and the thoracic markings. There is no material in the NHM but the agreement with Pinhey's description is excellent; this record is the first from mainland Africa.

Zygonyx F. flavicosta. – This species is common in the Bakossi lowlands of the Kumba-Kupe area wherever there are fast rocky streams; it does not seem to need forested surroundings and it is often found in farm bush. The material agrees well with NHM material from Nigeria (e.g. a male Oshun Bridge. Oshogbo I/II-1913, Dr. H. Strachan leg., and another male (Ibadam, 1960, Mound, leg., R.M. Gambles det).

Zygonyx S. speciosa. – This larger species is also found in similar habitats to *Z. flavicosta*. The females are magnificent insects with their large wings marked extensively with golden-yellow and dark brown; the males have only basal patches of brown in the wings. My males agree wiht NHM material from the 'Cameroons River' in the McLachlan Collection and labelled *Pseudomacromia speciosa* Karsch. The females agree perfectly with the female of *pretiosa* in the NHM. This is was described with a female holotype as a different species by Karsch although the locality is the same : Cameroons River; it is easy to see why the sexes were not believed to belong to the same species. Gambles has labelled the specimen of *pretiosa* 'this is a female of *speciosa*'. Pinhey (1964) has synonymised the two taxa.

Seasonality

With records available from three years' surveying, it is now possible to look for seasonal patterns in sightings of adults. For each species the months recorded were listed. There is a danger of an artefact here. If one believes a time to be good, one may go out to look-and-find; a time believed to be poor may mean little recording takes place. However, the following are put forward as tentative suggestions.

Generally, in this part of Cameroon all months are wet and most species can be found at all times of the year. However, the months from November to February are relatively dry and the sunshine figures are highest then. The usual pattern is for the numbers of each species to increase during these dry and sunny months; the numbers remain high in March and April, which although wetter, tend to be sunny. Number fall off as the wet season progresses but most species are still to be found, albeit in smaller numbers. There are some interesting exceptions to this pattern.

There is a widespread belief amongst local people that November is the time to see dragonflies. I believe that this is based upon the fact that the common libellulids, breeding in temporary pools which dry out at the end of the wet season, emerge in large numbers then. They are often seen in the villages at this time, before either moving to forests during the drier months, or migrating. When we arrived in March 1995, villagers expressed surprise that we had come at the 'wrong time'; an impression that was borne out by the absence of almost all Odonata, except *Anax tristis,* in the villages. However, in the montane streams in the shady forests numbers are high at at this time.

The rare species may only be recorded in the sunnier months owing to the lower numbers; they may be missed in the main wet season. For example, *Neurolestes trinervis* has not been found between June and November on Kupe, although the other, commoner, specialist montane stream species (*Stenocnemis, pachystigma, Pentaphleb stahli, Nubiolestes diotima*) have been found in all months.

There is a fairy strongly marked tendency for the chlorocyphids to be absent from records at the end of the wet and the early dry season (October to December), appearing in January, February and March (when many tenerals are found), but most records are for the part of the wet season, April to July. This pattern applies to *C. cancellata, glauca* and *curta,* although the commoner *selysi* has only been missed from records in December.

The gomphids are particularly difficult to record at all and several species are only known from a single female species (which cannot usually be named). The species diversity appears to be greatest between April and July, in the early wet seasons, none has been recorded in the early dry season (November and December) and numbers build up as the wet season approaches. Even the time of the heaviest rains produces interesting rarities: *Phyllogomphus montanus* and *P. corbetae* sp. n. are good examples.

Conservation

The species list of 180 is impressive when one considers the size of the area surveyed. Additions to this list are to be expected, especially in the Gomphidae which are so hard to collect; a total of at least 200 species can be confidently expected. For comparison, Belize which is of a similar area has 170 recorded species (Paulson, 1998). I suspect that few parts of Africa of equivalent area can match this. For example, Kenya which has an area of 580.000 km^2 or 24 times the area of the Province, has a recorded total of 194 spp. with a higher intensity of study. It is worth noting that many of the very common and widespread African taxa are absent from the list of species from the South-West Province owing to the lack of savanna habitat and this serves to stress even more the importance of the more specialised species which do occur.

In general, Africa has an impoverished tropical fauna compared with that of South-East Asia and South and Central America. However, in this part of the Continent, around the Bight of Biafra, the dragonfly fauna approaches the levels of richness of the best parts of these two areas. There is a rich mix of ancient relicts and endemics at genus and species levels, with the addition of a large number of more widespread Congolian species. Certain species, such as *Pentaphlebia sthalii* and *Nubiolestes diotima,* have their nearest relatives in South America, recalling ancient Gondwana land connections. Certain genera such as *Umma* appear to have their 'headquarters' here. The dragonflies therefore support the data obtained from many other groups that South-West Cameroon forms a curcial 'centre of biodiversity' of the highest level of importance from a conservation viewpoint. Certainly, the extensive well-forested area of the Bakossi Mountains, including Mount Kupe, are of great importance for conservation. Takamanda to the north, also appears to be very rich, with a possibly distinct fauna. Korup is insufficiently worked but can be expected also to be important. Mount Cameroon and the associated lowland are also very rich. It is to be hoped that efforts by such bodies at the World Wildlife Fund and Birdlife International will continue to safeguard these areas.

The World Wildlife Fund has recently instigated a programme of stream monitoring in the Bakossi area and Christopher Wild and Otto Mesumbe are Investigating which dragonflies can be used as water-quality indictors.

Acknowledgements

I am grateful for the help I have received from the other members of the Cameroon Dragonfly Project, and from the staffs of the Mount Kupe Forest Project and the Mount Cameroon Project. Otto Mesumbe and David Chelmick deserve special mention for their enthusiastic fieldwork. Christopher Wild (research coordinator with World Wildlife Fund) provided logistical support, invaluable help with habitat data, and read through a preliminary draft of the paper. I am greatly indebted to Raymond Andress for his excellent illustrations to Michael J. Parr for helpful discussion on the genus *Orthetrum*, and to Viola Clausnitzer for information about Kenyan Odonata.

References

Buchholz, K.F., 1958. Ein neuer Phyllogomphus (Gomphiade : Odonata) aus Kamerun. *Bonn.Zool. Beitr.* 9(2/4): 259-260.

Cammaerts, R. 1969. Paragomphus abnormis (Karsch, 1890) et deux especes precedemment confondues avee cc taxon (Odonata Gomphidae). *Bull.Annls.Socl.Ent.Belg.* 105(9/10): 234-248.

Corbet, S.A. 1977. Gomphids from Cameroon, West Africa (Anisoptera : Gomphidae). *Odonatologica,* 6(2): 55-68.

Corbet, S.A., Green, J., Griffiths, J. and Bentney, E. 1973. Ecological studies on crater lakes in West Cameroon. Lakes Kotto and Mboandong. *J. Zool. Lond.* 170 : 309-324.

D'andrea, M. and Carfl, S. 1997. Nuove recolte di odonati del Camerun con note su *Agriocnemis maclachlani* Selys, 1877 e. *A. aligulae* Pinhey, 1974 e descrizione di *Agriocnemis dissimilis* sp. nov. e *Trithemis osvaldae* sp. nov. Atti Soc. it. Sci. nat. Museo. civ. Stor. nat. *Milano.* 136 (2): 157-190.

Ejedepang-Koge, S.N. 1986. *The tradition of a people-Bakossi. A historical socio anthropological study of one of Cameroon's Bantu peoples.* ARC Publications, Alexandria, VA.

Fao-Undf-Onarest. 1977. Soil surveys and land evaluation for the second development programme of the Cameroon Development Corporation. Summary report. *Technical reports 7/8. Soil. Science Project,* Iraf-Onarest, Ekona, Cameroon.

Forster, F. 1906. Die Libellulidengattungen von Afrika und Madagascar. *Jber. Ver. Naturk. Mannheim, Mannheim.* 71-72: 1-71.

Forster, F. 1909. Beitrage zu den Gattungen und Arten der Libellen. *Jb. nass. Ver Nat* 62 : 211-235.

Forster, F. 1916. Beitrage zu den Gattungen und Arten der Libellen. IV. neue Beitr. syst. *Insektenk.* 1 : 23-25.

Fraser, F.C. 1956. Results from the Danish expendition to the French Cameroons 1949-1950: xiii-xiv Odonata. *Bull. Inst. fr. Afr. noire.* (A) 18:938-948:949-959.

Fraser, F.C. 1957. A revision of the genus *Phyllogomphus* Selys with descriptions of five new species. *Revue Zool. bot. afr.* 56 : 9-32.

Gambles, R.M. 1959. A new species of orthetrum (Odon., Libellulidae) from the Bamenda Highlands, British Cameroons. *Ent. mon. Mag.* 95 : 44-46.

Gambles, R.M. 1967. The type specimens of some West African dragonfly species. *Entomologist.* 100 : 196-201.

Gambles, R.M. 1975. A new species of Chlorocypha Fraser, 1928 (Odonata: chlorocyphidae) from Nigeria, and some new or little-known Nigerian subsspecies or forms better known from the Cameroons. *Ent. mon. Mag.* 110 : 105-121.

Gambles, R.M. 1979. West African species of Macromia (Odonata : Corduliidae) belonging to the picta and sophia groups. *Syst. Ent.* 4(4): 389-407.

Gambles, R.M. 1987. Hagen's tubercle, a largely overlooked but potentially useful character in gomphid taxonomy (Anisoptera : Gomphidae). *Adv. Odonatol.* 3 : 51-60.

Geze, B. 1943. Geographie physique et geologic du Cameroon Occidental. *Mem. Mus. natn Hist.nt.* 17, Paris.

Green, J. 1972. Ecological studies on crater lakes in West Cameroon, zooplankton of Barombi Mbo, Mboandong, Lake Kotto, and Lake Soden. *J. Zool. Lond.* 166 : 283-301.

Green, J., Corbet, S.A. and Betney, E. 1974. Ecological studies on crater lakes in West Cameroon. Debundhsha Lake. *J. Zool. Lond.* 173 : 199-233.

Grubb, P. 1982. Refuges and dispersal in the speciation of African forest mammals. In : G.T. Prance (ed.) *Biological Diversification in the Tropics,* Columbia Univ. Press.

Grunberg, K. 1914. Eine neue Calopterygidae aus Kamerun. *Ent. Rundsch.* 31(8) : 53-54.

Hamilton, A. 1976. The significance of patterns of distributions shown by forest plants and animals in tropical Africa for the reconstruction of Upper Pleistocene paleonvironments : a review. In : E.M. van Zinderen Bakker (Ed.) : *Paleoecology of Africa, the surrounding islands and Antartica.*

Hawkins, P. and Brunt, M. 1965. The soil and ecology of West Cameroon. Report to the Government of Cameroon, 2 vols. FAO Expanded programme of technical assistance 2083, Rome.

Karsch, F. 1890. Ueber Gomphidae. *Ent. Nachr.* 16: 370-382.

Karsch, F. 1891a. Die Libellulidengattungen Orthetrum Newm. (*Libellula* brauer) und Thermorthemis Kirby. *Ent. Nachr.* 17 : 58-63.

Karsch, F. 1891b. Uebersicht der von Herm. Dr. Paul Preuss in Deutsch-Westafrika 1890 gesammelten Odonaten *Ent. Nachr.* 119 : 193-195.

Karsch, F. 1982. Eine neue von Herm Dr. Paul Preuss bei Buea im Kamerunbirge entdeckte Odonate der Gattung Macromia. *Berl. ent. Z.* 37 : 15.

Karsch, F. 1893. Odonaten von Victoria in Kamerun, gesammelt von Herrn Dr. Paul Preuss, determiniert von Dr. F. Karsch. *Ent. Nachr.* 119 : 193-195.

Karsch, F. 1894. Libellen von der deutschen Forschungs-Station Yaunde im Hinterlande von Kamerun, gesammelt von Herm *G. Zenker : Berl. ent.* 39 : 11-16.

Karsch, F. 1896. Zwei neue Herm G. Zenker in Kamerun erbeutete Odonaten. *Ent. Nachr.* 22(2) : 17-21.

Karsch, F. 1899. Odonate von Johann-Alrecht-Hohe (Nord-Kamerun), gesammelt von Herm leopold (Conradt). *Ent. Nachr.* 25 : 161-176.

Legrand, J. 1986. Malgassophlebia westfalli spec. nov., nouveau Tetratheminae afrotrophical des forest du Gabon oriental : imagos, larves, notes biologiques et discussion sur le genre (Anisoptera : Libellulidae), *Odonatologica.* 15(1) : 97-105.

Leroi, O. 1915. Odonaten aus Aquatorial-Afrika. Ergebn. 2 dt. zentr. Afr. Exped. 1910-1911, *Fuehrung Adolf Fried. zu Mechlenberg.* 1 : 319-360.

Letouzey, R. 1968a. Etude phytogegraphique du Cameroun. *Ecyclopedie Biologique* 69. Lechevalier, Paris.

Letouzey, R. 1968b. Cameroun in conservation of vegetation in Africa south of the Sahara. *Acta phytogeogr. Suec.* 54. Uppsala, Sweden.

Livingstone, D.A. 1982. Quaternary geography of Africa and the refuge theory. In : G.T. Prance (Ed.), *Biological Diversification in the Tropics* Columbia Univ. Press.

Longfielf, C. 1936. Studies of African Odonata, with synonymy and descriptions on new species and subspecies. *Trans. R. ent. Lond.* 85: 467-498.

Longfield, C. 1951. A new African Micorgomphus and notes on some Asiatic types (Odonata). *Ent. News.* 62(3): 97-102.

May, M.L. 1997. Reconsideration of the status of the general Phyllomacromia and Macromia (Anisoptera : Cordullidae). *Odonatologica.* 26(4) : 405-414.

Mbuagbaw, T.E., Brain, R. and Palmer, R. 1987. A history of the Cameroon. (2nd edn.) Longman, Essex, U.K.

McLachlan, R. 1896. On some Odonata of the subfamily Aeschina. *Ann. mag. nat. Hist.* 6(17): 409-425.

Ministry of Planning and Regional Development. 1989. Republic of Cameroon : the Korup Project. Plan for developing the Korup National Park and its support zone. Yaounde, Cameroon.

Ndian Estate. 1987. Unpublished rainfall records quoted in Ministry of Planning and Regional Development, 1989. Republic of Cameroon : The Project, Yaounde, Cameroon.

Neba, A.S. 1987. *Modern geography of Cameroon.* Longman, Harlow, Essex, UK.

Paulson, D.R. 1998. Odonata of Middle America, by country. *Internet Comm.* (7) Jan. 1998.

Pinhey, E., 1961a. Dragonflies collected on an expedition from Rhodesia to Nigeria in 1958, pt. 1. *Ent. mon. Mag.* 96: 256-271.

Pinhey, E. 1961b. Dragonflies collected on an expedition from Rhodesia to Nigeria in 1958. pt. 2. *Ent. mon. Mag.* 97: 101-104.

Pinhey E. 1962a. some records of Odonata collected in tropical Africa. *J. ent. Soc. sth. Afr.* 25: 20-50.

Pinhey, E. 1962b. A description catalogue of the Odonata of the African Continent (up to December, 1959), pts 1 and 2, *Publocoes. cult. co. Diam. Angola.* 59 : 11-161, 163-322.

Pinhey, E. 1964. Dragonflies (Odonata) of the Angola-Congo borders of Rhodesia. *Publcoes cult. Co. Diam.* Angola. 63: 95-129.

Pinhey, E. 1971. Odonata of Fernando Po island and of neighbouring Cameroon Territory, *J. ent. Soc. sth. Afr.* 34: 215-230.

Pinhey, E. 1974. Odonata of the North-West Cameroons and particularly of the islands stretching southwards from the Guinea Gulf. *Bom. Zool. Beitr.* 25(1/3): 179-212.

Ris, F. 1909-1916. Libelluilnen monographisch bearbeitet. Collns Zool. Edm. de Selys Longchamps. 9/16: 1-1278 pls excl.

Schmidt, E. 1943. Bemerkungen uber Lestiden 2: eine neue Gattung aus Kamerun (Eolestes diotima). *Mitt. dt. ent. Ges.* 11: 102-111, 3 figs.

Schmidt, E. 1951. Uber neue and winiger bekannte afrikanische platycnemididen (odon.). *Mitt. Miinch. ent. Ges.* 41: 217-240.

Selys-Longchamps, M.E. de. 1879. Description of a new species of Phyllomaromia (Corduliina) from West Africa. *ent. mon. Mag.* 16: 103-104.

Selys-Longschmps, M.E. de, 1886. Revision du Synopsis des agrionines Premiere partie comprenant les legions Pesudostigma, Podagrion, Platyenemis, et Preotoneura. *mem. Cour. Acad. r. Sci. Belg.* 38(4): iv + 233 pp.

Sjosted, Y. 1899. Beitrage zur Kenntis der Insektenfauna von Kamerun, 6. Odonaten aus Kamerun, West-Africa. *Bih. K. svenska Vetenska Akad. Handl.* 25(4) 2: 1-62.

Sjosted T.Y. 1917. Odonaten aus Abessinian, Ost-und West-Afrika. *Ark. Zool.* 11(4): 27, 5pl.

Vick, G.S. 1996. Umma mesumbei spec. nov. with records of some other dragonfly species from the South West Province of Cameroon (Zygoptera: Calopterygidae). *Odonatologica.* 25(2): 167-178.

Vick, G.S. 1998. Notes on some damsefly larvae from Cameroon (Zygoptera : Perilestidae, Amphipterygidae, Playcnemididae). *Odonatologica.* 27(1): 87-98.

3

ENDOCRINOLOGY OF ODONATA

• *D.B. Tembhare*

Abstract

The advancement made in the field of endocrinology of Odonata is thoroughly reviewed since its inception. Hanström (1940), in true sense, laid the foundation of adonate endocrinology. The French workers (Cazal, 1947, 1948; Arvy and Gabe, 1952, 1953, 1962; Charlet, 1969, 1971, 1972a, b, c, 1974, 1977; Charlet and Schaller, 1974, 1975a, b, 1976; Charlet *et al*; 1975; Deroux-Stralla 1948a, b; Mouze, 1971; Mouze *et al;* 1973; Raabe and Provansal, 1972; Shcaller, 1959, 1960, 1962a, b, c, 1964, 1965; Shcaller and Charlet, 1970, 1980; Schaller and Defossez, 1974, Shcaller and Hoffmann, 1973, 1976; Schaller and Meunier, 1968; Schaller *et al;* 1974, 1975) as well as Indian workers (Tembhare, 1973, 1979, 1980, 1984, 1988, 1994, Thembhare and Thakare, 1976a, b, 1977; Thakare and Tembhare, 1975a, b, 1977; Tembhare and Andrew, 1991a, b, c, 1994) enriched the odonate endocrinology. As a result, now an extensive literature on the histology, histochemistry and ultrastructure of the endocrine system and hormonal regulation of metamorphosis, diapause, reproduction, osmotic, water and ionic balance and carbohydrate, lipid and protein metabolism is available.

Introduction

Although some information on the gross anatomy of cephalic endocrine glands is available from the classical description of the sympathetic nervous system of odonate larvae by earlier workers (Brandt 1836; Amans 1881; Tillyard 1917), it was perhaps, Baldus (1924) who first envisaged the true anatomical inter-relationship between the brain and the retrocerebral complex. He noticed the presence of two groups of large, glandular, chromophlic cells in the pars intercerebralis region of brain. Innervating the sympathetic ganglia, "ganglions stomatogastriques", by the well-defined nerves "nervi stomatogastrici", formed from their axons. Later, Hanström (1940) recognized the functional significance of the complex and introduced more authentic terms such as, the neurosectretory cells (NSC), the nervi coporis cardiaci (NCC) and the corpora cardiaca (CC) for the glandular nerve cells of pars intercerebralis, nervi stomatogastrici and ganglions stomatogastriques, respectively. He, therefore, truly laid the foundation of endocrinology of odonata. He also demostrated precisely the structure of the cephalic endocrine complex of four species of odonates, viz., *Aeschna juncea, Calopteryx virgo, Lestes sponsa* and *Coenagrion* sp.

During the past four decades, a lot of information became available on histomorphology of various endocrine components, histochemical analysis of the neurosecretory material (NSM), subcellular mechanisms of synthesis and release of NSM, cyclical activities in the various endocrine components in relation to physiological processes and hormonal control of the post-embryonic development, diapause, vitellogenesis, osmotic water and ionic balance, and intermediary metabolism. Although nearly two dozen species have been studied endocrinologically all over the world, yet the bulk of information hail mainly from *Aeschna cyanea* and *Orhtetrum chrysis* which have often been the objects of these studies.

In this paper an attempt is made to review the entire literature accumulated during past four decades to comprehend the advancement in our knowledge relating endocrinology of Odonata.

Structure of the Endocrine System in Odonata

The endocrine system of Odonata is built on the general plan of that of Insecta but it shows much primary type of organization in accordance with the primitive phylogenetic status of this group of insects. The endocrine system is integrated structurally with the central as well as sympathetic nervous system. The brain and ventral ganglia are embodying the neurosecretory cells and the latter are innervating the neighbouring endocrine glands. The aorta becomes the indistinct part of the endocrine system in the cephalic region.

The endocrine system of Odonata typicaly consists of (Fig. 3.1) :

(a) the neurosecretory cells in the brain and ventral ganglia,

(b) the corpora cardiaca and other neurohaemal organs,

(c) the corpora allata,

(d) the ventral glands,

(e) the gonads, and

(f) the midgut endocrine cells.

(a) The neurosecretory cells (NSC) in the brain (Fig. 3.2)

Since the report of Hanström (1940), several worker focused their attention to explore the cytomorphological and cytochemical features of the cerebral neurosecretory cells. The histological studies made so far suggest collectively the presence of distinct paired groups of NSC in various parts of the brain of Odonates and on the basis of their locations they are designated as follows :

(i) The medial neurosecretory cells : The presence of well-defined groups of the medial neurosecretory cells (MNC) in the mid-dorsal region of the pars intercerebralis on either side of median furrow and around the root of the medial ocellar nerve has been repeatedly reported in a large number of Odonata (Hanström, 1940; Cazal 1947; Arvy and Gabe, 1952; Schaller and Meunier, 1968; Gillott 1969; Charlet 1972a; Charlet *et al.*, 1974, Tembhare and Thakare, 1976a).

With the help of CHP and Azan staining techniques, Arvy and Gabe (1953) described for the first time the specific cytological characteristics of the MNC. The recognized two types of MNC, the chromophilic and phloxinophilic. Besides describing specialization of the MNC in their shape, size and cytological organelles, Arvy and Gabe (1953) have envisaged cyclic activity in synthesis and release of neurosecretory material (NSM) in these cell, mostly in relation to post-embryonic development.

Schaller and Meunier (1968) introduced distinct nomenclature, the A cells and B cells to the chromophilic and phloxinophilic MNC, respectively after examining thoroughly the CHP, Azan and newly employed Aldehyde fuchsin (AF) stained

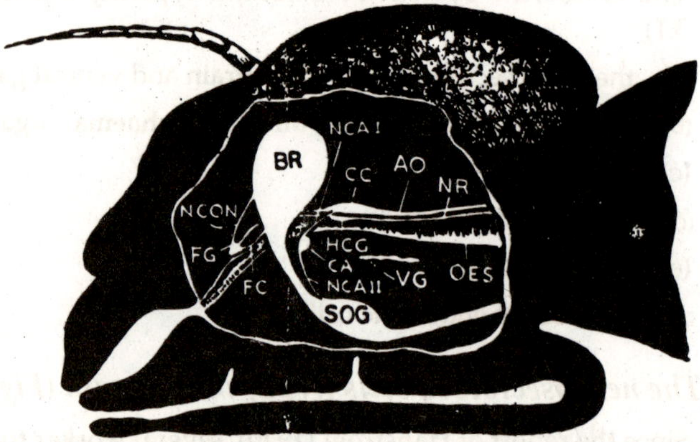

Fig. 3.1 : Cephalic neuroendocrine complex in lateral view (diagrammatic). AO = aorta; BR = brain; CA = corpus allatum; CC = corpus cardiacum; FC = frontal connective; FG = frontal ganglion; HGC = hypocerebral ganglion; NCA-I = nervi corporis allati-I; NCA-II = nergi corporis allati-II; NCC = nervi corporis cardiaci; NCON = nervous connectives; NR = recurrent nerve; OES = oesophagus; SOG = suboesophageal ganglion; VG = ventral gland.

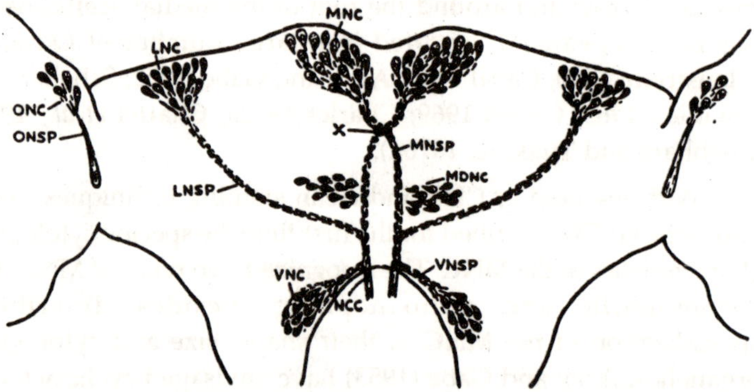

Fig. 3.2 : Location of neurosecretory groups and neurosecretory pathways in the brain (diagrammatic). LNC = lateral neurosecretory cells; LNSP = lateral neurosecretory pathways; MDNC = mid-dorsal neurosecretory cells; MNC = medial neurosecretory cells; MNSP = medial neurosecretory pathways; NCC = nervi corporis cardiaci; ONC = optic neurosecretory cells; ONSP = optic neurosecretory pathways; X = chiasma; VNC = ventral neurosecretory cells; VNSP = ventral neurosecretory pathways.

histological preparations. Gillott (1969) confirmed the classification of MNC proposed earlier by Schaller and Meunier (1968) in the damselfly, *Coenagrion angulatum*. He however, could not notice the cyclical secretory activity in the B cells and expressed doubt in recognizing them as the true NSC.

Charlet (1972a, b) made extensive histological studies and explored in detail, the tinctorial properties of the MNC of *Aeschna cynea* larvae after employing various selective techniques, such as CHP, AF, Azan and Alcian blue. He classified the MNC into three main types – the A, B and C cells and furthermore reclassified the A cells into two sub-types – A1 and A2 cells on the basis of size and cytological characteristics. On the basis of various tinctorial characteristics revealed by neurosecretion-selective staining techniques, such as CHP, AF, ABP (Alcian blue phloxine), ATH (aldehyde thionin), Azan and also *in situ* performic acid victoria blue (PAVB) and AF permanent whole mounts, Tembhare (1973) and Tembhare and Thakare (1976a) demonstrated two distinct groups of MNC in the pars intercerebralis composed of A1, A2, B and C cells in the larvae and only A1, B and C cells in the adults of *Orthetrum chrysis*.

While describing the cell-types, most of the earlier workers (Schaller and Meunier, 1968; Gillott, 1969; Charlet, 1972b), often realised the need of further work of resolve the dispute of assigning the nature of B cells as whether they are representing the negative secretory phase of the A cells or maintaining their independent *status quo*. Thakare and Tembhare (1975a) made an extensive histochemical survey by employing more than 30 tests to resolve the enigma. The found that the cerebral NSM is, no doubt, mucopreteinaceous in nature as reported earlier by Arvy and Gabe (1962) in *Aeschna cyanea*, but the critical analysis reveals that the A and C cells differ in their NSM from that of the B cells as the NSM of the first two cell-types is virtually characterised by the proteins containing cystine or cysteine aminot acids while that of B cells possesses the protein rich in basic amino acids.

The electron microscopic study of the neurosecretory pars intercerebralis-corpus cardiacum system in the larvae of *Aeschna cyanea* singularly supports the inference made earlier on the basis of histological and histochemical data that all the cells-types recognized in the dragonflies represent their distinct categories independently and differ in their cytomorphology, tinctorial affinities and chemical nature of their NSM (Tembhare, 1980). The ultrastructure of MNC strongly proves their true neurosecretory nature and classification in various types as these cells, besides possessing predominantly all essential cytoplasmic organelles for protenaceous secretion, are producing the neurosecretory granules (NSG) of quite variable sizes. Arvy Gabe (1953) and Tembhare (1980) discussed functional involvement of nucleus and cytoplasmic organelles in the process of NSM synthesis in MNC.

(ii) The lateral neurosecretory cells: In Odonata, the lateral neurosecretory cells (LNC) were described first by Arvy and Gabe (1953). They reported the presence of well-defined groups of LNC in lateral parts of the protocerebrum just above the globuli cells of the corpora peduculata, in both, the last instar larvae and adults of various species. Charlet (1972a) and Charlet *et al.* (1974) confirmed repeatedly an existence of LNC in the larvae of *Aeschna cyanea* and recognized them as the C cells. Tembhare and Thakare (1976a) conducted various histological staining techniques to display the cytological and tinctorial properties of these cells in the dragonfly, *O. chrysis* which revealed their differentiation into the A2, B and C cells in larvae but only B and C cells in the adults. The presence of A2 cells in the larvae and their absence from the adults signifies their functional confinement with physiological activities limited to the larval life only, mostly the metamorphosis and other developmental processes (Tembhare and Andrew, 1994).

(iii) The paramedial (mid-dorsal) neurosecretory cells: Schaller and Meunier (1986) made an elegant discovery of the presence of some mediodorsal and paramedial NSC in the protocerebral region of the brain of some anisopterans. According to these workers, the mediodorsal cells are characterized by possessing quite a large number of fine NSG, giant sized and pyriform in shape, scattered in the homogeneous cytoplasm and exhibiting themselves as the A cells. The axons of these cells, after running for a short distance, join the axonal tracts of MNC after the formation of chiasma.

The paramedial cells are of two types, the A and B cells. These cells extend their axons to the suboesphageal ganglion. They mostly occupy ventral region of the anterior part of protocerebrum, i.e., similar region but opposite side of that of mediodorsal cells. The A cells of both the groups resemble each other in shape, size and tinctorial affinities. The transport of the NSM from the paramedial A cells to the suboesophageal ganglion via circumoesophageal connectives has been demonstrated *in vitro* preparations (Schaller and Meunier, 1967; Charlet *et al.*, 1974).

(iv) The ventral neurosecretory cells: The presence of the ventral neurosecretory cells (VNC) in the triticerebral region of the brain has been reported in the dragonfly, *A. cyanea* (Charlet, 1972; Charlet *et al.*, 1974) and *O. chryss* (Thembhare and Thakare, 1976a). Charlet (1972a) did not give classification of VNC while Tembhare and Thakare (1976a) recognized them as the B and C cells in both the larvae and adults. Raabe (1964) reported earlier the existence of VNC in the tritocerebral region of the brain and formation of the third pair of nerves, the NCC III, innervating the corpora cardiaca in the adult dragonfly. *Aeschna cyanea* but Schaller and Meunier (1968) could not trace either the VNC or NCC III. Charlet and Schaller (1974) could moreover observe only the VNC. Tembhare (1973) supported the observations of Charlet (1972a) after confirming the presence of VNC but at the same time the absence of NCC III in both the larvae and adults.

(v) The optic neurosecretory cells: Tembhare (1973) noticed the presence of a well-defined group of NSC in the anterodorsal region of the root of optic lobes representing the B and C cells in both the larvae and adults of *O. chrysis.* An accumulation of NSM in the axonal endings of ONC at the internal optic chiasmatic region indicated, to some extent, its functional interrelationship with the physiology of eye which is yet to be confirmed (Thembhare and Thakare, 1976a).

(b) The neurosecretory cells inteh ventral gangalia (Fig. 3.3)

Arvy and Gabe (1952, 1953) reported first the existence of a pair of NSC containing chromalus-haematoxyline positive cytoplasmic inclusion in the mid-ventral region of the suboesophageal ganglion of various species of Odonata. Though these workers strongly emphasized the innervation of ventral

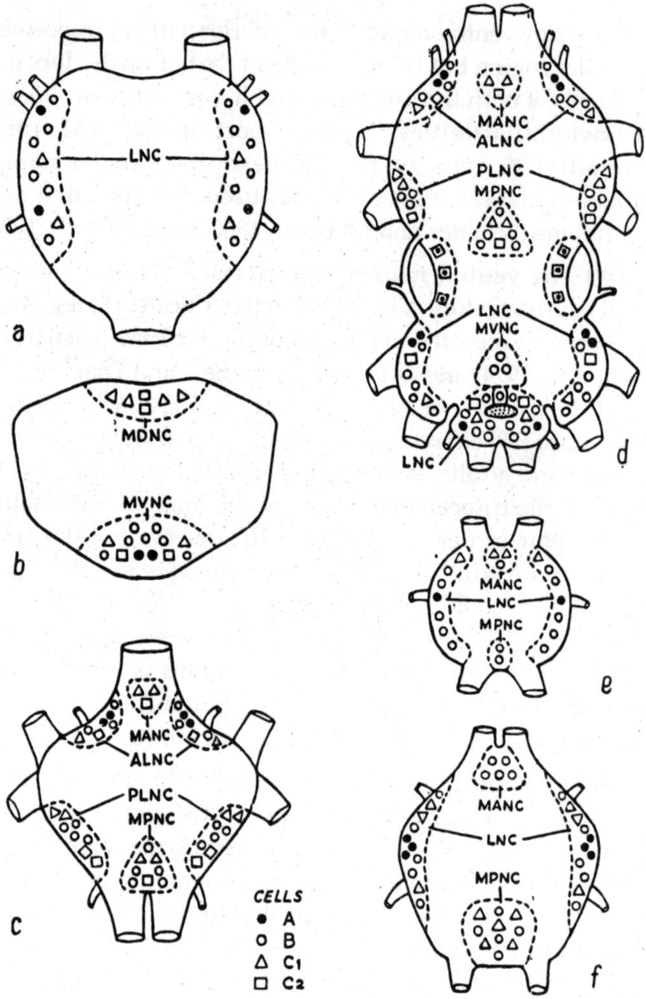

Fig. 3.3 : a-f : Digrammatic representation showing distribution of the neurosecretory cells in (a) and (b) subesophageal (a) longitudinal and (b) transverse plane; (c) prothoracic, (d) meso, meta+first abdominal, (e) abdominal 2-7 and (f) last abdominal ganglia of *Orthetrum chrysis*. ALNC = antero-lateral neurosecretory cells; LNC = lateral neurosecretory cells; PLNC = postero-lateral neurosecretory cells; MDNC = mid-dorsal neurosecretory cells; MVNC = mid-ventral neurosecretory cells; MANC = mid-anterior neurosecretory cells; MPNC = mid posterior neurosecretory cells.

glands by NSC of the suboesophageal ganglion, nobody else supported their observations. Schaller and Meunier (1968) confirmed the observations of these workers in some species of Anisoptera and, in addition, noticed two more pairs of similar types of cells in the lateral regions of the ganglion. These cells, they have designated as the A cells which differ in size, i.e., the A cells are large in the lateral but small in the mid-ventral region of the ganglion, particularly, in *A. cyanea.* The also reported the presence of a large number of B cells intermingled often with the A cells.

Charlet (1969) described NSC in the suboesophageal and rest of the ganglia of the ventral nerve cord of *A. grandis* with cytomorphology, staining properties and classification. He recognized the A, B and C cells and reclassified the A cells into five subtypes; (1) the principal A cells, (2) the compact A cells, (3) the paramedial A cells, (4) the abdominal A cells and (5) the last abdominal A cells, on the basis of tinctorial affinities and distribution. Charlet (1972a-c) demostrated similar organization of NSC in the ventral ganglia of *A. cyanea* and found differentiation of the A cells since the fourth instar larae and higher activity in the NSC prior to the moulting in each larval instar and throughout the life in the adult.

Tembhare and Thakare (1977) gave detailed account on distribution, staining affinities and classification of NSC, histochemistry of NSM of various types of NSC, the axonal pathways and symphathetic neurohaemal organs in *O. chrysis.* Recently, ultrastructure of the thoracic A cells of *A. cyanea* larvae has been discussed in relation to their role in osmotic and ionic balance regulation (Thembhare, 1979a).

In *O. chrysis,* NSC were classified into A, B, C1 and C2 cells. The NSC in the ventral ganglia are situated in various regions; the mid-dorsal, mid-ventral and lateral regions of the suboesophageal ganglion, in the mid-anterior, antero-lateral and postero-lateral regions of the thoracic ganglia and in the mid-anterior, mid-posterior and lateral regions of the abdominal ganglia. The A cells are confined to the mid-ventral and lateral regions of the suboesophageal ganglion, in the antero-lateral region of thoracic ganglia and in the lateral regions of the abdominal ganglia. The B cells occur in all regions of the ventral ganglia, except the mid-dorsal region of the suboesophageal ganglion. The C1 cells lie in all regions of the ganglia except the metathoracic gangia where they are confined to the lateral regions only. The C2 cells are located in the mid-dorsal and mid-ventral regions of the suboesophageal ganglion, in all regions of the pro- and mesothoracic ganglia, anterior, posterior, lateral regions of metathoracic ganglion and in the lateral region of the first abdominal ganglion only. The number of NSC differs from one ganglion to another.

The NSC of ventral ganglia produce the NSM composed of PAS-positive material with 1,2-glycol groups and some proteins. The proteins with cystine or cysteine occur pre-dominantly in the NSM of A cells, moderately in cells and negligible in the B cells, while proteins containing arginine occur in the NSM of A and B cells only and those containing basic amino acids are confined to the C2 cells. Histiochemical demonstration, thus supports the classification of these cells which show homology with the mid-ventral cells in the subesophageal ganglion. All the rest of sub-types of A cells of Charlet (1969) represent merely various secretory phases of the principal A cells in *O. chrysis*.

(c) The neurosecretory pathways

(i) The cerebral neurosecretory pathways : The axons of various groups of NSC run together in the form of well-defined axonal tracts or bundles and transport independently the NSM from the perikarya of their cells to the respective neurohaemal organs. These neurosecretory axonal tracts are referred to as nervi corpori cardiaci (NCC) by a large number of workers except Tembhare and Thakare (1676a), according to whom, these axonal tracts are still naked inside the brain and they retain the form of nerves only after coming out of the brain. These worker have, therefore, referred the term 'neurosecretory pathways' to the intracerebral axonal bundles while NCC to the 'true nerves innervating the CC' after coming our of the brain. Since the time of Baldus (1924), anatomical organization of the axon MNC and medial neurosecretory pathways (MNSP) have been described in a large number of odonates. They run first mid-ventrally, form a chiasma at the middle, thereafter run ventero-posteriorly and, finally, leave the brain as the nervi corporis cardiaci or NCC-I innervating the CC. Most of the workers have observed NSM along the MNSP in histological, histochemical and also ultrastructural preparations in the dragonflies studies so far. Arvy and Gabe (1953) described the course of lateral neurosecretory pathways (axons of LNC) as running independently in antero-posterior direction and finally forming a pair of fine nerves, the nervi corporis cardiaci externi (NCC-II) after leaving the brain in the last instar larvae and adults. Although some workers have reported the innervation of CC by two pairs of nerves, the NCC-I and II (Cazal, 1947; Deroux-Stralla, 1948a), large number of investigators have observed only a single pair of nerves, the NCC, (Hanström, 1940, Schaller and Meunier 1968; Gillott, 1969; Charlet, 1972; Charlet *et al.*, 1974, Tembhare and Thakare, 1976a, Tembhare, 1980).

Schaller and Meunier (1968), Charlet (1972 a) and Tembhare and Thakare (1976a) reported the fusion of LNSP with MNSP in various anisopteran species. In *O. chrysis,* the LNSP fuses the MNSP in prechiasmatic region in the larvae while in postchismatic one, in the adults. Other workers, however, reported the fusion of LNSP with the MNSP after formation of chiasma. Challer and Meunier (1968) and Charlet *et al.* (1974) noticed the union of axons of the anterior mediodorsal cells with the MNSP in postchiasmatic region and, in addition, the axons of paramedial cells running through the circumoesophageal connectives and finally terminating into the suboesophageal ganglion.

Raabe (1964) first repoted the formation of NCC-III from the axons of VNC and the innervation of each CC by three nerves, the NCC-I, II and III in *A cyanea.* The studies of Charlet (1972a) and Tembhare and Thakare (1976a) revealed clearly the fusion of the VNSP (axons of VNC) with the MNSP before their emergence from the brain in the dragonflies, *A. cyanea* and *O. chrysis.*

Almost all NSC, MNC, LNC, VNC and anterior mediodorsal cells appear to discharge their NSM into the CC except the anterior paramedial and optic NSC. The definite site of release of NSM of the paramedial cells is still and enigma (Charlet *et al.*, 1974). The ONC extend their axons towards the internal optic chiasmata region where they get terminated suggesting the release of NSM from these cells in the brain itself without running to the definite neurohaemal organs. Such type of release of NSM, though appears unusual, is not uncommon, in arthropods (Maddrell, 1970). Although Gabe (1972) reported in a large number of pterygote insects the progressive increase in the amount of PAS-positive substance, mostly the 1,2-glycol group, during proximodistal axonal transport of NSM, yet no favourable evidence is available in odonates.

(ii) The aorta : *In vitro* experimental studies on the cephalic neuroendocrine system in the larvae (Thakare and Tembhare, 1975b) and adults (Tembhare and Thakare, 1976a) showed the innervation of aorta wall by some cerebral neurosecretory axons after by passing the CC in *O. chrysis* suggesting their secondary neurohaemal function. Ultrastructural studies also revealed the neurohaemal functicn of aorta in *A. cyanea* (Tembhare, 1980).

(iii) The ventral neurosecretory pathways : The axons of various pathways of NSC in vetral ganglia have been traced with success to some extent in *O. chrysis* only. They terminate exclusively into the neighbouring NHO. The sympathetic neurohaemal organs in the dragonflies have been described by some workers (Charlet, 1971; Grillot *et.al.*, 1971; Raabe and Provansal, 1972; Tembhare and Thakare, 1977).

(d) The corpora cardiaca

Although the corpora cardiaca (CC) were described and illustrated precisely as paired ganglia of the sympathetic nervous system (Brandt, 1836; Amans 1881; Tillyard, 1971); detailed account was given for the first time by Baldus (1924) on structural organization and their association with the pars intercerebralis (PI) region of the brain. Hanström (1940) established the functional inter-relationship of these structures in accordance with the activity of NSC of the PI and recognized finally their functional status as the 'neurohaemal organs' in the odonata. Cazal (1947) described histomorphology of the CC of various species of Odonata. He reported the presence of paired elongated CC behind the brain encircling the aorta dorsolaterally and after being fused with each other dorsally the are separated from the ventral side by the hypocerebral ganglia in both the larvae and adults. He noticed the presence of to types of cells, the chromophobes and the glandular cells or chromophilis in the substance of the CC, besides a large mass of axonal endings of the cerebral NSC. He demonstrated various cytological characteristics of the chromophilis in the favour of their secretory nature, such as large glandular cells body, mero-holocrine secretory activity, accumulation of acidophilic neurosecretory granules, abundant mitochondria in the cytoplasm and the pseudopodal axonal process running towards the aorta wall.

Arvy and Gabe (1952, 1953) confirmed histomorphological organization of the CC described by earlier workers. With the help of CHP staining technique, they observed the stainable cerebral NSM in varying quantities inside the body of the CC and in addition, noticed two types of NSM in the chromophils; phloxinophilic in two species of Zygoptera and an anisopteran *Symetrum vulgatum,* while on the contrary, CAH-positive in *A. cyane.* After extensive histological and histochemical analysis of the cephalic endocrine system in relation to the post-embryonic development of odonates, these workers finally proposed homology of the CC of Odonata with the sinus glands of crustacean decapods and isopods.

Gillott (1969) extended singularly an agreement with the histomorphological description of the CC given by earlier workers for other dragonflies including the presence of chromophobic and chromophilic cells. Gillott (1969) in *Coenagrion angulatum* and Tembhare and Thakare (1976a) in *O. chrysis* observed fuchsinophilic or CAH-positive neurosecretory granules in the chromophilic cells and varying quanitites of stained fuchsinophilic colloidal NSM in the intercellular spaces.

In *O. chrysis,* histological investigations revealed the presence of two nervous connections, the nervi cardio-stomatogastrici connecting both the CC with the hypocerebral ganglion ventrally and the CC in the middle region by a dorsal commissure, the commissure corporis cardiaci, (Tembhare and Thakare, 1976a). Recently, Tembhare (1980) studied ultrastructue of the CC of *A. cyanea* larvae and noticed two types of chromorphilic or intrinsic neurosecretory cells (INC), the INC1 and INC2 containing the electron dense granules of 3000 Å and 2000 Å in diameter, respectively. Both types of INC represent all signs of being secretory cells although they differ from one another in their location and gross cytomorphology. Ultrastructural studies demostrated varying sizes of the NSG in different extrinsic axon endings in the CC suggesting the CC as a primary store house for the NSM of almost all types of NSC of the brain. The exocytotic mode or release of NSG from the CC is well-evident.

(e) The corpora allata

The corpora allata in Odonata were first described by Hanström (1940, 1942) notwithstanding their position as ventro-dorsal to CC as the spherical or oval bodies innervated by the nerve, nervi corporis allati (NCA), in the adults only.

Cazal (1947) could judge their location from the serial sections as it was quite difficult for him to see them in dissections due to their close association with the circumoesophageal connective, oesiophagus and tentorium. He pointed out the modification of the shape of CA from oval in young larvae to the elongated one in the adults. He confirmed the observation of Hanström (1940) regarding innervation of CA by NCA 1 in dadults and also succeeded in tracing the NCA1 in the larvae. On the basis of histological structure, he compared the CA of Odonata to the lymphoid tissue of vertebrates.

Deroux-Stralla (1948a) confirmed the topography and histomorphology of these glands in *A. grandis,* reported earlier in various species of Odonata by Cazal (1947). He, in addition, noticed the innervation of CA by two nerves, the NCA1 from CC and the NCA2 from the suboesophageal ganglia. Arvy and Gabe (1952, 1953) emphasised the formation of NCA1 from the axons of NSC of the pars intrcerebralis after bypassing the CC and that of NCA2 from teh axons o NSC of suboesophageal ganglion, and suggested the regulation of the acitivity of CA by the NSC of both the ganglia. They observed increase in size of the CA in relation to gradual advancement in development of the last instar larva and considerable fall in their volume with the degeneration of the ventral glands in the newly emerged imago. Gillot (1969) did not agree with the observation of earlier workers regarding an innervation of the CA by NCA1 and NCA2 in the damselfly,

C. angulatum. He however, disagreed with the observations of earlier workers regarding transformation of the CA from oval or spherical shape to elongated tubular one. He, however, mentioned manifold elevated volume of the CA in the matured adults.

Tembhare and Thakare (1976a) noticed the CA in *O. chrysis* as cylindrical paired bodies occupying the similar position reported by earlier workers and exhibiting alike histomorphological structure described in other odonata studied so far. They have further reported that the NCA1 bifurcates at the anteriror end of the CA and both the branches innervate the superficially and again reunite at the other side forming the NCA2. In *O. chrysis* thus the NCA1 and NCA2, both are merely axonal extensions of the cerebral neurosecretory cells which arise initially from the brain and lastly terminate into the suboesophageal ganglion. Our observations, therefore, contradict those of Arvy and Gabe (1952, 1953) in relation to the origin of NCA2 from the NSC of suboesophageal ganglion and also of Gillott (1969) who could not observe innervation of the CA either from the brain or the suoesophageal ganglion.

The CA of *O. chrysis* basically represent the similar histological structure both in the larvae and adults, but sometimes cyncytial organization of the CA can be seen. In the female dragonfly, particularly, some giant cells become well evident at the time of vitelogenesis with modification in volume of the glands as well as in that of size of cell bodies and nuclei. Up till now, there is no report mentioning an accumulation of NSM in the CA of Odonata.

(f) The ventral glands

Since the initial report of Pflugfelder (1947), the histomorphology, of the ventral glands has been described in a considerable number of odonates (Cazal, 1947, 1948; Deroux-Stralla, 1948a, b, Arvy and Gabe, 1952; Schaller, 1960; Gillot, 1969; Tembhare and Andrew, 1994). Cazal (1947), contemporary of Pflugfelder, independently studied the structure of the ventral glands of some species of odonata. He unexceptionally observed in all species of odonata the presence of well-defined lobulated structures, mostly attached with the submentum and richly supplied with the trachea in the posteroventral region of the head after being shifted from the prothoracic segment. He described the histology of the ventral glands as composed of the large number of small sized cells having large nuclei at the centre but scantly cytoplasm. He noticed maximum size of the ventral glands in the last instar larvae prior to commencement of moulting and, sometimes, retaining the cyncytial structure. He further suggested that the ventral glands are histomorphologically analogous with the CA or lymphoid tissue of the vertebrates. He noticed degeneration of these structures due to pycnosis in the imago just after moulting and proposed functional resemblance with the thymus glands of the vertebrates. He reported their role in the metabolism of nucleoproteins. Deroux-Stralla (1948a, b) confirmed the findings of earlier workers pertaining to the developement, structure, location and metamorphic role of the vental glands in the dragonfly, *A. grandis.* He described in detail the tracheating of the glands and after performing elegant experiments, advocated their role in the moulting and, metamorphosis.

Arvy and Gabe (1952) gave much relevant impetus on the structure and function of ventral glands as a part of their versatile studies on the endocrinology of Odonata. These workers extensively explored cytological as well as cytochemical aspects of the cells of ventral glands and noted the synthesis of RNA predominantly in the cells at the time of active phase of the glands. These workers reported the innervation of the ventral glands by the axons of NSC. Some adjoining neurons of mid-ventral region of the suboesophageal ganglion. Since then their observation has been repeatedly cited in the reviews on the insect endocrinology but none of the latter researchers on Odonata has supported this much acclaimed investigation. In fact, Arvy and Gabe (1952) themselves did not provide even a single photomicrograph in support of their statement but on the other hands, mentioned that "Les glandes ventrales se trouvent au contact de volumineuses trachees et des trachooles se ramifient a leur interieur. Leur innervation, que Pflugfelder ne signale pas, est assuree oar un nerf venu du ganglion sousoèsophagien."

Besides Arvy and Gabe (1952), recently Tembhare and Andrew (1994) studied correlative histological changes in the ventral glands, MNC and CC-CA in *Tramea virginea* in relation to the moulting and metamorphosis, Gillott (1969) displayed the histomorphology and the cyclical activity in the ventral glands in accordance with the one set of metamorphosis and moulting in *C. angulatum.*

(g) Midgut endocrine cells

The midgut is now considered as the large endocrine organ in insects. The TEM and immunocytochemical techniques revealed the presence of endocrine cells in the midgut epithelium and positive reaction for Gastrin/Cholecystochinin, Pancreatic polypeptide, Substance - P, Enkephalin - like vertebrate peptide hormones in *Aeschna cynea* (Andries and Tramu, 1984).

Endocrine Regulatory Mechanism in Odonata

In Odonata the following endocrine regulatory mechanisms are extensively studied:

(a) post-mebryonic development,

(b) diapause,

(c) vitellogenesis,

(d) osmotic and ionic balance and

(e) intermediary metabolism.

(a) The post-embryonic development

In Odonata, the physiological approach towards the endocrine system was made first in order to assess the mechanism of moulting and metamorphosis. In the year 1947, two pioneer workers Pflugfelder and Cazal had independently recorded the cytological signs in the ventral glands indicating their involvement in the process of moulting and metamorphosis. These workers observed enhanced activity of the ventral glands during the period of imaginal moult and their degeneration immediately after the emergence of the imago. The active role of the ventral glands in the process of moulting was experimentally confirmed by Deroux-Stralla (1948b) in the dragonfly, *A. grandis.* (Fig. 3.4).

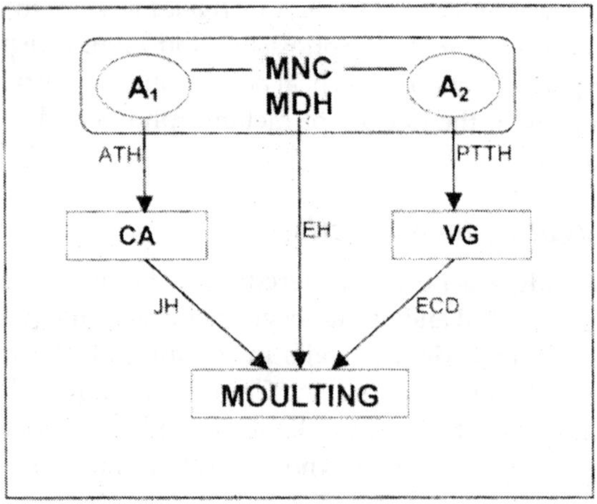

Fig. 3.4 : Endocrine Control of Development. A1 = A2-MNC, A1, A2 = medial neurosecretory cells; ATH = allatotropic hormone; CA = corpora allata; ECD = ecdysone; EH = eclosion hormone; JH = juvenile hormone; MDC = mid-dorsal cellos; PTTH = prothoracicotropic hormone; VG = ventral glands.

Besides the ventral glands, involvement of other endocrine components, such as the NSC of pars intercerebralis (PI), suboesphageal ganglion, the corpora cardiaca and the coropora allata was investigated in various species of Odonata (Arvy and Gabe, 1953). Maximum acitivty in the NSC of pars intercerebralis and consecutive enlargement in the size of CC and CA with the onset of metamorphosis of ultimate larva to imago was noticed on one hand and correlative activity among the NSC of suboesophageal ganglion and the ventral glands, on the other.

On similar lines, Schaller and co-workers made very valuable contribution (1959-1980). In *A. cyanea* it is now well established that the moulting cycle is controlled mainly by two hormonal factors, the moulting hormone (Ecdysone) secreted by the ventral glands and the prothoracicotropic hormone (PTTH) produced by the NSC of the pars intercerebralis. Besides these two endocrine centres, the corpora allata play a key role in metamorphosis. Whenever, normal activity of these endocrine component is inhibited (Schaller *et al.*, 1974) or disturbed by the external environmental factors, the developing larvae undergo diapause (Charlet, 1977).

The ventral glands : Extirpation of ventral glands results into the formation of 'permanenet larva' due to total inhibition of the process of moulting. Implantation of the ventral glands in even 200 days old permanent larva switches on the moulting cycle suggesting univocally a key role of the ventral glands in facilitating the process of moulting (Schaller, 1959,1960a-c). It is further pointed out that the extirpation of the ventral glands within 24 hours since emergence of last instar larva can only prevent the moulting and if it is carried out after 24 hours does not cause any harm to the normal process of development. It suggests that the ventral glands secrete a sufficient quantity of ecdysone within a period of initial 24 hours and, after exercising series of events throughout intermoult period of 25 days, the hormone brings the transformation of last instar larva to the adult dragonfly (Schaller, 1962c). Whenever, the permanent larva (resulted due to ventralectomy) is injected with synthetic ecdysone, the moulting is induced and the larva simultaneously exhibits the adultoid characters in *A. cyanea* (Mouse *et al.*, 1973). It is also revealed that ecdysone acts mostly on the epidermis initiating apolysis and subsequently, synthesis of a new cuticle. With the help of radiolabelled ecdysone administration, Schaller and Hoffman (1973) observed the activity of ecdysone initiating on the 16th day and reaching to the maximum on the 20th day during the intermoult period in the last instar larva. Schaller and Charlet (1980) noticed high titer of α- and β-ecdysones, particularly 20-hydroxy-edysone (β-ecdysone) during the last third duration of intermoult period.

The first period is moreover related to the conjugation and oxidation in order to produce 3-dehydro-ecdysone (α-ecdysone). During the second period of 16th to 25th days, intense hydroxylation takes place in order to produce 20-hydroecdysone. It is further observe that the 3-dehydrogenation takes place in the midgut and Malpighian tubules, while the 20-hydroxylation in the fat body and Malpidhian tubules. Schaller *et al.*, (1975), however, noticed no adverse effect of extirpation of the ventral glands on the rate of 20-hydroxylation. *In vitro* studies made with radioimmunoassy (RIA) and high pressure liquid chromatography (HPLC), Schaller and Charlet (1980) confirmed the results of earlier studies provoking two peaks of ecdysone activity during intermoult period in the last instar larvae of the dragonfly, *A cyanea*. Schaller and Hoffman (1976) noticed that the resumption of moulting in the permanent larvae after implantation of ventral glands depends mostly on the number of ventral gland-pairs implanted and the age of the donor from which they are removed. The higher moulting activity occurs when the donor is near the end of the intermoult period. Schaller and Charlet (1980) reported that the amount of ecdysone produced by a single pair of glands is sufficient to cause the whole effect. The critical period for extirpation of ventral glands from donors is the 16th to 21st day and the ventral glands taken out during this period are always able to restore efficiently the moult of the recipients.

(ii) The NSC of pars intercerebralis: Schaller and Charlet (1970) noticed accumulation of NSM in the A cells of pars intercerebralis, paramedial cells of protocerebrum and ventral ganglia in the permanent larvae after ventralectomy following 102 days and the release of NSM from the NSC after implantation of the active ventral glands on the second day. They also observed rapid accumulation of NSM in these NSC following degeneration of the ventral glands in the newly moulted imago. They reported that the pseudodiapause can be induced by performing ventralectomy. On the strength of these observations, the feedback mechanism in between the NSC and ventral gland activity has been established.

Charlet and Schaller (1975a, b) observed complete block of moulting due to electric coagulation while resumption of the same was made out after implantation of pars intercerebralis-corpora cardiaca (PI-CC) complex. The electric coagulation of PI block moulting if it is performed within one to two days in the penultimate and two to three days in the ultimate larvae since their emergence suggesting these respective periods as the critical one in the last two instars of *A. cyanea*. If the electric coagulation is performed after the critical period, then the moulting is simply delayed. A single PI-CC complex or 5 to 6 PI required to exert sufficient prothoracicotropic effect on the inactive ventral glands in the larvae from which electric coagulation of PI was made earlier. Schaller and Hoffman (1976) reported that PI-CC complex taken from the donor for implantation in the permanent larva exerts prothoracicotropic effect regardless of the number of the complexes or the age of the donor.

Schaller and Charlet (1980) have recently, emphasized that the prothoracicotropic hormone (PTTH) of *Aeschna* probably has a less degree of specificity and may affect, in thfirst instance, general metabolism before activating the ventral glands. The brain-CC, after implantation in the PI-cauterised larvae or diapausing larvae switch on the process of moulting regardless its physiological state, as they are ever capable of stimulating the ventral glands. Tembhare and Andrew (1994) noticed that in *Tramea virginia* last instar larva, the cerebral A1 and A2 cells secrete allatotropic (ATH) and prothoracicotropic (PTTH) hormones respectively. Both, the A2 cells and ventral glands degenerate completely in the last moult and are lacking in the adult dragonflies. This is perhaps, the first report signifying the role of medial A1 and A2 cells in Odonata.

(iii) The corpora allata : It is perhaps difficult to perform surgical operation such as extirpation of the CA, in the dragonfly larvae and adults due to their much inconvenient location in the head. The direct evidence, therefore, could not be collected for a long time (Schaller *et al.*, 1974).

Implantation of CA into the last instar larvae yielded two types of unusual forms, the supernumerary larvae and adultoids as a result of total or partial inhibition of metamorphosis respectively (Schaller, 1959, 1960, 1962c). These results confirm, through indirectly, that the CA remain inactive throughout the life of last instar larvae and imaginal moulting takes place in the absence of a juvenile hormone. Administration of FME in the last instar larvae also produced the supernumerary and adultoid forms. The action of FME found to be directly dependent on the doses and age of the experimental larvae (Schaller and Defossez, 1974).

It has been pointed out that an administration of FME at 700 μl prevents the metamorphosis completely producing supernumerary larvae while that of 200 to 400 μl quantity of FME inhibits the metamorphosis only partially producing the adultoids. It is further suggested that the higher dose creates the juvenile and the lower dose signifies the prothoracicotropic effects of the FME. More than 700 to 800 µl, quantity of FME exerts the lethal effects. Whenever FME is injected at the range of 200 to 400 µl, the individual larva moults in several transitional forms in accordance with the dose and thus confirms the determination of successive metamorphic changes occurring normally during the post-embryonic development of the dragonfly by the juvenile hormone. The action of the ecdysone, however, is reported to be dose-dependent (Mouze, 1971; Defossez and Schaller, 1973).

Schaller and Defossez (1974) observed that the FME can exert juvenile effects if injected before the 6th days since the emergence of last instar larve more or less at the same time of as activity of PTTH. The synchronization of the activity of PTTH- ecdysone and ATH-JH is established first so that the entire programme of the proess of moulting and metamorphosis is computerized at the beginning of the intermoult period and the required morphogenetic and bio-chemical process are exercised accordingly.

(iv) The paramedial NSC of protocerebrum : Charlet and Schaller (1976) noticed active role of paramedial A cells during the process of ecdysis. They observed that after the electric coagulation of these cells, the last instar pharate larvae became incapable of casting off their exuviae suggesting the neuroendocrine control of exuciation. It is, therefore, recommended that these cells might be producing the eclosion hormone.

(b) The diapause

The last instar larva of the dragonfly undergoes diapause for a longer period mostly due to totally different environmental conditions, it faces primarily photoperiod and secondarily temperature (Corbet, 1956; Schaller, 1960, 1965). (Fig. 3.5).

The photoperiod seems to be stimulating the acitivity of the neurosecretory cells of pars intercerebralis directly and, in the absence of sufficient photoperiodic stimuli these cells become inactive inducing the diapause. During diapause, the ventral glands are also inactive due to an absence of PTTH. If the larvae are transferred to the long photoperiod in the laboratory, them the diapause is terminated and larvae resume successive development rapidly. Whenever, active brain-CC complex is implanted from non-diapausing into diapausing larvae, the diapause is terminated but on the contrary, the active ventral glands are incapable of doing so (Schaller, 1960, 1962a, b).

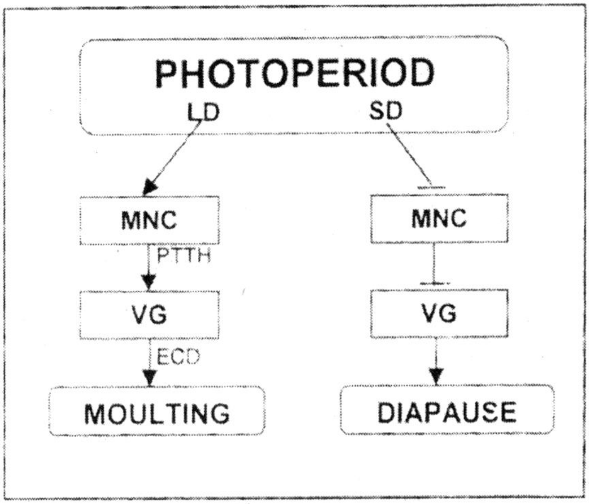

Fig. 3.5 : Endocrine Control of Diapause. ECD = ecdysone; LD = long day; MNC = medial neurosecretory cells; PTTH = prothoracicotropic hormone; SD = short day; VG = ventral glands; —▶ = stimulatory; —| = inhibitory.

The ecdysone efficiently induces apolysis, and exuviation in the diapausing larvae suggesting role of prothoracicotropic hormone not only in stimulating the ventral glands but also in controlling their activity for a long time so that sufficient amount of ecdysone is secreted (Schaller, 1962a, b, 1964).

(c) *Vitellogenesis*

Neuroendocrine control of vitellogenesis in Odonata has been explored in the dragonfly, *Orthetrum chrysis* only (Tembhare, 1973). The ovaries in the dragonfly, *O. chrysis* are paired structures and show gradual increase in their size, weight and cytological changes in relation to oogenesis and vitellogenesis since emergence of the female dragonfly till the oviposition occurring first on the 8th day and thus furnishing the first cycle of egg maturation. Each ovary is composed of a large number of panoistic ovarioles and the process of vitellogenesis takes place only in teh terminal oocytes of the ovarioles. During vitellogenesis, the follicular epithelium and ooctyte both undergo successive cytoarchitectonic modifications in order to facilitate the incorporation of yolk material. The carbohydrate, protein and lipid yolk are exogenous and their accumulation takes place in a specific manner, firstly carbohydrates followed by proteins and lastly the lipids through the follicular epithelium (Tembhare and Thakare, 1975). (Fig. 3.6).

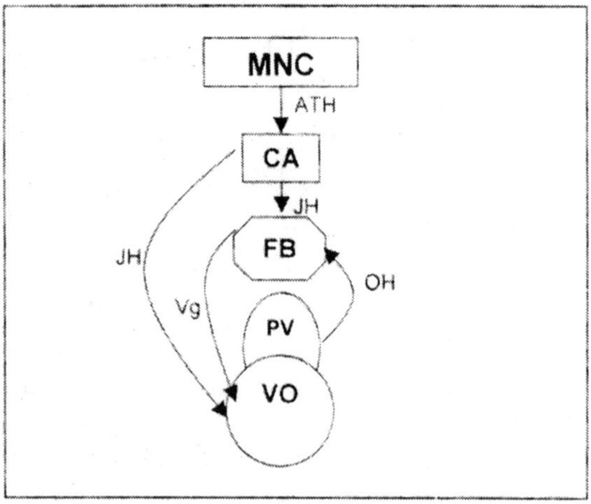

Fig. 3.6 : Endocrine Control of Vitellogenesis. ATH = allatotropin hormone; CA = corpora allata; FB = fat body; JH = juvenile hormone; MNC = medial neurosecretory cells; OH = ovarian hormone; PV = previtellogenic oocyte; Vg = vitellogenin; VO = vitellogenic oocyte.

The correlative histological, histochemical, biochemical and experimental studies demostrated that the deposition of yolk material initiates on the fifth day and continues till the beginning of chorion formation on the seventh day. The A-MNC of the brain and the CA exhibit concomitant enhanced secretory activity during the period of vitellogenesis suggesting allatotropic role of MNC and gonadotroic role of CA during yolk synthesis and deposition (Tembhare and Thakare, 1976b). Injections of farnesyl methyl ether (FME) elevated the haemolymph lipid level significantly confirming gonadotropic role of CA in yolk lipid synthesis. The MNC become inactive after the 5th day, while the CA remain most acitive till the 7th day, i.e. during the period of protein and lipid yolk deposition in the terminal oocytes suggesting the role of CA in performing the process of transport and incorporation of yolk also.

Besides these two endocrine component, synthesis of steroids in the follicular epithelium during the early vitellogenic phase can be demostrated histochemically after performing relevant tests for cholesterol and some enzymes like Δ5, 3β -hydroxysteroid dehydrogene, Glucose-phosphate dehydrogenase and nicotinamide adenine dinucleatide diphorase, suggesting perhaps, steroidal nature of the ovarian hormone (OH) (Thakare and Tembhare, 1977). It is now well-established that the OH in insects is the β-ecdysone stimulating synthesis of vitellogenins in the fat body. Treatment of Precocen II, the anti-JH compound on *Orthetrum chrysis* confirms the role of JH in lipd yolk synthesis and deposition (Thembhare, 1994).

(d) Osmotic, water and ionic balance

Charlet (1974) demonstrated the function of the anterior paramedial and mediodorsal A cells in the regulation of water balance in the larvae of *A. cyanea* by releasing mostly the diuretic hormonal factor. Thakare *et al.* (1977) reported that the medial neurosecretory cells control osmoregulation in the adult dragonfly, *O. chrysis* by producing an antidiuretic factor (Fig. 3.7).

Recently, regulation of osmotic and ionic balance in the haemolymph of the larvae of dragonfly, *A. cyanea* has been demonstrated much adequately (Tembhare, 1979a). Effects of removal of pars intercerebralis-CC, neck ligation and cauterization of the lateral NSC of thoracic ganglia on the haemolymph osmolarity, sodium and potassium concentrations yielded significant data to suggest the thoracic ganglia as a primary source of a hormone playing active role in regulation of osmotic and ionic balance in the haemolymph.

The histological and electron microscopic studies indicated an involvement of the A cells of the thoracic ganglia in order to control osmoticandionic balance in the larvae of the dragonfly, *A. cyanea.* Experimental studies on the larvae of *O. chrysis* also suggest singularly the control of ionic balance of haemolymph by the diuretic hormone released from the A cells of thoracic ganglia (Tembhare, 1984).

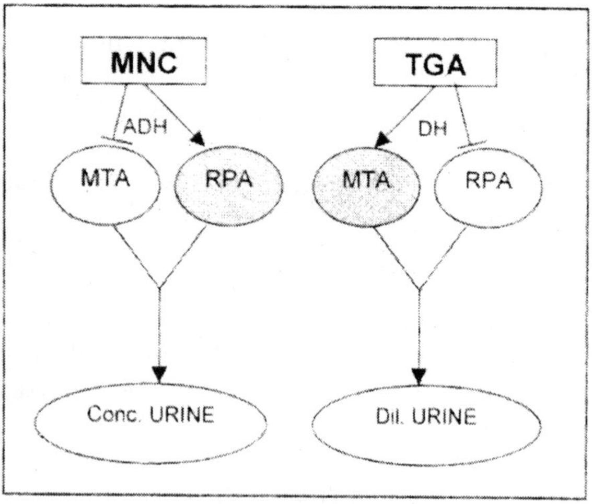

Fig. 3.7 : Endocrine Control of Water and Ionic Balance. ADH = anti-diuretic hormone; DH = diuretic hormone; MNC = medial neurosecretory cells; MTA = Malpighian tibules absorption; RPA = rectal pad absorption; TGA = thoracic ganglionic A-cells.

(d) Hormonal control of metabolism

In *Tramea virginea* the haemolymph protein concentration increases gradually during the intermoult period and diminishes with the onset of moulting. Cauterization of the MNC causes depletion while administration of the CC aqueous extract elevates the haemolymph protein level. The cauterization of MNC and removal of CC reduce the midgut protease activity suggesting control of haemolymph protein concentration by the MNC neurohormone (Tembhare and Andrew, 1991a). Similar to the haemolymph protein concentration, the haemolymph trehalose level increases gradually and as the larva approaches next moulting it falls down greatly. Cauterization of MNC and treatment of FME did not affect the haemolymph trehalose concentration significantly but administration of the CC extract elevates to a great extent suggesting hypertrehalosemic effect of the C extract. As the

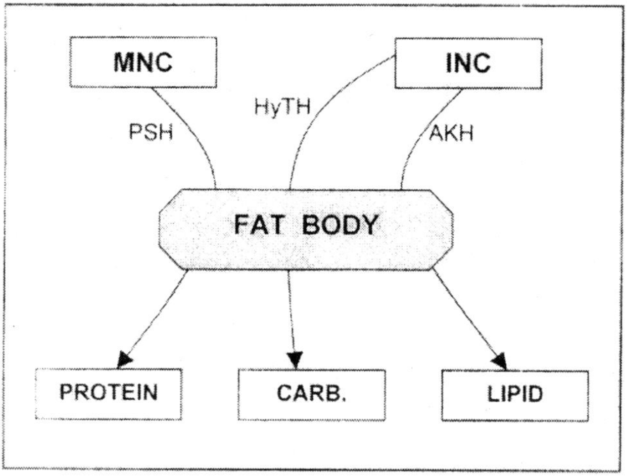

Fig. 3.8 : Endocrine Control of Metabolism. AKH = adipokinetic hormone; HyTH = hypertrehalosemic hormone; INC = intrinsic neurosecretory cells; MNC = medial neurosecretory cells; PSH = protein stimulating hormone.

MNC exert no significant effect, the CC extract seems to contain intrinsic trehalosemic hormone similar to that in other insects (Tembhare, 1991b, 1995, 1997, 1998, 1999). The haemolymph and fatbody lipid concentration rise gradualy till the beginning of succeeding moulting cycle during which it is consumed in adult tissue formation. The CC extract cause hyperlipemia in haemolymph and hypolimia in the fatbody suggesting a presence of the adipokinetic factor in the CC extract (Tembhare and Andrew, 1991c). The adipokinetic hormone (AKH) is isolated and the peptide is characterised in some species of Odonata (Tembhare 1995) (Fig. 3.8).

References

Amans, P. 1881. Recherches anatomiques et physiologiques sur la larve de I' *Aeschna grandis Rev. Sc. Nat. Montpelier,* 3: 63-74.

Andries, J. and G. Tramu, 1984. Detection immunohistochemique de substances apperenteesa des hormones peptidiques de Mammifers d' *Aeshna cynea* (Insecte, Odonate). *C.R. Academie Science Paris,* 299 : 181-184.

Arvy, L. and M. Gabe, 1952. Donnees histophysiologigqes sur les formations endocrines retrocerebralis de quelques Odonates. *Annls Sci. Nat. Zool.* 14 : 345-374.

Arvy. L. and M. Gabe, 1953. Donnees histophysiologique sur la neurosecretion chez les paleopteres (Ephermeroptereset Odonates). *Z. Zellforsch. Mikrosk. Anant* 38: 591-610.

Arvy, L. and M. Game, 1962. Histochemistry of the neurosecretory product of the pars intercerebralis of pterygote insects. *Mem. Soc Endocrinol.* 12: 131-344.

Baldus, K. 1924 Untersuchungen uber Bau and Funktion des Gehirnes der Larven und Imago von Libellen. *Z Wiss. Zool.* 121 : 557-620.

Brandt, J.F. 1836. Bemerkungen uber die Mund Magen. Oder Eingeweidenerven der Evertebraten. *Mam Acad St. Peterbourg* 2.

Cazal, P. 1947. Recherches sur les glandes endocrines retrocerebrales des insects.2. *Odonates Arch. Zool Exp. Gen.* 85.: 55-82.

Cazal, P., 1948. les glands endocrine retrocerebrales des insectes (etude morphologique) *Bull Biol. (Suppl.)* 32: 1-227.

Charlet, M., 1969. Etude des cellules neurosecrtrices dans la chaine nerveuse ventrale d' *Aeschna grandis* L. (Insecte Odonate) *C.R. Acad. Sc. Paris* 269: 1554-1557.

Charlet, M. 1971. Presence d' organes neurohemaux associes a la chaine nerveuse ventrale d' *Aeschna cyane* Mull. *C.R. Acad. Sc. Paris.* 272: 2910-2912.

Charlet, M. 1972b. Etude du system neurosecreteur d' *Aeschna cynea* Mull. These 3e cycle. Strasbourg.

Charlet, M. 1972b. Etude histologique de la pairs intercerebralis de la larve d' *Aeschna cyane* Mull. *C.R. Acad Sci. Paris* 275: 1047-1050.

Charlet, M. 1972c. Evoulation dusysteme neurosecreteus de la pars intercerebralis d' *Aeschna cyanea* Mull. *C.R. Acad Sc. Paris.* 275 : 2255-2258.

Charlet, M. 1974 Mise en evidence d'un centre peurosecreteur protocerebral intervenant dans I' equilibre hydrique de la larve d' *Aeschna cyanea* Mull. Charlet, M, 1979. *C.R. Acad. Sc. Psris.* 279 : 835-838.

Charlet, M. 1977. These do Doctorat d' Etat Univ. Strasbourg, A.O. CNRS.

Charlet, M. and F. Schaller, 1974. Etude des cellules neurosecretrices cerebrales non-associees a da pars intercerebralis chez la larvae d' *Aeschna cyanea* Mull.

Charlet, M. and F. Schaller, 1975a. Resauration de la mue chez des larves d' *Aeschna cyanea* Mull rendues permanents par electrocoagulation de la pars intercerebralis. *C.R Acad Sc. Paris* 281 : 1617-1619.

Charlet, M. and F. Schaller, 1975b. Effects de la destruction de la pars intercerebralis sur la meu chez la larve d' *Aeschna cyanea* Mull. *C.R. Acad Sc. Paris.* 281 : 831-834.

Charlet, M. and F. Schaller, 1976. Blocage de I' exuviation chez la larve d' *Aeschna cyanea* apres electrocoagulation d'un centre neurasecreteur du protocerebron anterieur. *C.R. Acad. Sc. Paris* 283 : 1539-1541.

Charlet, M., S. Schaler and P. Joly, 1975. Donnees surles phenomenes de neurosecretion chez les Odonates. *Zool. J. Physiol.* 78 : 279-288.

Corbet, P.S. 1956. Environmental factors influencing the iduction and termination of diapause in the Emperor dragonfly, *Anax imperator* Leach. *J. Exp. Bio.* 33 : 1-14.

Defossez, A and F. Schaller, 1973. Inhibition de la metamorphose de I' appareil copulateur males des larves d' *Aeschna cyanea* Mull. Par injection d' un mimetique de I'hormone juvenile. *C.R. Acad. Sc. Paris (D).* 276 : 2051-2054.

Deroux-Stralla, D. 1948a. Recherches anatomohistologique preliminaires a une etude des mechanismes endocerines chez les Odonates. *Bull. Soc. Zoll. France.* 73: 31-36.

Deroux-Stralla, D. 1948b. Recherches experimentales sur le role des glandes vent rales dans la mue et la metamorphose chez *Aeschn cyanea* Mull. *C.R. Acad Sc. Pris.* 227 : 1277-1278.

Gabe, M. 1972. Donnees histochemiques sur I' evolution due produit de neurosecretion protochephalique des Insects Pterygotes au cours de son cheminement axonal Acta. *Histochem* 43 : 163-183.

Gillott, C. 1969. Morphology and histology of the cephalic endocrine glands of the damselfly, *Coenagrion angulatum* Walker. *Can. J. Zool.* 47 : 1187-1192.

Gillott J.P., A Provansal, N. Baudry, and M. Raabe, 1971. Les organs perisymphathiques des insctes Pterygotes. Les Prinipaux types morphologiques. *C.R. Acad. Sci. Paris.* 273 : 2129.

Hanstrom, B. 1940. Inkretorische Organe, Sinnersorgane und Nervensystem des kopfes einiger niderer Insektenordsugen. K. Svenska Vetensk. *Akad Handl.* 18: 1-265.

Hanstrom, B. 1942. Die corpora cardiaca und corpora allta der insekten. *Biol. Gen.* 15 : 485-531.

Maddrell, S.H.P. 1970. Neurosecretory control system in insect. In : *Insect Ultrastructure* (A.C. Neville, ed). Blackwell Sci. Pub Oxford/ Edinburgh.

Mouze, M. 1971. Role de I'hormone juvenile dans la metamorphose oculaire de larves d' *Aeschna cyanea* Mull. *C.P. Acad. Sc. Paris.* 273 : 2316-2319.

Mouze, M., J.C. andies and F. Schaller, 1973. Effects sur la mue et la metamorphoe de I' injection de'ecdysone a des larves paramenents *d'Aeschna cyanea* Mull. *C.R. Acad. Sc. Paris,* 277 : 2509-2512.

Pflugfeder, O. 1947. Uber die Ventral drusen and einige andere inkretorische organe des Insekfenkopfes, *Biol Zbl.* 66 : 211-235.

Raabe, M. 1964. Nouvelles recherches sur laneurosecretionchez les Insects. *Ann. Endocr.* (Supp.) 25: 107-112.

Raabe, M. and A. Provansal, 1972. Les organs perisympathiques des Paleopteres. *C.R. Acad. Sci. Paris* 275 : 925-928.

Schaller, F. 1959. Control humoral due development postembryonnaire *d'Aeschna cyane* Mull. *C.R., Acad Sc. Paris.* 248 : 2525-2527.

Schaller, F. 1960. Etude du development posteembryonnaire *d'Aeschna cyanea* Mull. *Ann. Sc. Nat. Zool. Fr.* 12 : 751-868.

Schaller, f. 1962a. Role de la diapause et de la vitesse du development dans le cycle biologique d' *Aeschna cyanea* Mull. *Bull. Assn Philomath. Als Lorr* 11 : 128-137.

Schaller, F. 1962b. Rupture dè la diapause larvaire chez *Aeschna cyanea* Mull. Para implantation de complexes cerveaucorps cadiaquesde larves activees. *C.R. Soc. Biol.* 156 : 658-661.

Schaller, F. 1962c. Phenomenes d'inhibition de la metamorphose chez des larves agees *d'Aescha cyanea* Mull. *Bull. Soc. Zool. Fr.* 87 : 582-600.

Schaller, F. 1964. Action de I'ecdysone sur des larves *d'Aeschna cyane* Mull. en diapause. *Mem. Soc. Nat. Math. Cherbourg* 6: 135-140.

Schaller, F. 1965. Action de la photoperiode croissante sur des larves an diapause, d'*Aeschna cyanea* Mull. Maintenues a basse temperature. *C.R. Sco. Biol.* 159 : 846-849.

Schaller, F. and M. Charlet, 1970. Evolution du systeme neurosecreteur de larves d' *Aeschna cyanea* Mull. Privees d' ecdysone. *C.R. Acad. Sci. Paris.* 271 : 2004-2007.

Schaller, F. and M. Charlet, 1980. Neuroendocrine control and rate of ecdysone biosynthesis in larvae of a paleopteran insect, *Aeschna cyanea* Mull. In : *Progress in Ecdysone Research* (J.A. Hoffmann, ed) pp. 99-110. Elsevier/North-Holland, Biochemical Press, Amsterdam.

Schaller, F. and J.A. Defossez, 1974. Inhibition de la metamorphose, plasrve d' *Aeschna cyanea* Mull. Par un mimetique de I' hormone juvenile. *Withelm Roux' Arch Entw Mech Org* 172 : 20-32.

Schaller F. and J.A. Hoffman, 1973. Evolution due taux d' ecdysone au cours du dernier stade larvaire d' *Aeschna cyanea* Mull. *C.R. Acad. Sci Paris (D).* 277 : 741-743.

Schaller, F. and J.A. Hoffman, 1976. Metabolism de L' a-Ecdysone in vivo et in vitro chez *Aeschna cyanea*. Coll. Internat. *CRNS Atualite's sur les hormones d' invertebrates*. 261 : 393-401.

Schaller, F. and J. Meunier, 1968. Etude systeme neurosecreteur cephalique des insects Odonates. *Bull. Soc. Zool. France* 93 : 233-249.

Schaller, F., J.C. Andries, M. Mouze and Z.A. Defosse 1974. Neuveaux aspects due controle hormonal du cycle biologique des Odonates : researches sur la larve d' *Aeschna cyanea* Mull. *Odonatologica* 3 : 49-62.

Schaller, F., J.A. Hoffmsn and J. Koolman, 1975. Metabolism de I' ecdysone chez active a des animaux normaux et a' des larves au cycle de mue bloque. *C.R. Acad Sc. Paris*. 2801 : 1813-1816.

Thembhare, D.B. 1973. Studies on the neuroendocrine system in dragondly, *Orthetrum chrysis* (Selys) Ph.D. Thesis, Nagpur University.

Thembhare, D.B. 1979a. Hormonal regulation of osmotic and ionic balance in the haemolymph of the larvae of drangonfly, *Aeschnacyanea* Mull. *Arc. Internal, Physiol. Bioch.* 87 : 557-563.

Tembhare, D.B. 1980. An electron microscopic study of the neurosecretory pars intercerebralis-corpus cardiacum system in larvae of the dragonfly. *Aeschna cyanea* Mull. *Z. Mikrosk-anat Forsch. Leipzig*. 93 : 60-72.

Tembhare, D.B. 1984. Hormonal regulation of the ionic balance in the haemolymph of the larvae of the dragonfly, *Orthetrum chrysis* (Selys). *Odonatologica* 13 : 565-572.

Tembhare, D.B., 1994. Effect of precocene II on yolk lipid depostion in the dragonfly, *Orthetrum chrysis*. *Fraseria* (N.S.) 1: 15-18.

Tembhare, D.B. 1995. Molecular chemistry of insect hormones. In *Recent advances in insect endocrine research* (Eds. D. Mulaleedharan and M. Jacob). Association Adv. Entomol. Triandrum, pp. 39-67.

Tembhare, D.B. 1997. *Modern Entomology*. Himalaya Pub. House, Mumbai.

Tembhare, D.B. 1998. Odonate ovary and vitellogenesis. *Fraseria* (N.S.) 5 : 15-18.

Tembhare, D.B. 1999. Recent development in insect endocrinology. In *Advances in Insect endocrinology and Physiology*. (Ed. R.K. Bhola) Kalyani Publ. Ludiana.

Tembhare, D.B. and V.K. Thakare, 1975. The histological and histochemical studies on the ovary in relation to vitellogenesis in the dragonfly, *Orthetrum chrysis* (Selys) Z. mikrosk-anat. *Forsch. Leipzig* 89 : 108-127.

Tembhare, D.B. and V.K. Thakare, 1976a Cephalic neuroendocrine system of the dragonfly, *Orthetrum chrysis* (Selys) *Odonatologica* 5 : 355-370.

Tembhare, D.B. and V.K. Thakare, 1976b. Neuroendocrine control of vitellogenesis in the dragonfly, *Orthetrum chrysis* (Selys) Z. mikrosk-anat. *Forsch Leipzing* 90 : 691-704.

Tembhare, D.B. and V.K. Thakare, 1977. Neurosecretory system of the ventral ganglia in the dragonfly, *Orthetrum chrysis* (Selys), *Cell. Tiss. Res.* 177. 269-280.

Thakare, V.K. and D.B. Tembhare, 1975a. Histochemistry of the heorusecretory cells of the pars intercerebralis of the dragonfly, *Orthetrum chrysis* (Selys) *Odonatologica* 4 : 225-235.

Thakare, V.K. and D.B. Tembhare, 1975b. The effect of electrical stimulation on the cephaic neurosecretory system in the last instar nymph of the dragonfly. *Orthetum chrysis* (Selys). *Experimentia* 31 : 693-694.

Thakare, V.K. and D.B. Tembhare, 1977. Light microscopic structure and steroid histochemistry of the ovary in relation to steroidogenesis in the dragonfly *Orthetrumchrysis* (Selys). *ZoologicaPoloniae.* 26 : 556-572.

Thakare, V.K., B.S. Thakare, and D.B. Tembhare, 1977. Neurosecretory control of osmoregulation in the adult dragonfly, *Orthetrum chrysis* (Selys) *Zool. Beit.* 23 : 311-321.

Thakare, V.K., D.B. Tembhare, M.W. Khan, and V. Pendharkar, 1980. Chemical composition of haemolymph in the larvae of the dragonfly, *Orthetrum chrysis* (Sylys) *Odonatologica* 9 : 321-324.

Tembhare, D.B. and R.J. Andrew, 1991a. Neuroendocrine influence on haemolymph protein concentration and midgut protease activity in the final instar larva of *Tramea virginia* (Rambur). *Odonatologica* 20 : 91-95.

Tembhare, D.B. and R.J. Andrew, 1991b. Hormonal influence on the haemolymph trehalose concentration in the dragonfly, *Tramea virginia* (Rambur) larva. *Entomon* 16 : 165-166.

Tembhare, D.B. and R.J. Andrew, 1991c. Hormonal influence on the haemolymph and fat body lipid concentration during development and moulting of the ultimate nymph of the dragonfly, *Tramea virginia* (Rambur). Indian J. *Cmop. anim. Physio.* 9 : 74-80.

Tembhare, D.B. and R.J. Andrew, 1994. Activity of the cephalic neuroendocrine system during development of the last instar larva of *Tramea virginia* (Rambur), *Adv. Oriental Odonatol.* pp. 31-40.

Tillyard, R.J., 1917. *The Biology of Dragonflies.* Cambridge University Press.

4

Ecology of Larval Odonates in Lentic Freshwater Ecosystems

• *Arvind Kumar*

Abstract

The present investigation deals with the seasonality, species diversity, richness of species and secondary productivity of larval odonates as well as the various abiotic factors of the surroundings from January to December, 2000. The anisopteran larvae showed trimodal peaks wheareas zygopteran larvae exhibited bimodal in their seasonality during the period of study. The species diveristy varied between 0.332 to 1.918 bits and between 0.672 and 1.902 bits in anisopteran and zygopteran larvae respectively.

The biomass of the larvae was recorded maximum in September (0.553 g/haul) whereas minimum was in April (0.212 g/haul). The net growth of the population varied between 0.0087 g. dry wt./cub. met./day to 0.0005 g.dry wt./cub.met./day having positive growth in the population. The product moment correlation coefficient for biomass of larval population with different abiotic factors of water were also computed and dissolved oxygen (DO) showed marked correlation with the biomass of the odonata larvae ($r=0.898$; $p>0.05$).

Introduction

Odonates are usually the most conspicuous insect group found near and body of inland water and commonly called as "**devil's darning needle**", "**snake feeder**", "**snake doctor**" and "**horse stinger**". The common name "**mosquito hawk**" refers to their value in destroying mosquitoes, gnats and other pests (Smith and Pritchard, 1956). The nymphal odoantes are predaceous like the adults and spending their entire larval period in the inland water.

These larval population may often be a predominant feature of pond ecosystems and they are the most abundant fauna of benthic and littoral zones. Their role in the structure and function of the littoral zone communities appears to be much more important than has been previously recognised. As dominant members of the benthic and littoral fauna, they are considered promising organisms for use in the organic pollution biomonitoring because of their ease of collection, large numbr of species and also sensitivity to the water quality. Thus, the larval odonates, which are invariably aquatic and inhabit in almost all types of aquatic habitats, such as fast and slow running streams, rivers, lakes, ponds etc., show functional-morphological adaptation to the environment in which they inhabit. Odonates had evolved and flourished during the carboniferous period and had predatory mode of feeding. These insects are preying upon the fish spawn, fish fry and even the fingerlings and are troublesome for pisciculturists since long (Wilson, 1923 and 1924). Since odonates are at the top of predatory food chan, their mere presence in any habitat shows the conditions are varied, congenial and healthy; supporting diversified faunal components, which in turn supported by complex flora. It is said that the complex flora allows for a greater diversity of faunal components and more complex food webs (Boyd, 1971).

The available literature suggests that odonate larvae are top carnivores (Lawton, 1970; Benke, 1972, 1976; Hassan 1974, Roy, 1984; Kumar, 1996), and thus play a significant role in limiting the numbers of organisms present at the lower trophic levels of the food chain. These larvae are feeding on various zooplankton, aquatic insects, fish spawns, fish fry and fingerlings and are secondary and tertiary consumers. It is well known fact that in the absence of these top canivores, no balanced ecosytem could continue to persist. Therefore, the ecosystem can be manipulated and monitored with the help of organisms occupying at the apex of the food chain. As evident from Lotka (1924) and Volterra (1925) equations, a predator population is prerequisite in keeping the ecosystem in a state of balance.

These larvae may be distinguished from all other aquatic insect larvae by two strong characters: the huge grasping labium and the form of tracheal gills which serve for respiration. Snodgrass (1954) had given an account of the comparative morphology of the odonate larvae. The labium of the adult and larval odonates was studied by the Butler (1904). Corbet (1953) established the terminology for the labium of the larval odonates. Corbet (1962) has also discussed the morphological and ecological adaptations shown by larvae associated with respiration, feeding and concealment. Kumar (1973a and b) also studied such adaptations in relation to their habitat. In Anisoptera, the inner surface of rectum has become modified and is richly supplied with tracheae and called rectal gills. This serves for respiration as well as jet-propulsive device throughout the larval life.

In Zygoptera, however, the epiproct and paraproct have beome specialised to form foliaceous lamellae. The function of these lamellae has been a subject of much controversy for sometime, but there is no doubt that the lamellae have evolved as a supplement to other means of respiration in larve (Macneill, 1960; Corbet, 1962). Furthermore, the shape and size of the lamellae can be said to be generally in inverse proportion to the availability of oxygen in a normal habitat. The lamellae of zygopterous species which are found in slow-running marshy streams or ponds are foliaceous, mostly duplex type and show protrusive growth and have well developed tracheation, thus, being able to breathe maximum amount of oxygen. On the other hand, those zygopterous larvae which inhabit in hill streams (in which oxygen concentration is quite high), the larvae show reduction both in the surface area and the extent of the tracheation of the lamellae and these are simplex. Thus, great adaptive rediation has been shown by these larve in colonizing different habitats and their body form has been modified accordingly. These larvae exhibit some sort of behavioural adaptation to their environment. Protection from predators, like carnivorous (Corbet, 1962) has also greatly influenced their concealment behaviour. Some carnivorous fishes such as *Wallago attu* have been recorded as a major predator on odonate larvae (Corbet, 1961, 1962; Data Munshi *et al.*, 1991).

Niche specialization has also been encountered in these larvae. If they live in relatively fine sediment, they may borrow into it, cover themselve with it or hide among debris which accumulate near the bottom, so-called "burrowers" or "hiders". Such dragonflies usually lay eggs exophytically. If, on the other hand, they live among large stable objects, such as stone, or upright plants, they grasp these tightly and do not attempt to cover themselves. Dragonflies of the second type usually lay eggs endophytically (Kumar and Khanna, 1983). These larvae are found in coarse sand near banks of wetlands and are found more or less covered with it. In the Libellulidae, some larvae also have tendency to live at the bottom of the ponds burried superficially among debris and mud. Another type of larvae which are almost akin to shallow burrowers can, according to Corbet (1962) be called "hidders". These larvae have a tendency to remain lying flat in the decomposing leaves at the bottom of the ponds. Their abdoman is noticeably flattened dorsoventrally and is oval in shape. The cordullid larvae (Kumar, 1973b) which are also found in the same type of habitat, have been found to have unusually long spider-like legs. The abdomen of this species is almost flattened and is oval in shape. The larvae have a tendency to fold their legs tightly against the abdomen and to keep the abdomen closely attached to the bottom of the pond. Some of the Libellulids live among stable objects are weed-dwellers. These larvae grow quickly and are active in habits. Another type of larvae in this group are those which are found in fast flowing hill-streams. They are dorso-ventrally flattened and are adhered tightly against the stones to which they cling. The flattened shape of these species is assumed to offered least resistance to fast flowing water, and permits the current to pass unempeded over their bodies. The vertical component of the current thereby presses their bodies more tightly against the substrates.

Thus, odonate larvae comprise one of the most important groups of fresh water invertebrates. Kiauta (1975) has given the following characterisation of odonata, "dragonflies comprise one of the most important groups of aquatic animals. Being voracious predators in both immature (aquatic) and adult (aerial) stages, they are important elements of all, except the drier (or high alpine) environment in the temperate and tropical regions, occupying a position at the apex of the food chain of invertabrate life. In addition to their biological importance in such environments, these relationships render them the ultimate accumulator of persistent compounds in polluted water. Thus, they are the ideal organisms to be used as indicator of water pollution and contamination. Because of their unique morphology and physiology, dragonflies are used extensively in the study of many biological phenomena. Thus, the morphological and physiological adaptations of these larvae are of ecological interest because they pertain to their seasonal occurrence and abundance. These larvae live in a habitat in which such physical factors as temperature, water current etc. are the major regulating sources and they play an imortant role in their growth, distribution, emergence as well as in their other physiological attributes. They can undergo quick physiological changes under the conditions of rapid environmental changes.

Considerable works have also been made on various aspects of odonates of which the following deserve special mention (Fraser, 1919; Laidlaw, 1920; Needham, 1930; Asahina, 1942; Snodgrass, 1954; Nicholson, 1958; Corbet, 1962; Benke and Benke, 1975; Wolf and deWolf, 1977; Poyyamoli and Pandian, 1984; Roy, 1984 and 1986; Kumar, 1989; Kaushik *et al.*, 1990; Sandsberg and Bonsdroff, 1992; Tonczyk, 1994; Hanel, 1995; Suri Babu *et al.*, 1996; Lombardo, 1997; Khan and Tembhare, 1998; Chelmick, 1999; Vick, 1999 and Zhu and Zhang, 1999). Thus, available information suggests that very little work has been done on the odonate ecology in relation to their physiological parameters of these larval inhabiting fresh water systems of this north-eastern part of the country except the work of Kumar (1993, 1994a, b, c, 1995a b, c, 1996, 1997, 1998, 1999 and 2000) and Kumar and Kumari (1999). In view of such paucity of information, an ecological study of odonate larvae inhabiting in certain wetlands of Santal Pargana has been undertaken in the present dissertation with special reference to periodicity and secondary productivity.

Materials and Methods

(A) Seasonality of Larvae

The larval odonates were collected at fortnightly intervals from January to December, 2000 between 7.30 to 10.30 a.m. by D-shaped aquatic insect net, made up of nylon cloth. Its mesh size was 40 to 90/cm^2. The D-frame aquatic insect net was made according to Merritt and Cummins (1978). The samples were taken by the method used by Allan (1975). The density of the larval odonates was determined from the five samples taken randomly. The smaples were cleaned, sorted and preserved in 70% alcohol, in the laboratory. Then the larvae were counted and identified. The same practices were done throughout the period of study.

The total number of individuals/haul and number of species (S) were counted and then, the species diversity (H) and evenness of species (J) were computed from the following mathematical formula:

$$H = -\sum_{i=1}^{S} n_i/N \log_2 n_i/N \text{ (Shannon and Weaver, 1964)}$$

where, N is the total number of individuals in S species, n_i is the number of individual in the ith species, and the information content of diversity is expressed as number of bits. The greatest possible value of H_{max} was calculated as follows (after Pielou, 1969).

$$H_{max} = \log_2 1/S$$

where, S species are present, then the evenness (equatibility) was calculated by,

$$J = H/H_{max}$$

(B) Secondary Productivity

The productivity of odonate larve was determined by biomass method. The biomass of larvae were determined according to the method suggested by Winberg (1971). The samples of each hauling were sorted in the laboratory and they were placed in the oven at 105°C for 24 hr and its weight was recorded until a constant weight of the material was recorded. The weight loss at this temperature is considered to represent the ash-free dry weight of the larvae and is used as a measure for their biomass (Beukema, 1976). The biomass g/haul (494.0 litres of water has been passed in each hauling through the net was converted into g/cubic metre of water). Biomass (g/cubic metre day) and net monthly growth (g/cubic metre/month) were also calculated.

(C) Physico-chemical Analysis of Water

The seasonal variations of the physico-chemical factors of the waterbodies have also been studied from January to December, 2000 following the standard methods as suggested by APHA (1985) and Trivedy and Goel (1986).

Results

Seasonality of Larvae

The observations made for the period of one year on larval odonate reveals some kind of interesting seasonal patterns in their abundance. Numbers of both sub-orders, viz., Anisoptera and Zygoptera of Odonates have shown some interesting periodicity in their emergence. The population density of Anisoptera and Zygoptera have been depicted in Figs. 4.1 and 4.2.

Sub-order : Anisoptera

The maximum population density of Anisoptera was recorded in the month of December 2000. In this sub-order the members of the families Gomphidae, Libellulidae and Cordulegasteriadae were recorded. In Gomphidae, *Mesogomphus lineatus* selys was recorded throughout the year except January-February. The maximum population was recorded in September. In Cordulegasteridae, *Cordulegaster* sp. was recorded throughout the year. Therefore, it is considered most abundant larvae among Anisoptera. In Libellulidae, *Potamarcha obscura* was recorded and its maximum population was in August. *Zyxomma petiolatum* was comparatively less abundant larvae and its maximum population density was in March and October.

The niche specialisation and habitat selection has also been observed in these larvae. *Mesogomphus lineatus* was recorded buried in mud, therefore, it is considered as benthic insects. *Cordulegaster* sp. was also found in the muddy buttom, therefore, it is also benthic fauna. *Potamarcha obscura* and *Zyxomm petiolatum* were recorded amongst the aquatic weeds at the bottom.

The total number of larvae/haul (N) number of species (S), species diversity (H) and evenness of species (J) are depicted in

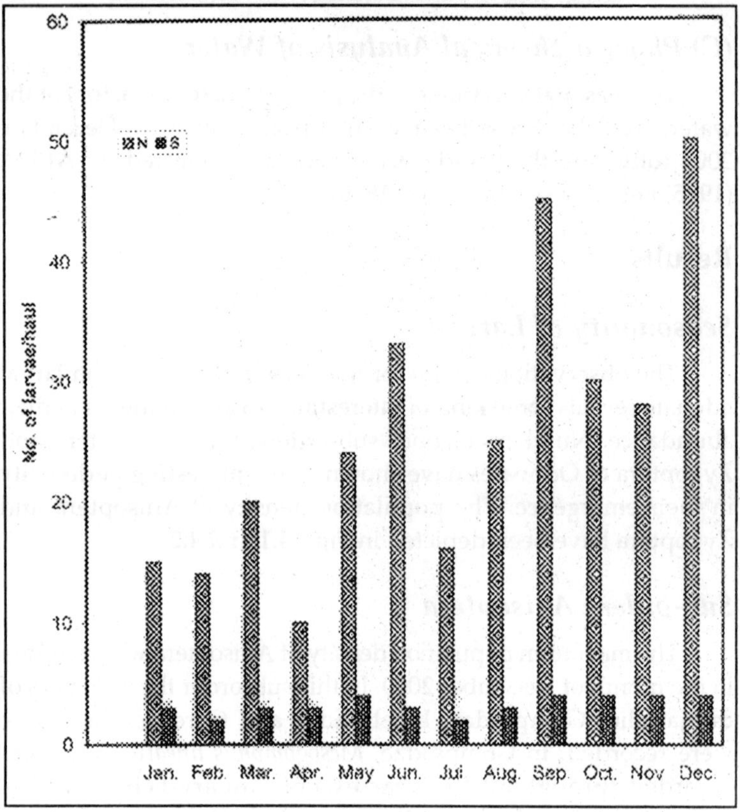

Fig. 4.1 : Monthly variations in population density of Anisopteran larvae.

Fig. 4.3. The species diversity varied from 0.332 to 1.918 bits with minimum and maximum value in July and November. The evenness of the varied from 0.338 to 0.995 with minimum and maximum values in July and March.

Sub-order : Zygoptera

The seasonality of the members of family Coenagriidae and Lestidae have been recorded. Among Coenagriidae, *Ischnura delicata* was the most abundant and outstanding damsel fly larvae contributing subsequently odonate larvae fauna of the water body. It is recorded throughout the year with the maximum number in September and minimum in June. *Ischnura senegalensis* was

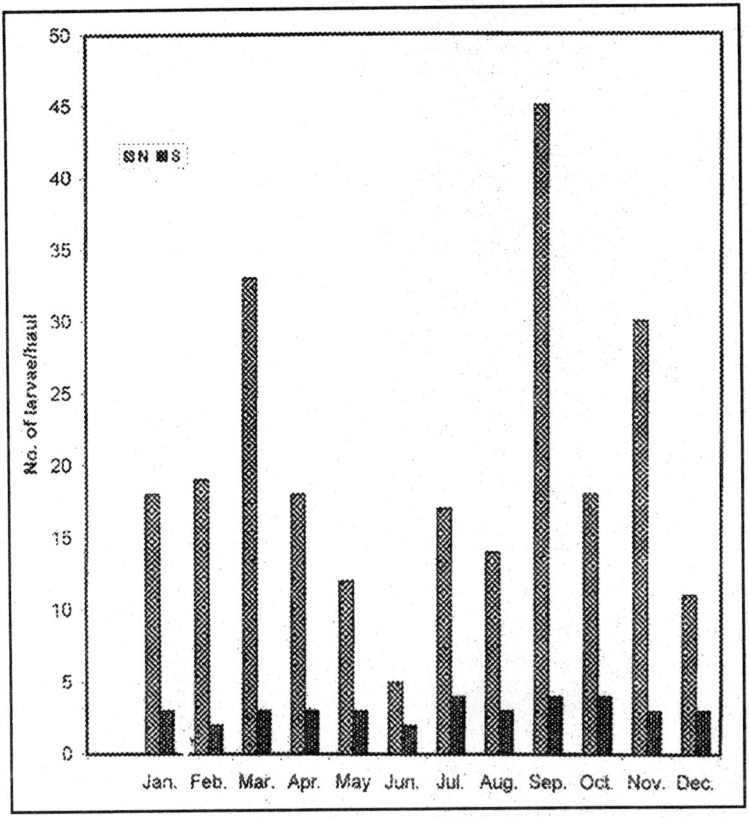

Fig. 4.2 : Monthly variations in population density of Zygoptoran larvae.

recorded regularly from September to December but in decreasing order. *Rhodischnuranursei* was also recorded during monsoon period.

In family Lestidae, *Lestes* sp. was recorded throughout the year except in June. The damsel fly larvae also show niche specialisation and habitat preferences. *Ischnura delicata* was recorded abundantly among the aquatic weeds and algal mat of the water body. *Ischnura senegalensis* was recorded burried in and where profuse algal network was present. *Rhodischnura nursei* was also recorded in aquatic weeds. *Lestes* sp. was recorded from pond bottom where aquatic vegetations were found in abundance.

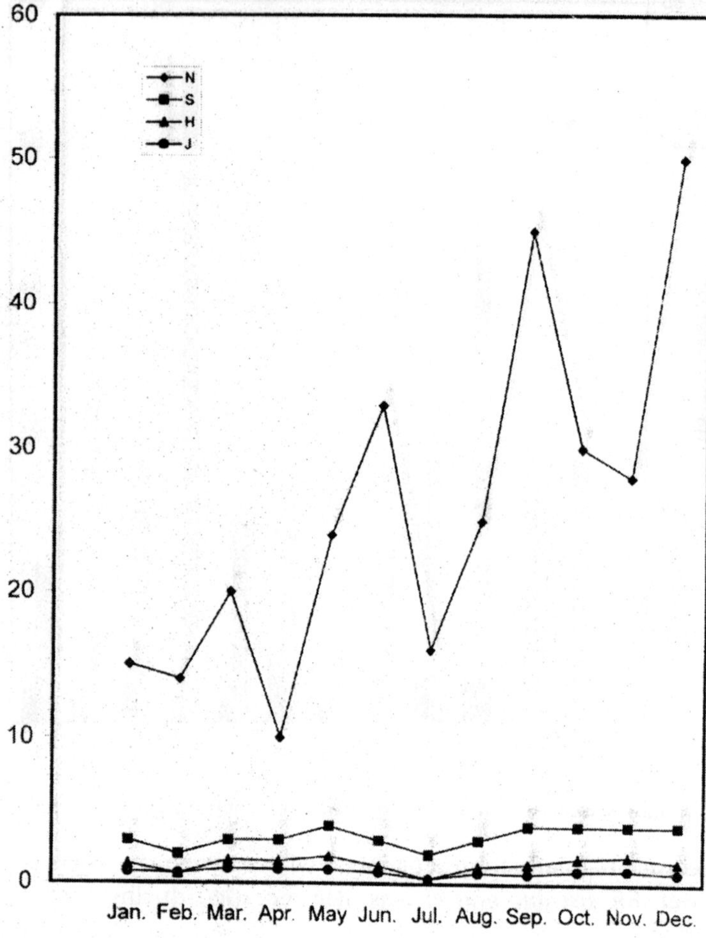

Fig. 4.3 : Monthly variations in tolal no. of larvae/haul (N), no. of species (S), species diversity (H) and richness of species (J) of Anisopteran larvae.

The total number of larvae/haul (N), number of species (S), species diversity (H) and evenness of species (J) are presented in Fig. 4.4. The species diversity varied from 0.672 to 1.902 bits with minimum and maximum values in February and October. The evenness of species varied from 0.673 to 0.955 with minimum and maximum values in February and October.

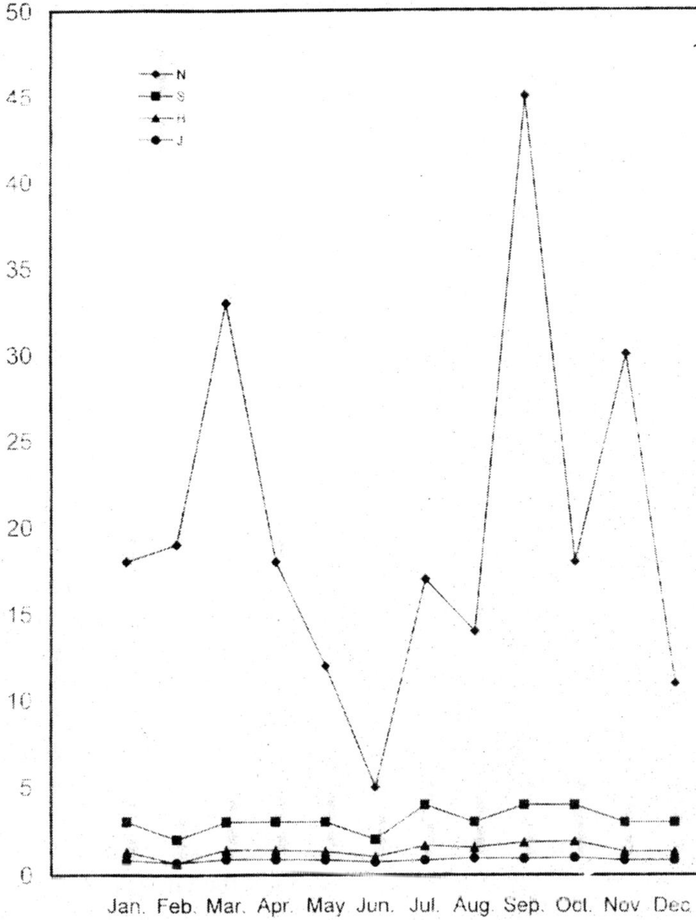

Fig. 4.4 : Monthly variations in tolal no. of larvae/haul (N), no. of species (S), species diversity (H) and richness of species (J) of Zygopteran larvae.

(B) Secondary Productivity

The following larval odonate populations were collected for the estimation of secondary productivity:

1. *Sub-order - Anisoptera :*

(a) Family : **Gomphidae**

(i) *Mesogomphus lineatus*, Selys.

(b) Family : **Libellulidae**

(i) *Potamarcha obscura* Rambur

(ii) *Zyxomma petiolatum* Rambur

(c) Family : **Cordulegasteridae**

(i) *Cordulegaster* sp.

2. *Suborder - Zygoptera :*

(a) Family : **Coenagriidae**

(i) *Ischnura delicata* Hagen

(ii) *Ischnura senegalensis* Rambur

(iii) *Rhodischnura nursei* Morton

(b) Family : **Lestidae**

(i) *Lestes* sp.

The biomass, net montly and net daily growth of the larval population have been presented in Figs. 4.5, 4.6 and 4.7. The maximum biomass (g. dry wt/haul) was recorded in September (0.553 g/haul) whereas minimum biomass was recorded in April (0.212 g/haul). The maximum and minimum biomass g. dry wt. cub met was also calculated which followed the same trend. The biomass of odonate larvae (g. dry wt/cub. met./day) was obtained after dividing monthly net growth of the population (g. dry wt./cub. met./month) by the days of intervals between two successive dates of sampling. The maximum net monthly growth of the larval population was found in summer and monsoon months when usually positive increase in the biomass of the larval population was recorded. However, negative growth of the population was found in winter season. The net daily growth of the population varied between 0.0087 g. dry wt./cub.met./day to 0.0005 g.dry wt./cub. met./day having positive growth in the population. The annual net production come to 1.89 g. dry wt./cub. met./year with montly productivity being 0.098 g.dry wt./cub. met/month.

Physico-chemical Characteristics of the Water

Various abiotic factors of water are depicted in Figs. 4.8 to 4.10. The surface water temperature was recorded between 20.0

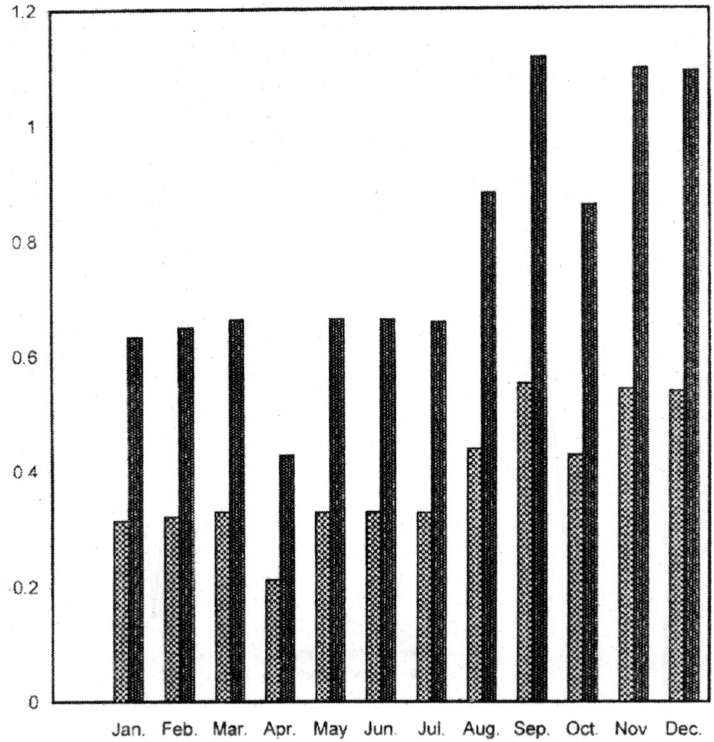

Fig. 4.5 : Monthly variations in biomass (g. dry wt./haul) and biomass (g. dry wt./cub./met.) of Odonate larvae.

and 39.0°C with minimum and maximum values observed in January and June respectively. From the middle of March till the end of September, it remains more or less constant. The pH of water was interestingly observed alkaline throughout the period of investigation and the value ranged from 7.4 to 8.8.

The dissolved oxygen content of the water was comparatively low in summer than the winter and varied form 3.00 to 9.5 ppm. Its minimum and maximum values were recorded in January and November. Free carbon dioxide was observed in the months of April, May, June, November and December and its concentration varied between 7.00 to 12.00 ppm with minimum and maximum recorded in April and December respectively.

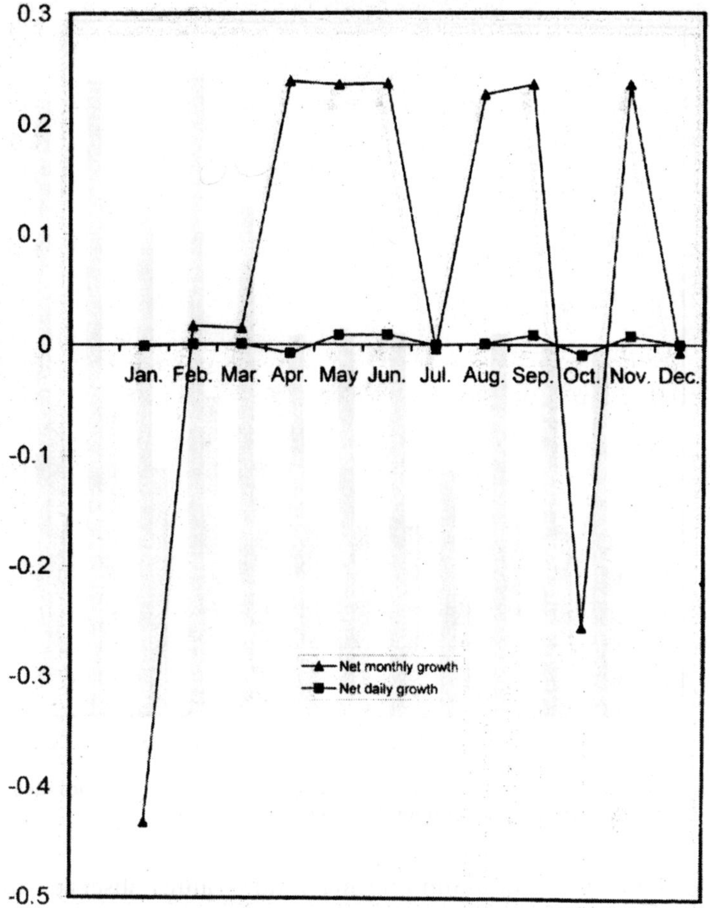

Fig. 4.6 : Monthly variations in Net monthly growth (g.dry wt./cub met/ month) and Net daily growth (g. dry wt./cub. met./day) of Odonate larvae.

The content of carbonate alkalinity (Phenolophthalein) ranged from nil to 9.0 ppm with maximum value in August whereas the concentration of bicarbonate alkalinity (Methyl orange) varied from 140.0 to 280.0 ppm with minimum and maximum values in January and July respectively. It was usually low in winter and high in summer and rainy months. The concentration of chloride content of water was found low and it varied between 0.45 to 0.90 ppm with minimum and maximum values in January and May respectively. Total hardness was quite higher during the present investigation and ranged between 79.0 and 98.0 ppm with minimum and maximum values in the month of January and December respectively.

Discussion

Seasonality of Larvae

The data callected on larval odonate population during the present investigation in a shallow tropical freshwater body of Santhal Parganas (Dumka), show some interesting seasonal fluctuations in their population density. They exhibit some kind of periodicity and seasonality in their abundance. From the present study certain conclusion could be made regarding the seasonality of the odonate larvae in the wetlands.

In the suborder Anisoptera the families Gomphidae, Aeschnidae, Cordulegasteridae and Libellulidae and among suborder Zygoptera, families Coengriidae and Lestidae are well represented by their larval stages. The general trend of population fluctuation of these families revealed that from June to December, there was maximum number of individuals in the wetland in the remaining months of the year relatively low numerical abundance of population ws recorded.

Anisoptera

The observation made on the larval population of dragonflies (Anisoptera) showed somewhat trimodal peaks in their abundance, one in June, second in September and third in December. This observation indicated that some species lay their eggs in spring, i.e. in March and thereafter larval development takes place until they become fully subceptible to sampling in May-June and produced the peaks of larvae in these months. The presence of low population densities from January to April in the period of study in due to low temperature in winter season, because in these months no proper egg laying occurred. Therefore, comparatively lower population density was recorded in these months. But when temperature gets increased from March onwards, egg-laying takes place in these waterbodies with subsequent increase in larval population in May-June and thereafter, the low population density in July-August is due to monsoon flooding and high water discharge from the habitat. The majority of these nymphs are benthic and sedentary. The high water discharge in the monsoon and flooding play an important role in the dispersal of these sedentary larvae. Therefore, the drastic reduction of population of these larvae in June-August of study period is due to monsoon flooding. In October-November, comparatively lower population density was recorded which might be due to emergence larvae from the waterbody. In these months high population of these nymphs were recorded. In September, the nymphs have attained their final instar stage having wing pads in them.

Zygoptera

The observations made on the larval damsel flies (Zygoptera) show somewhat bimodal peaks in their abundance, one in March-April and another in September. In winter months comparatively lower population density was recorded. In the winter months, low temperature plays an important role in egg laying and also in the larval development.

In general the odonate population are both univoltine and bivoltine. Some of the species lay their eggs in March which becomes fully susceptible to sampling in September and produce the peak of larvae in these months. Some of the species lay their eggs in November which owing to the winter period do not reach up to the final instar and remain in water up to January, giving a small peak of larve in February-March of the present study.

The availability of food also affects the rate of larval developments of odonate. The larval development and the duration of the developmental period is directly related with the environmental conditions of the habitat. In a seminatural enclosure, vegetations provided hiding places for the prey of these larvae (Hassan, 1974). In such conditions these larvae have very hard time to find their specific food prey. It has been demonstrated by Hassan (1974) that this probably accounts for the much longer larval duration (58 days) of *Palpopleura lucia lucia* (Anisoptera) in nature than in the laboratory (39.8 days). He concluded that the former figure is probably close to what happens in the eutrophic waters. Therefore, due to unavailability of proper food during the active period of development or body growth retards the emergence of adults in such water bodies.

Therefore, in the present investigation, high abundance of macrophytic vegetations (emergent type) in winter and summer months are responsible for low population densities of these larvae in the pond. Since the odonate larval population generally attack only moving prey and vegetation provides hiding places for their prey, therefore, in the active period of life, no proper food is available to them (Benke, 1972 and 1976). Therefore, in the periods of high abundance of macro-vegetations are the periods of low abundance of larval odonata population.

The other causes for the population dynamics of odonate larvae are the presence of predators in the waterbody, mortality and emergence of these larvae. The fish were a significant cause of mortality (Benke, 1972). In the present investigation Carnivore fish *Wallago attu* was recorded in abundance from January to May. This fish is at the top of the food chain and regarded as a greatest predator for these larvae (Datta Munshi *et al.*, 1991). The constant mortality in a population with over-lapping generations indicate that those factors affecting the population are density independent or density-proportional (Odum, 1971). However, such an interpretation does not hold for a synchronously developing populations with non-overlapping generations, specially if increase in size of individuals results in an increase in populations biomass. Larger individuals usually require a greater share of potentially limiting resources (such as food or space) and as individuals grow, the population may be increasing rather than decreasing its use of resources. It was suggested by Benke and Benke (1975) that *Anax junius* grow faster when food was more plantiful and consumed more as its size increases. Pianka (1967) had studies lizard species diversity and suggested that three most important niche dimension among potentially competing species are food, time and space. Thus, interspecific relationships are dominant factors in species abundance and are important from the view point of community perspective.

The emergence of larvae is also an important factor in their seasonality Corbet (1964) considered that members of Placoptera, Ephemeroptera, Odonata, Diptera and Trichoptera are exceptionally well studied for population studies at the time of emergence, as individuals of a population pass through the water surface and thus become unusually accessible to the observer, whereas, insect like odonata undergo at critical final moult on an emergence support on the shore (Mathavan and Pandian, 1977) others like Dipterns emerged directly from the water surface. The various workers as Corbet and Danks (1973), Harper (1973) and other estimated the biomass of the emerging adults per unit area per unit time. Thus, the emergence patterns in the odonate larvae in the various seasons of the year play an important role in the seasonal variations of the populations.

Species Diversity and Evenness

Figures 4.3 and 4.4 indicate that the species diversity (H) and evenness of species (J) of these odonate larvae show fluctuations in their range. A measure of the species diversity at any instant of time is based on Shanon-Weaver Theory (1964) indicates that the more complex the community is, the greater will be the values of its species diversity and stability (MacArthur, 1975). A crude measure of the stability of a community over a given time can be obtained from that range of values of species diversity calculated within time period. Margalaf (1968) suggested that the higher values of species diversity is correlated with longer food chains and more stable community. MacArthur (1975) hopotesized that community stability is a function of food web complexity. The greater the links in any food web the more stable the community. Paine (1966) found that the lack of stability in community is due to a unique set of interactions among the species and between the species and the physical environment. Thus, evenness of the species in a community and diversity of species is an area are two sides of the same coin. Diversity tends to be high in older community and low in newly established ones.

In the present investigation, it was observed that in those months when the total individuals was found to be maximum even then the species diversity as well as evenness values as calculated was found slightly lower than the months where total number of individuals/haul was found lower than those months. In these months evenness of species were also calculated to be minimum. It was mainly due to the fact that though total number of individuals was more in a month, the number of species remained more or less same in the successive months. It was observed that increase in the total number of species resulted in the decrease in species diveristy and also in evenness. In case of Anisoptera, individuals/haul (N) was found to be maximum in the month of December, however, the values of the species diversity (1.513 bits) is found lower than in the months where total number of species was found much than this month (November : 28/haul). It was due to increase in the total number individuals (50/haul) and also in the number of species (S) as compared to the successive months. It was due to abundance of many species of Anisoptera in the habitat such as *Mesogomphus lineatus* (25/haul), *Cordulegaster* sp. (15/haul), *Potamarcha obscura* (4/haul), *Zyxomma petiolatum* (2/haul) etc. On the contrary, individuals/haul (10/haul) was found to be low in the month of April, but species diversity in higher (1.545).

In case of Zygoptera, individuals/haul (N) was found to be higher in the month of September (45/haul) but the species diversity (H) is found to be lower (1.820). The number of individuals/haul (N) was lower in the month of October (18/haul), but the species diversity was higher (1.902).

High diversity occurs when there was no numerically dominant species in the habitat. Periods of high diversity are times of rapid changes and transition, during which the aquatic odonate larval community is composed of some species that are increasing and others are declining in number. When all the species in a population was more or less equally distributed, the evenness was found maximum. Species diversity (H) is thus, significantly correlated with the evenness (J) or the distribution of individuals among species of odonate larvae. It is not correlated with the number or with the sample size (Sagar and Hassler, 1969).

Thus, the species diversity concept points out that natural biotic communities typically characterised by the presence of a few species with many individuals and many species with few individuals. The species diversity index gives equal value to each individual in a population (Sagar and Hassle, 1969). As the sepcies diversity relates to then information theory, more account of information is contained in the natural community than in a polluted community. A polluted system is simplified and those species that could survive encounter less competition and they increase in numbers. Therefore, in more natural community will have higher diversity index and lesser fluctuation in its range. Therefore, a measure of the pollution of a particular system over a given time can be obtained form the range of values of species diversity calculated within the time period. Howell and Gentry (1973) observed that thermal effluents from a nuclear power plant decreased the diversity of aquatic insects in the Savannah river of South Carolina.

Pollutants will generally cause adjustment and alterations in species abundance and community species composition in an aquatic system. These adjustments should be detectable through use of an appropriate diversity measure. The present study on species abundance and diversity of aquatic odonate larvae revealed that these derelict ponds are not contaminated with any type of pollutants because species diversity (H) and evenness of species (J) of these larvae show no wide fluctuations in its range. From the above observations, it may be postulated that a measure of the pollution of a particular system over a given time can be obtained form the range of the values of species diversity and evenness calculated within the time period.

Secondary Productivity

In the present investigation, production losses between the sampling periods are estimated as the product of numbers lost in the interval and the mean weight of individuals. The sum of these losses throughout the life cycle is equal to total production since some odonate larvae are either consumed by aquatic predators especially carnivorous fish and aquatic birds or leave the system through emergence. From the Figs. 4.5, 4.6, 4.7 and 4.8 losses for a

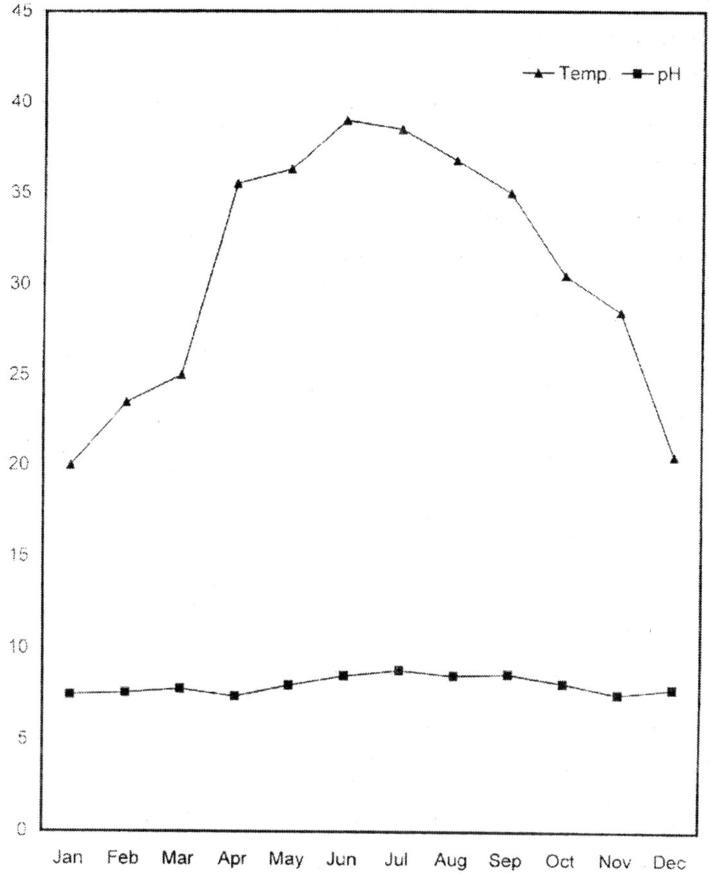

Fig. 4.7 : Monthly variations in temperature (°C) and pH

given time interval (October-December) are calculated as the product of number or weight loss. Since the sum of the weight loss column accounts for all losses throughout the life cycle of the population, it is equal to annual production. Although, the sum of the weight loss column is equal to annual production, the weight loss during given intervals is not equal to total production during that interval. This total new production is, in fact, equal to the weight loss calculation plus increase in standing stock biomass. Therefore, this indicates that during the periods of October to

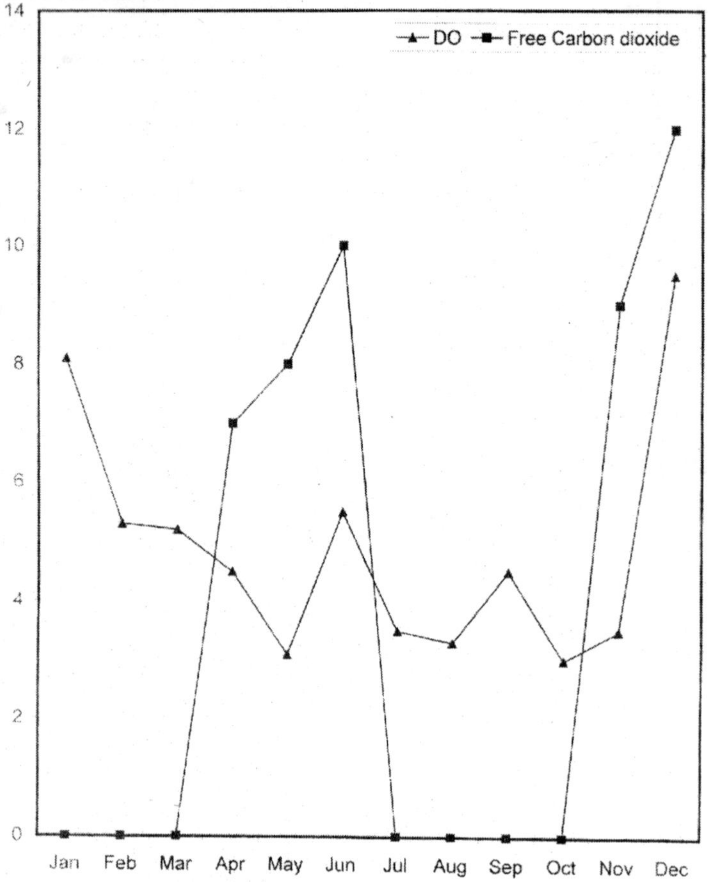

Fig. 4.8 : Monthly variations in DO and free Carbon dioxide values (mg/l).

December, very little production is going on in the aquatic system. This little production may be due to heavy mortality of the larvae, unavailability of their food substances, migration and adverse abiotic factors. The biomass of odonate larval population depends on several factors such as active competition, intrinsic life history patterns, competition for food and interspecific predation. According to Winberg (1971) predation stimulates the gorwth of the consumed species by decreasing competitions.

In the present investigation, the low production of these larvae in winter may be due to the fact that they enter into non-growing phase and as such there was no consistent increase in their biomass during this season. In the spring season, when increase in the environmental temperature occurs, the development resumed and biomass once again got increased. Thus, it seems that the environmental temperature appeared to be one of the important factors responsible for the fluctuations of biomass.

In the present study, both positive and negative growth of the odonate larval population was recorded. According to Boughey (1968), the rate of increase may have a negative as well as positive value. At any instant of time, a population may be declining by the loss of in individuals rather than growing by their addition. The change could be due to emigration or immigration and the birth of the new individuals or natality and mortality alongwith population dispersal, i.e., emigration and immigration play an important role in oscillations of overall biomass production of these larvae.

The life cycle pattern of these odoante larvae may be responsible for the fluctuations of biomass during differnt seasons of the year. The phenological and developmental patterns differ from species to species while individual of certain species emerge, others do not. As a result, there is oscillation in the overall biomass production of these aquatic insects at a given time. The quantitative information concerning the seasonal abundance and biomass production of the insects in the habitat is very limited, however, information from the literature indicates that these larval population are univoltine, bivoltine and multivoltine, importing no any definite trend of larval population in the wetlands.

During monsoon, i.e. in July-August, comparatively low biomass was recorded. Monsoon rains and flooding are unfavourable to sedentary larvae and benthic dragon fly nymphs. Therefore, the rapid water discharge and monsoon flooding during this season cause the drastic reduction in the biomass production of these insects. In winter season, the larval population undergo diapause or over-wintering. As such in the winter season, specially in December large number of larval population was recorded since

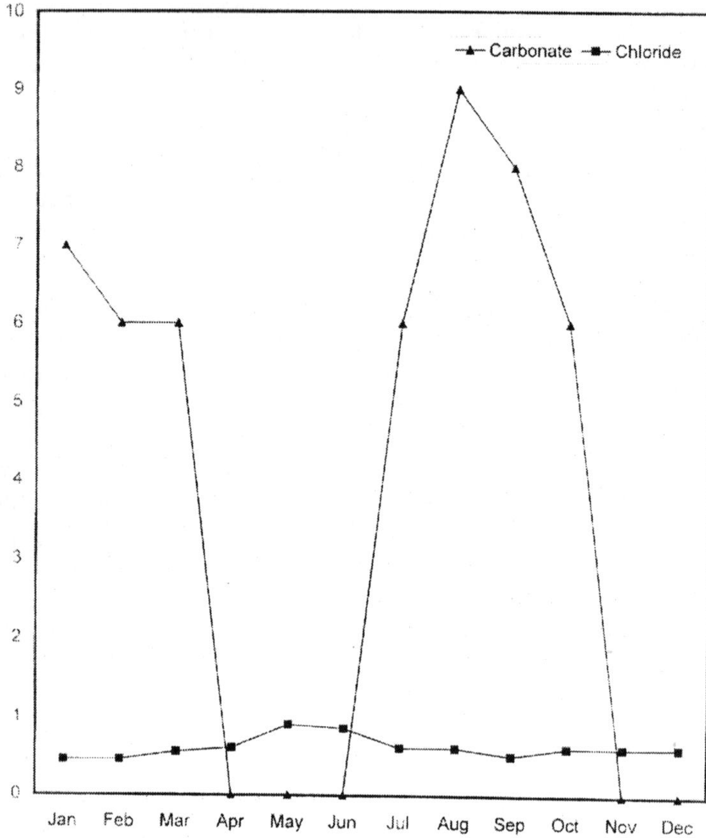

Fig. 4.9 : Monthly variations in Carbonate Alkalinity and Chloride contents (mg/l).

no emergence occurs due to the low temperature. Thus, it may be concluded that phenologial divergence and climatic regime may be responsible for their biomass fluctuation. These insects enter into the non-growing phase in the winter as such there was no consistant increase in their biomass during this season. In the spring, when development resumes, biomass once got again inreased. The biomas of these insects dropped off rapidly as some of them emerged or attempted to emerge in remaining seasons of the year. Our findings support views of Benke (1972 and 1976)

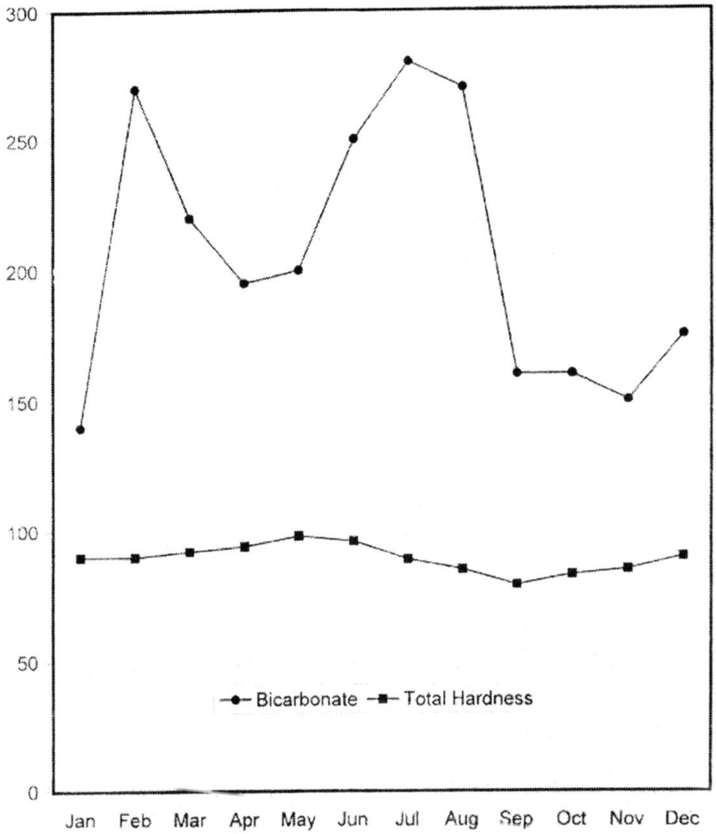

Fig. 4.10 : Monthly variations in Bicarbonate alkalinity and total hardness (mg/l).

who studied dragon fly production and prey turn over rates in Allanta, Georgia.

It was found by Benke (1976) that highest production occurred from June to September. This finding of Benke (1976) supports the present observation in which maximum production was recorded from June to September in this wetland. Little production occurred during the winter months. There are a number of workers who have estimated the larval production in various geographical regions of the world. Lawton (1970) found the damsel fly, *Pyrrhosoma nymobula* in an English pond to have an annual production of 0.7 g. met. sq. Benke (1972) strongly suggested that their annual production is directly related to the availability of food prey. The abundance of vegetation provide hiding places for their prey. The abundance of prey is one of the major factors in their biomass oscillation and which in turn depends upon the seasonal appearance and disappearance of macro-vegetations of the wetlands. Thus, the production of odonate larvae is directly controlled by the prey turn over. By larval odonata it is cleared that a very small standing stock of primary consumers must be turning over at an enormous rate to support these predatory insects in aquatic system.

Physico-chemical Characteristics of Water

The physico-chemical environment of the water functions in one or more ways in exerting the influence upon biotic components. So, it gives a pictures of environmental suitabilities of water for maintaining normal life. As such the abiotic factor is important in elucidating some causal influences for the faunal composition and abundance of aquatic insects, therefore, the knowledge, of physico-chemical characteristics of water is quite imperative for proper ecological studies of odonate larvae.

Temperature is the most important factor of water in controlling the developmental rates of aquatic odonata and is also responsible for the release of nutrients in the waterbodies by initiating the decomposition of organic matters. The water and air temperatures are correlated to each other because the water temperature varied with rise and fall of atmospheric temperature. It is probably brought about due to the fact that smaller the mass of water, greater the response to all temperatures (Welch, 1952). The observation of Munawar (1970), Nasar and Munshi (1974) and Rai (1980) have also confirmed the same situations on shallow tropical water bodies present in different parts of the country.

It is clear from the present investigation that the value of pH was slightly higher during the summer months than the other months. According to Hutchinson (1957), pH is regulated by the CO_2-carbonate system. The variations in the pH values of pond water takes place due to the changes in carbons in carbonate and bicarbonate contents. Both the biological phenomena, photosynthesis and respiration run together simultaneously in the aquatic medium. It produces entirely different products. The metabolic products affect the buffering system of the medium, hence controls the pH values of water. The present water is alkaline due to presence of sufficient quantity of carbonates. The pH tolerance of organisms varies from species to species but the greatest diversity and the highest productivity occur within the range 5.6 to 8.5 (Wiebe *et al.*, 1934).

A considerable variation in the dissolved oxygen content was observed in this waterbody. According to law of solubility of gases, periods of high temperature should be the time of low oxygen content and *vice versa.* As such dissolved oxygen varied inversely with temperature. But it is apparent from the present investigation that photosynthetic activity of phytoplankton and growth of macrophytes also matter much. It is clear cut from the results that high dissolved oxygen values were obtained in December. This high dissolved oxygen content of water may be due to the high photosynthetic activity of macrophytes and phytoplankton bloom in the wetland. The minimum record of oxygen in May was due to consumption of oxygen by the respiration of autotrops and heterotrops. It may be due to presence of comparatively less growth of macrophytes in this period. In general, the disolved oxygen content of water was found to be low during summer and post-monsoon months, because of the decomposition activity at higher temperature during this month.

Water and air temperature collectively affect the solubility of atmospheric oxygen into the water. Rai (1980) has recognised the air and water temperature as one of the factors controlling the concentration of dissolved oxygen into water but these are not the only factors. The present study does not support this view as there were several occasions when levels of dissolved O_2 content have not strictly followed the fluctuations of air and water temperatures. The other factors responsible for reduction of dissolved oxygen content are decomposition of the organic matter of the bottom, agitation of water during fishing activity, and also due to excessive human interferences. Thus, the oxygen content of fresh water shallow bodies actually depends upon several other factors, viz., temperature, intensity of light penetration, photosynthetic activities of aquatic flora, respiration by plant and animal communities, wind action as well as certain anthropogenic activities.

The free carbon dioxide was found more in the months of April, May, June November and December. But in the rest months, there is depletion of carbon dioxide and in these months carbonate alkalinity was present. The main sources of carbon dioxide in certain months of the year seem to be the decomposition of organic matters and respiration by the aquatic organisms. During the months of October to December, the waterbody remained packed up with huge growth of several kinds of aquatic macrophytes specially emergent types. Such conditions probably develop a barrier between water and atmosphere for free gaseous exchange between these two media. Green plants are responsible only for the production of oxygen but respiration takes place throughout the whole day and night time. Due to respiration, over production of CO_2 usually gets diffused into the atmosphere which lowers its concentration in the water. Maitland (1978) has explained this phenomenon that the rain water usually percolates through the soil, it may pick up even more CO_2 and ultimately adding into the waterbody.

The carbonate alkalinity was recorded in the months where free CO_2 was absent and it was in low concentration. It varied from nil to 8.0 ppm. The presence of cabonate alkalinity is an indication of higher rate of carbon assimilation. Its content is high during monsoon months.

The present investigation clearly reveals that the concentration of bicarbonate was greater than the carbonate throughout the year. The bicarbonate contents of the bottom layer of the pond were usually higher than that of the surface due to prophylatic activities releasing CO_2. There is an evident fluctuation in the content of bicarbonate alkalinity which is more or less inversely proportional to the carbonate alkalinity. The rainfall is also one of the factors to decrease the values of bicarbonate.

The chloride content of the wetlands was very meagre and its concentration ranged from 0.45 to 0.90. According to Thresh *et al.* (1944) the chloride content is an index of pollution of animal origin. Hence, it is clear that this wetland in not so contaminated by sewage of animal origin.

Most inland waters are dilute solutions of calcium and sodium carbonates but variations contributed by lithology and climate are very wide. According to Rai (1974), the total hardness is an index of water quality and is of considerable significance in connection with the discharge of organic pollutants. During the present investigation, total hardness showed its higher value during summer which might be due to high temperature and low water levels as well as the addition of calcium and magnesium salts from detergents and soap used for bathing and cloth washing. It's lower values were recorded during rainy months which might be due to dilution on account of high rains as precipitation causes in lowering of values of hardness (Patil, 1982).

The product moment correlation coefficient for biomass of larval population with water temperature, pH, dissolved oxygen, free carbon dioxide, alkalinity and chloride contents were computed and established in the present study and shown in Table 4.1. Of all the physico-chemical factors studied, dissolved O_2 show marked correlation with the biomass g/cub. met. of the

Table 4.1 : Correlation coefficient between biomass g.dry wt./haul/ with various physico-chemical parameters of water

Parameters	Correlation coefficient	Regression coefficient	Intercept	T for R
Temperature	-0.163	-2.575-03	0.465	0.525
pH	0.206	4.652-03	1.360-03	0.666
DO	0.898*	4.803	0.362	0.282
Free CO_2	0.151	3.402-03	0.375	0.483
Carbonate	0.078	2.363-03	0.376	0.248
Bicarbonate	-0.238	-1.348-04	0.419	0.776
Chloride	-0.152	-0.121	0.460	0.488
Total hardness	-0.158	-0.159	0.448	0.385

larval population (r = 0.898; p>0.05). These larvae live in such environment in which dissolved oxygen content are the most important limiting factor. These larvae take dissolved oxygen of water in their respiration. The environment in which dissolved oxygen is meagre then the larval population becomes less represented.

Thus, the overall picture emerged from the above discussion, it can be concluded that the nymphal odonata usually form a high proportion of biomass in freshwater and also occupy an important position in its energy flow pathway. They are frequently dominant aquatic insects in the littoral zones and their ecological adaptations are of great significance in understanding the structure and function of fresh water lentic ecosystem.

References

Allan, J.D. 1975. The distributional ecology and diversity of benthic insects in Cement Creck, Calorodo. *Ecology*. 56 : 1040-1053.

APHA. 1985. *Standard methods for the examination of water and wastewater.* American Public Health Association., Washington, D.C.

Asahina, S. 1942. Odonater gesammelt Von. Dr. S. Kwayama in Manchoukuo nebst einer Listeder bekanntes Odnatenartes ausder Mandschurei and den angrenzenden Gebieten. *Kontyu*. 16 : 62-82.

Benke, A.C. 1972. An experimental field study on the ecology of co-existing larval odonates. Ph. D. dissertation. Atheens, Univ. of Georgia. p. 122.

Benke, A.C. 1976. Dragan fly production and prey turn over. *Ecology.* 57 : 915-927.

Benke, A.C. and Benke, S.S. 1975. Comparative dynamics and life histories of eo-existing dragon fly population. *Ecology*. 56 : 302-317.

Beukema, J.J. 1976. Biomass and species richness of the macrobenthic animal living on the tidal flats of the Dutch Wodden Sea. Netherland. *J. of Sea Research*. 10 : 236-261.

Boughey, A.S. 1968. *Ecology of Population*. The Macmillan Company, New York. 133. p.

Boyd. C.E. 1971. The limnological role of aquatic macrophytes and their relationship with reservoir management. Reser. *Fisch. Limn. Spl. Pub.* 8 : 153-165.

Butler, H. 1904. The labium of odonata. *Trans. Amer. nt. Soc.* 45 : 335-396.

Chelmick, D.G. 1999. Larvae of the genus *Anax* in Africa (Anisoptera : Aeschni). *Odonatologica.* 28 : 209-218.

Corbet, P.S. 1953. A terminology for the labium of larval odonata. *Entomologist.* 86 : 191-196.

Corbet, P.S. 1961. The food on non-chalcid fishes in the Lake Victoria basin, with remarks on their evolution and adaptation to lacustrine conditions. *Proc. Zool. Soc. Londan* 136 : 1-101.

Corbet, P.S. 1962. *A Biology of Dragonflies.* Witherby Ltd., London.

Corbet, P.S. 1964. Temporal pattern of emergence in aquatic insects. *Canad. Ent.* 96 : 264-279.

Corbet, P.S. and Danks, H.V. 1973. Sex ratios at emergence of two species of high arctic *Aedes* (Diptera : Culcidae). *Can. Ent.* 105 : 647-651.

Datta Munshi, J.S., Datta Munshi, J., Singh, D.K., Singh, O.N. and Poddar, T. 1991. Graphic representation of the food and feeding habits of certain animals of river Ganga : Mode of feeding and Dimension of food items. *J. Fresh Water Biol.* 3 : 45-58.

Fraser, F.C. 1919. Description of New Indian odonata larvae and excuviae. *Rec. Indian. Mus.* 16 : 459-467.

Hanel, L. 1995. The contribution to the knowledge of dragonflies (odonata) in podblanicko region (Ceutra Bohemia). *Bohemia Centralis.* 24 : 129-149.

Hannon, C.E. and Weaver, N. 1964. *The Mathematical Theory of Communication* , University of Illinios Press, Urbana. p. 125.

Harper, P.P. 1973. Emergence, reproduction and growth of setipalpian plecoptera in Southern Ontario. *Oikos.* 24 : 94-107.

Hassan, A.T. 1974. Studies on the ecology, behaviour and life history of libelluline dragon flies. Thesis, Univ. Ibadan, Nigeria.

Howell, F.G. and Gentry, J.B. 1973. Effect of thermal effluents from nuclear reactors on species diversity of aquatic insects. In thermal ecology. U.S. Atomic Energy Commision.

Hutchinson, G.E. 1957. *A Treatise on Limnology.*, Willey Publ., New York. p. 1015.

Kaushik, S., Sharma, S., Saxena M.N. and Saksena, D.N. 1990. Habitat Ecology of Anisopteran Nymphs in Certain water bodies of Madhya Pradesh. Acta. hydrochim. *Hydrobiol.* 18 : 443-447.

Kiauta, B. 1975. Cytotaxonomy of Dragonflies with special reference to the Nepalese fauna. Lectures delivered at the Tribhuvan

University, Kathmandu, Nepal Research Centre, Kathmandu. pp. 1-77.

Kumar, A. 1973a. Description of the last instar larvae of odonata from the Dehradun Valley (India) with notes on biology. I. (Suborder-Zygoptera). *Oriental Insects.* 7 : 83-118.

Kumar, A. 1973b. Descriptions of the last instar larvae of odonata from the Dehradun Valley (India) with notes on biology. II. Suborder-Anisoptera. *Oriental Insects.* J. 2 : 291-331.

Kumar, A. 1989. Studies on the life history of Indian dragonflies : *Orthetrum sabina sabina* (Drury) (Odonata : Libellulidae). *Rec. Zool. Surv. India.* 85 : 573-581.

Kumar, A. 1993. Estimation of calorific contents in the macroinvertebrates of the river Mayurakshi at Dumka. *The Indian Zool.* 17: 55-58.

Kumar, A. 1994a. Analysis of gut content and dimension of food items of certain dragonfly nymphs of the river Mayurakshi in Santal Pargana, Bihar. *Proc. Nat. Aca. Sci.* B 64 : 269-273.

Kumar, A. 1994b. Population on density, species diversity and evenness of dragonfly larvae in a freshwater pond of Santal Pargana, India. *J. Freshwater Biol.* 6: 231-235.

Kumar, A. 1994c. Role of species diversity of aquatic insects in assessment of pollution in the wetlands of Santal Pargna, Bihar. *J. Environ. Pollut.* 1 : 117-121.

Kumar, A. 1995a. Species composition and population dynamics of odonate larvae in a fish farming pond of Santal Pargana, Bihar. *Proc. Nat. Aca. Sci.* B65 : 401-411.

Kumar, A. 1995b. Estimation of productivity of odonate larvae of a tribal village pond of Santal Pargana, Bihar. *Environ. and Ecol.* 13 : 313-316.

Kumar, A. 1995c. O_2 uptake by odonate larvae in a tropical wetland of Santal Pargana, Bihar. *J. Freshwat. Biol.* 7: 177-181.

Kumar, A. 1996. Comparative studies on stomach content and forage ratio of zygopteran and anisopteran larvae in a fish pond of Santal Pargana, Bihar. *Proc. Nat. Aca. Sci.* B66: 315-323.

Kumar, A. 1997. Energy contents of dragonfly nymphs (Anisoptera : Odonata) in a tropcal freshwater ecosystem of Santal Pargana, Bihar. *Environ. and Ecol.* 15 : 190-193.

Kumar, A. 1998. Biomonitoring of pollution by aquatic insect community and micro-organisms in different freshwater ecosystems. In : *Ecotechnology of Pollution Control and Environmental*

Management (Eds. A. Kumar and R.K. Trivedy), Enviro Media, Karad. pp. 139-144.

Kumar, A. 1999. Qualitative and quantitative evaluation of predatory aquatic insects in two ecologically different wetlands in Santal Pargana, Bihar. *J. Environ. Pollut.* 5: 37-43.

Kumar, A. 2000. Assessment of nutrient budget of benthosin lentic freshwater ecosystems. In : *Current Topics of Environmental Biology* (Eds. G. Tripathi and G.G. Pandey), ABD Publishers, Jaipur. pp. 128-139.

Kumar, A. and Khanna, V. 1983. A review of the taxonomy and ecology of odonata larvae from India. *Oriental Insects.* 17 : 127-157.

Kumar, A. and Kumari, P. 1999. Seasonal population dynamics of zygopteran larvae in a fish pond. *Fraseria.* 6 : 3-8.

Laidlaw, F.F.1920. Notes on some interesting larvae of dragonflies in the collection of Indian Museum. *Rec. Indian Mus.* 19 : 185-187.

Lawton, J.H. 1970. Feeding and food energy assimilation in larvae of the damsel fly *Pyrhosoma nymphula* (Sulz.). (Odonata : Zygoptera). *J. Anim. Ecol.* 39 : 669-689.

Lombardo, P. 1997. Predation by *Enallagma* nymphs (Odonata, Zygoptera) under different conditions of spatial heterogenecity. *Hydrobiologia.* 356 : 1-9.

Lotka, A.J. 1924. *Elements of Physical Biology.* Williams and Wilkins, Baltimore. Reprinted by Dover, Publ., New York, p. 460.

Macneill, N. 1960. A study of caudal gills of dragonfly larvae of the suborder Zygoptera. *Proc. R. Irish Acad.* 61 : 115-140.

MacArthur, R. H. 1955. Fluctuations of animal populations and a measure of community stability. *Ecology.* 36 : 533-536.

Matiland, P.S. 1978. Quantitative studies on the invertebrate fauna of sand and stony substrates in the river Endrick, Scotland. *Proc.R. Soc. Edinb.* B68: 277-301.

Margalaf, R. 1968. *Perspectives in Ecological Theory.* Univ. of Chicago Press, Chicago. p. 122.

Mathavan, S. and Pandian, T.J. 1977. Patterns of emergence, import of egg energy and energy export via emerging Dragonfly populations in a tropical pond. *Hydrobiol.* 54 : 257-272.

Merritt, R. W. and Cummins, K.W. 1978. An Introduction to the aquatic insects of North America. Kendall/Hunt Publishing Company, IUWA-52001, U.S.A. pp. 427.

Munawar, M. 1970. Limnological studies on freshwater ponds of Hyderabad, India. *Hydrobiologia.* 31 : 101-128.

Nasar, S. A. K. and Datta Munshi, J. S. 1974. Seasonal variations in physico-chemical and biological properties of a tropical shallow water. *Japanese J. Ecol.* 24 : 255-259.

Needham, J.G. 1930. A manual of the dragonflies of China. *Zool. Sin* (A). 11 : 114.

Nicholson, A.J. 1958. Dynamics of insect population. *Annual Rev. Ent.* 3 : 107-136.

Odum, E.P. 1971. *Fundamentals of Ecology.* W.B. Saunders Company, Philadelphia. p. 557.

Paine, R. T. 1966. Food web complexity and species diversity. *Am. Nat.* 100 : 65-76.

Patil, P. 1982. An ecological study of the algal flora of the lakes of Bhopal. Ph.D. Thesis, Bhopal University, Bhopal. p. 310.

Pianka, E.R. 1967. Lizard species diversity. *Ecology.* 48 : 333-351.

Pielou, E.C. 1969. *An Introduction to Mathematical Ecology.* John Wiley and Sons, New York.

Poyyamoli, G. and Pandian, T.J. 1984. Diel emergence patterns of some Indian odonates. Abs. 26, First Indian Symposium of Odonatolgoy, Madurai.

Rai, D. N. 1980. Ecological structure of certain swamps of Darbhanga, Bihar, Ph. D. Thesis, Bhagalpur University. p. 267.

Roy, S.P. 1980. Population dynamics and diversity of Mayfly numphs (Insecta : Ephemeroptera) in a fresh water pond of Bhagalpur. *Biol. Bul. India.* 2: 11-14.

Roy, S. P. 1984. Studies on gut content analysis of odonate nymphs in a fresh water fish pond at Bhagalpur (Bihar). *Entomon,* 1984. 1 : 25-29.

Roy, S. P. 1986. Studies on the seasonal variations in the biomass and production of aquatic insects in a fresh water pond. *Biol. Bull. India.* 8 : 169-173.

Sagar, P.E. and Hassler, A.D. 1969. Species diversity in lacustrine phytoplankton. I. The components of the index of diversity from Shannon's formula. *Am. Nat.* 103 : 51-59.

Sandsberg, E. and Bonsdroff, E. 1992. Competition for food between predator of marine and limnic origin : Experiments with saduria entomon (Isopoda) and Libellula quadrimaculata (odonata). Proc. 12th Baltic Marine Biol. Symp. pp. 141-144. Shannon, C. E. and Weaver, N. 1964. The mathematical theory of communication. Univ. Illinois Pres, Urbana. pp. 125 : 1-11.

Smith, R.F. and Pritchard, A.E. 1956. Odonata. In : *Aquatic Insects of Cālifornia with keys to North American Genera and Californis Species* : (ed. R. L. Usinger), University of California Press, Berkeley and Los Angeles. pp. 106-153.

Snodgrass, R.E. 1954. The dragonfly larva. *Smiths. Misc. Coll.* 123 : 1-138.

Suri Babu, B.; Srivastava, B.K. and Srivastava, V.K. 1996. Morphology and biologyof the final instar larva of *Trithemis pallidinervis* (Kirby) (Odonata : Libellulidae). *Bioved.* 7 : 107-110.

Tonczyk, C. 1994. Some odonatological suggestions, as how to collect, preserve specimens, bread and work out the dragonflies. I. Collecting the adults. *Bul. Ent.* 2 : 2-4.

Trivedy, R. K. and Goel, P.K. 1986. Chemical and biological methods for water pollution studies. *Enviro Media., Karad.* p. 251.

Vick, G.S. 1999. A checklist of the odonata of the South-West province of Cameroon, with the description of *Phyllogomphus corbetae* Spec. Nov. (Anisoptera: Gomphidae). *Odonatologica.* 28 : 219-256.

Volterra, V. 1925. Variations and fluctuations of the number of individuals in animal species living together. In : *Animal Ecology* (R. N. Chapman, Ed.) McGraw-Hill Book Co. INC, New York. pp. 409-448.

Volterra, V.S. 1925. Variations and fluctuations of the number of individuals in animal species living together. In : *Animal Ecology* (Ed. R.N. Chapman), McGraw-Hill Book Co. Inc., New York, London. pp. 409-448.

Welch, P.S. 1948. *Limnological Methods*, McGraw-Hill Book. Co. INC., New York, London.

Welch, P.S. 1952. *Limnology*. 2n. Ed. McGraw-Hill Book. Co. INÇ., New York.

Wiebe, A.H.; Gorock, A.M.M.; Fuller, A.C. and Markus, A.C. 1934. The ability of freshwater fish to extract oxygen at different hydrogen-ion-concentration. *Physiol. Zool.* 7 : 435-448.

Wilson, C. B. 1923. Life history of the scavanger water beetle *Hydrophilus triangularis* and its economic relation to fish breeding. *Bull. Bureau Fish.*, 3 : 9-38.

Wilson, C.B. 1924. Water beetles in relation to pond fish culture with the life histories of those found in fish ponds at Fairport. IONA. From *Bulletin of the Bureau of Fisheries.* 39 : 231-345.

Winberg, G.G. 1956. Rate of metabolism and food requirements of fishes (Original in Russian). Translation Series no. 194. (*J. Fish Res.* Ed. Can.).

Wolf, W.J. and deWolf, L. 1977. Biomass and production of zoobenthos in the Gravelington estuary. *The Netherland Estuarine and Coastal marine service.* 5 : 1-24.

Zhu, H. Q. and Zhang, X.X. 1999. *Somayochlora shanxiensis* : A new dragonfly from Sanxi, China (Anisoptera : Corduliidae). *Odonatologica.* 28 : 289-292.

5

Biology of Odonata of Indian Sub-continent: A Review

• *B. Suri Babu and Arvind Kumar*

Abstract

A review on the biological studies on Odonates during 1985-2001 from Indian subcontinent is provided. The information on the egg stage, larval growth, larval habit and habitat, emergence, reproductive behaviour of adults, parasites and predators of adults have been dealt in detail in the text.

Keywords : *Odonata, Anisoptera, Zygoptera, Larvae, Emergence, Rendezous, Territoriality, Wheel position, Oviposition.*

Introduction

The odonates or dragonflies constitute a small, well known, widely distributed order of an ancient group of hemimetabolous insects. Primitively terrestrial, they became largely aquatic during permian period (Corbet, 1962). In animal texa, odonate larvae are at the top of aquatic food chain (Kumar, 1996). This order includes three suborders: the Anisozygoptera, containing only two species; the zygoptera and the Anisoptera; all of which have aquatic egg and larval stages constituting greater part of life span, whereas the terrestrial winged adults are comparatively short-lived, paradoxically many aspects of ododate biology have been incompletely studied, eventhough odonates are well known taxonomically. The biology of Odonate have been studied by Tillyard (1917) who emphasizes systematic and functional morphology and Corbet (1962 and 1980) who described ecology and behaviour; further studies on biology of Indian odonata have failed to attract the attention of Indian entomologists.

The Odonata of Indian sub-continent have been studied for the past 250 years. Fraser (1933-1936) published three volumes on odonata in the form of *Fauna of British Indian Subcontinent and Abroad* have studied the odonata of Indian suncontinent. The larvae and biology of several species of odonata have been studied in recent years (Kumar, 1972, 1973, 1985; Prasad and Varshney, 1995; Suri Babu, 1998a and b). This article is a revised and an updated review on the biology of odonata of this region and can be regarded as a supplement to the work already published. We have also included the work which are not mentioned in the above cited reviews.

As freshwater insects, dragonflies are exceptionally vulnerable to urban and agricultural expansion, which commonly entail the draining of ponds and marshes (Corbet, 1980). This group of insects is, therefore, directly associated with the welfare of mankind particularly, concerned with agriculture, medical and veterinary as they are active feeders of various pests of paddy, cotton, sunflowers and swarms of termites. They feed on adult mosquitoes, sand flies and stable flies of medical and veterinary importance. The odonates are known to be one of the best biological indicators of ecological degradation and pollution in the water bodies (Tembhare, 1998).

The Egg Stage

The eggs of many tropical and temperate species of odonates exhibit direct development and hatch after 5-40 days, whereas those of certain other temperate species undergo delayed development, overwinter and hatch after about 80-230 days and eggs laid by a single female on the same day can vary widely in the date of hatching (Corbet, 1980). Kumar (1985) described the egg structure of *I. rapex*. He found that each egg has a tightly coiled very large slender filament at its posterior pole which uncoils and become a long cord when the egg released form the body of female and comes in contact of the water. The eggs get entangled with surrounding objects in water and thus serve as a grasping hook to arrest movement of the egg. He reported hatching of the eggs in 16 days at 30°C in Dehradun.

Kumari and Nair (1985) reported the hatching of egg of *P. flavescens* in 6 days in Trivendrum, Kerala. Kumar (1988) recorded hatching of egg of *N. t. tullia* in 11 days at 28°C and Kumar (1989) reported hatching of eggs of *Tramea virginia* in 24 days at 24°C at Dehradun. Thomas and Gladstone (1989) reported the hatching period of Begum *et al.* (1990b) reported the incubation period for eggs of *N.t. tullia* 7-9 days and prolarva hatched out through a longitudinal slit in the eggs, whereas Begum *et al.* (1990a) reported hatching for the eggs of *B. rufa* in 12-15 days at 30±2°C at Dhaka, Bangladesh.

The Larval Stage

Generally, larvae occupy conventional aquatic habitat, but certain Sylvian zygoptera develop in water that accumulates in the leaf bases of plants and megapodagrinoid *Podopteryx selvsi* occupies water containing trees holes. The larva of the aeshnid, *Antipodophlebia asthenes* in habits subtropical rain forest in Australia, is probably terrestrial at least in later instars (Corbet, 1980). He also stated that the larvae of odonatea selects the site within the aquatic environment that has implications for resources partitioning and for concealment from both predators and prey. Chowdhury and Mohiuddin (1994) recorded that majority of species like *Tholymis tillarga, Brachydiplex chalybea* and *Orthetrum glaucum* breed in the streams. Some species like *Rhodothemis rufa, B. plutonia and Hydrobasileus croceus* breed in permanent stagnant waterbodies. They stated that a good number of species like *Diplocodes nebulosa, D. trivialis* and *Neurothemis fulvia* exploit temporary monsoon pools whereas certain species like *D. trivialis, P. flavescenes and N. t. tulia* breeds in rice fields. They also reported that some species such as *Ictinogomphus rapax* make use of concrete cisterns. Kaushik *et al.* (1990) observed the habitat ecology of anisopteran nymphs in certain waterbodies of Madhya Pradesh.

Suri Babu *et al.* (1996) observed that larvae of *Trithemis pallidinervis* are found among the aquatic weeds like *Eichhornia crassipes, Hydrilla verticillata, Potamogeton crispus, P. natans, Nelumbo nucifera* and the cultivating hydrophytes of economic value like *Trappa bispinosa* of littoral zone of Sagar lake. Suri Babu (1998a) reported that the larvae of *Ischnura aurora aurora* found in narrow slow running streams and remain attached to the submerged vegetation. He (1998b) also reported that the larvae of *P. decorum* flourished in a wide variety of biotopes like the littoral zone of Sagar lake, perennial ponds and temporary monsoon ponds of Sagar district, M.P., where larvae remain attached to the submerged vegetation. Suri Babu (2000) reported the larva of *N.i. intermedia* are slugghish weed dwellers and are stagnicolous and they found among aquatic vegetation within algal mats of the littoral zone of Sagar lake, perennial ponds and temporary monsoon pools in Sagar.

Biswas *et al.* (1994) recorded the larvae of *B. contaminata, Zyxomma petiolatum, T. tillara* and *Ictinus* sp. from Ramna lake, Bangladesh. They stated that the variations of population density of these species were perhaps due to the variation of submerged rooted vegetation, availability of food, sun light and nature of bottom bed of the water body. Begum *et al.* (1986) reported that dirty water with good vegetation produced abundant population of odonate larvae whereas it was observed less abundant in clear water with poor vegetation. Gangulee (1979) reported that the vegetation is one of the factors which affect the larval population of dragonflies in aquatic environment.

Roy (1998) studied the energetics and trophic biology of larval odonates with special reference to their role in the management of aquatic ecosystem of a fish pond of Bhagalpur (Bihar). He advocated that due to the utilization of food present at various trophic levels of the food chains, the larval odonates have regulatory impact in the management of the aquatic ecosystems as well as their mere presence indicates healthy and non-contaminated environments.

Kumar (1994) recorded the population dynamics, species diversity and evenness of species of dragonfly nymphs in a freshwater ecosystem of Bihar. Kumar (1995) estimated the seasonality in the secondary productivity of odonate larvae in aquatic ecosystem. He further prepared an exhausive check list of odonate larvae in ponds, reservoirs and lakes of Santal Pargana, Jharkhand. Kumar (1996) also collected teh larvae of Odonates and showed the comparative food preferences between anisopteran and zygopteran larvae. Kumar and Kumari (1999) worked in a tribal village fish pond of Santal Pargana, Dumka (Jharkhand) and opined that damselfly larvae showed niche specialisation and habitat preferences. They recorded larvae of *Ischnura delicata* were abudantly among the aquatic weeds and algal mass, the larva of *I. senegalensis* were found burried in the profused algal network, those of *Rhodischnura nurei* were inhabiting among aquatic weeds. *Lestes umbrins* larvae were mostly lying on pond bottom rich in aquatic vegetation. *Agriocnemis* sp. larve were observed in certain stony bottom of the monsoon pond whereas those of *Rhincocypha bisignata* were concealed in narrow spaces beneath the pebbles. The larvae of *Neurobasis chinensis chinensis* were attached to the submerged vegetation.

Kumar and Kumari (1999) stated that the larval damselflies show somewhat bimodal peaks in their abundance, one in March-April and another in September. In winter months, lower population density was recorded due to low temperature, delaying in egg laying and larval development. Bimodal abundance seems to be due to their habit of egg laying in March and November.

Suri Babu and Srivastava (2001) reported that the larvae of *I. rapex* occurred in fine sediment of littoral zone of Sagar lake and in sandy bottom of perennial streams and larvae of *Anax guttatus* and *A. indicus* inhabit the littoral aquatic vegetation of Sagar lake, Sagar (M.P.). Whcreas *Hemianax ephippiger* larval cling to grasses and Vallisnaria in a stagnant stream and larvae of *Anax immaculifrons* found in small stream and ornamental cements tanks often concealed among algal mats. They observed the larvae of *O. s. sabina* and *Potamarcha congener* live among aquatic vegetation in oxbows, larvae of *B. geminata* colcealed among algal mats in septic lavatory tanks under construction, where as *P. flavescence* and *Zyxomma petiolatum* found in cemented ornamental tanks.

Larval Growth

It is usual for species in permanent lowland habitats in tropical and warm temperate latitudes to complete one or more generations per year and for species in warm-to-cool-temperature latitudes to be univoltine or semivoltine, but the simplest pattern of larvae growth, in which the rate seems to be primarily as a result of temperature and food availability is found in tropics (Corbet, 1980). Several species of Zygoptera and Anisoptera in tropics routinely complete larval development in about two months (Kumar, 1972). Kumar (1985) reared *Ictinogomphus rapax* and reported 13 instars excluding prolarva at 18°-35°C which completed its cycle in 248 days at Dehradun. Kumari and Nair (1985) recorded 11 instars and took 104 days for *Pantala flavenscens* to attain adulthood in Trivandrum. Kumar (1988) described *N. t. tllia* as univoltine species in Dehradun. He observed that in the laboratory larval development completed in 179 days with 11 instar with one prolarve. He stated that prolonged larval development in some Indian species is due to low temprature as in the case of larval developement of *N. t. tullia* is rapid till December and no further moulting occur till February. Suri Babu (1988) observed that *P. flavescens* could complete three generations in one year in a small cemented ornamental garden tank at Sagar (M.P.)

Begum *et al.* (1990) recorded 10 instars including a prolarva in *B. rufa* which completed its life cycle in 132 days at 30±2°C and Begum *et al.* (1990b) also reported 9 instars for *N. t. tullia* and total time required to complete its life cycle in 81.5 days at 31±1°C in Dhaka, Bangladesh. Begum *et al.* (1990b) that the variations in time for development of *N. t. tullia* in Dehradun (Kumar, 1989) and Begum *et al.* (1990b) have happened due to the climatic variations prevailing in two different regions of the subcontinent.

Suri Babu and Srivastava (1990) studied the life history of certain *Ceriagrion coromandelianum* in the artificial cemented tank of Sagar (M.P.), India. There are four broods, out of which two are confined to the summer, one occurred in monsoon and the last brood extended over the winter season. They stated that the water temperature and seasonal rain have influenced the multivoltine nature and larval growth of this species.

Chowdhury and Mohiuddin (1994) also reported that *P. flavescens* breeding in artificial concrete tanks and adults were found throughout the year with no migration was seen in Chittagong, Bangladesh.

Emergence

It is rather a well-known phenomenon that the odonates emerge in the early hours of morning and also that most or majority of them emerge almost at the same time after first rains. Poyyamoli and Pandian (1985) reported the emergence patterns of four tropical odonate species in relation to certain environmental factors in a pond situated 20 km away from Madurai city, Tamil Nadu. They observed the emergence of *Orthetrum sabina, Pantala flavescens, Ischnura senegalensis* and *Ceriagrion ceromandelianum* exclusively between 0500 and 0700 hr and they exhibited highly synchronized emergence and the sex ratio ranged from 0.5 to 1.6. These authors indicated that the direct of light intensity in combination with the indirect effect of relative humidity played a dominant role in dictating diel emergence patterns of the odonates. Kumari and Nair (1985) recorded the emergence of *Pantala flavescens* in the months of December and January in the zone which receives northeast monsoon whereas Nasiruddin and Begum (1985) reported that dawn is the preferable period of emergence of *Tramea virginia* takes place from weedy tanks and during the months of April and May at Dehrudun. He also reported the emergence of *P. flavescens* in August and September in the areas of south-west monsoon.

Begum *et al.* (1992) observed that the emergence in some species of dragonflies takes place from very early in the middle of the day in Ramna lake of Dhaka city, Bangladesh Chowdhury and Mohiuddin (1994) reported that first category of larvae of *Potomarcha obscura, Tramea basilaris burmeisteri, Orthetrum sabina, Anax guttatus, Ryothemis plutonia, Hydrobasileus croceus* and *Tramea limbata* emerge over a single relatively shorter period of 1½ to 3½ months, where the second category species like *Ictingomphus rapax, Palpopleura sexumaculata sexumaculata* and *Acisoma panorpoides panorpoides, Rhyothemis variegata variegata, Orthetrum pruinosum neglectum* and *Neurothemis intermedia atalanta* emerged for six months. Some species like *Brachethemis contaminata, Tholymis tillarga* and *Neurothemis tullia tulia* have two long emergence periods in a year. Species like *Diplocodes nebulosa, D. trivialis* and *Rhodothemis rufa* have one long and one short period and species like *Pantala flavescens, Trithemis pallidinervis* and *Brachydiplax sobrina* have two short emergence period in Chittagong, Bangladesh. Srivastava and Suri Babu (1994) reported the emergence periods in Chittagong, Bangladesh. Srivastava and Suri Babu (1994) reported the emergence of *Enallagma parvgum* in March, then June and again in the post-monsoon period, i.e. September and October. Suri Babu *et al.* (1995) reported that the mature larvae of *Ictinogomphus angulosus* crawled up one the submerged stones, sides of anchored boats and flating leaves of aquatic plants for emergence. The emergence occurred during 1900 to 2400 hr, thus avoiding predation by birds. The emergence started in the month of May and continued up to the last week of July at Sagar lake (M.P.).

Suri Babu *et al.* (1996) also reported that the mature larvae and emergence of *Trithemis pallidinervis* (Kirby) two times in every year, i.e. in the months of March and April and later in September and October. Suri Babu (1998a) observed the emergence of *Ischnura aurora aurora* twice in a year, once in early summer (April) and again towards the decline of the monsoon (September and October) in a slow running stream in Sagar (M.P.). Suri Babu (1998b) recorded the emergence of *P. decorum* throughout the year except in extreme winter at water temperature of lake 17-18°C, i.e. during the months of December-February. Suri Babu (2000) reported the emergence of *Nurothemis intermedia intermedia* (Rambur) during. April to mid-May and again from mid-October to November in Sagar lake.

Biswas *et al.* (1994) reported that males and females of dragonflies do not emege at equal rate throughout the emergence period, they also noted that males emerge significantly earlier than females and presumed that early emergence of males perhaps assures their presence when females return to the water for mating.

Reproductive Behaviour

In a typical condition mature males arrive at the rendezvous earlier in the season and in the day than femals, and place themselves out before the female arrive and the rendezous serve as the focus or departure point for population and subsequent oviposition and therefore, as the arena for sexual selection (Corbet, 1980).

The Rendezvous

The different sex of odonates of certain species sometimes or normally meet and copulate elsewhere, for the great majority the rendezvous is at or near the oviposition site (Corbet, 1980). Srivastava and Suri Babu (1985a) reported an artificial cemented tank at Sagar (M.P.) as rendezvous for *C. ceromandelianum*. They observed that males used to arrive at rendezvous usually in the morning between 8.30 and 9.00 hr and frequently eruised over without maintaining individual territory. The males used more than 5 sites as a base perhaps before acquired tandem position which also indicated lack of territoriality in the this species.

Chowdhury and Karim (1994) reported the *Copera annualata* has chosen a shaded place not far from the waterbody as rendezvous and males arrived the breeding place earlier than females.

Territoriality

The males of some odonates localize in a more or less fixed area within which they behaves in a way that evokes escape and avoidance among intruders, and this area properly termed as territory according to criteria recognised for other animals, specially birds (Corbet, 1980).

Srivastava and Suri Babu (1985b) reported territorial area for *Disparoneura quadrimaculata* as a circular area of 1.5-2.0 m. Chowdhury and Karim (1994) observed territorial area of *C. annulata* as a circular area with a radius of 20-30 cm. Srivastava *et al.* (1994) stated that males of *P. decorum* establized base perch generally at a high spot on *Eleocharis* and *Potamogeton* plants and defending territories within a radius of 0.5-1.0 m. Prasad (1991) reported that males of *B. contaminata* selects a small portion of the pond about 0.8 to 2.0 m in diameter as its territory. Prasad (1990a and b) reported that males *C. s. servilia*, *C. coromadelianum* and *Pseudagrion rubriceps* demonstrated territorial behaviour.

Mathavan (1985) reported that males of *B. contaminata* did not establize a settled base but maintained a territory of defence covering a maximum shore line of 2.0 to 4.0 m whereas males of *O. sabina* establize a settled base but not a territory of defence.

Sangal *et al.* (1994) reported that males of *N.i. intermedia* demonstrate strong territorial behaviour whereas *O. s. sabina* exhibited this behaviour occasionally aggressive behaviour.

The aggressive behaviour exhibited by mature males of odonates at the rendezvous is directed predominately towards conspecific males and nature and intensity of aggressive behaviour vary among species, and also within species according to density of males and distance from the centre of the territory (Corbet, 1980).

Srivastava and Suri Babu (1985b) stated that males of *D. quadrimaculata* showed aggressive behaviour similar to flight towards intruder behaviour of Bick and Bick (1965). The males of this species never chased the intruder beyond the territory and harmed physically unlike the behaviour observed in *Pseudagrion decorum* which express aggresive behaviour not only against conspecific males but also against heterospecific males to the extent that an intruding male of *Enallagma parvum* was captured and preyed upon by a resident male (Srivastava *et al.*, 1994). Chowdhury and Karim (1944) stated that the males of *C. annulata* showed four types of behaviour towards conspecific male:

(a) Approach without aggressiveness.

(b) Approach threat display.

(c) Chase display.

(d) Confrontation flash display.

Pre-Copulatory Courtship

Corbet (1962) stated that the pre-copulatory courtship was shown by males of *Caloptervgid* sp. before forming a tandem. Srivastava and Suri Babu (1985a) and Chowdhury and Karim (1994) reported precopulatory courtship found to be lacking in *C. ceromadelianum* and *C. annulata* respectively. Srivastava *et al.* (1994) reported no specific courtship except the male flew along side the female some distance before to secure a tandem link in *P. decorum* whereas Srivastava and Suri Babu (1985b) observed courtship in *D. quadrimaculta* was exhibited by males who prominantly displayed the blackish brown spots of their wings and performed flights above, below and laterally to the female damselfly.

Intra-Male Sperm Transfer

After a brief period in pre-copulatory flight, the male alighted on a terrestrial object an aquatic plant while the female remain hanging. The male then curved his abdomen ventrally and forward in order to bring gonopore near the secondary copulatory apparatus. Intra-male sperm translocation took place from ninth abodominal segment to the penis vesicle on the second abdominal segment (Srivastava and Suri Babu, 1985a and b). These authors reported that intra-male sperm translocation occurred after the seizure of the female interruptions in perched conditions in *C. ceromandelianum* and *D. quadrimaculata*. Srivastava *et al.* (1994) reported this process took place while male maintaining the tandem grip and settled on the perch.

Copulatory Wheel Position

Anisopteran dragonflies mostly assume the wheel position during the flight (Robert, 1958) whereas zygopteran damselflies usually assume it when perched (Rowe, 1978). Srivastava and Suri Babu (1985a) reported copulation in *C. coromandelianum* took place without any interruptions for a long duration of 18.4 mintues. They also reported (1985b) that the copulatory wheel was formed in the perched condition and took place for 4.1 minutes in *D. quadrimaculata* and copulation occurred under two conditions, settled and flight.

On the basis of dual function of the damselfly penis sperm removal and sperm transfer (Waage, 1979). It has been further argued by Corbet (1980) that sperm removal by the penis from spermatheca needed a very long duration of the wheel position and sperm transfer was carried out in short duration. Srivastava and Suri Babu (1985b) reported short duration of copulation observed in *D. quadrimaculata* may be explained if it involved mating only with virgin females. Mathavan (1985) observed copulation of *B. contaminata* in flight which lasted for 5-45 seconds. He reported that copulation in *O. sabina* took place when male settled on a site like tree top or bush and the duration was 2-28 mintues. Prasad (1990) observed wheel position for *P. rubriceps* for 3-8 minutes and for *C. ceromandelianum* for 33.6 minutes. In 1991 he observed copulation of *B. contaminata* for 4-10 seconds only.

Chowdhury and Karim (1994) observed copulation in *C. annulat* for 20-30 minutes with intercepted by rests. Srivastava *et al.* (1994) observed that in *P. decorum* in copulatory wheel the male displayed the aggressive behaviour by vigorously vibrating wings until the intruder flew away when any conspecific male approached.

Post-Copulatory Exploratory Flight

After insemination, the copulatory wheel was broken but the tandem link maintained. The dragonflies then performed post-copulatory flights in tandem. Srivastava and Suri Babu (1985a) observed post-copulatory flight in tandem for 66.5 minutes for *C. ceromeandelianum*. They observed (1985b) the female of *D. quadrimaculata* explored among the fallen rotting leaves, twings and algal mat on the water surface by touching these objects with her ovipositor. They observed the presence of fine setose at the tip of the styli of the ovipositor valvulae and the setae forming as a special sensory organ.

Oviposition

The females of odonates either lay egg "endophytically" within or among plant tissues or similar material or "Exophytically" by releasing them above or upon a surface (Corbet, 1980).

Srivastava and Suri Babu (1984) observed that females of *Ischnura aurora aurora* oviposited on grass stems usually on the water surface without gugarded by male. The eggs were laid in a linear pattern below epidermis of the stem of grass. They observed (1985b) that females of *C. coromandelianum* oviposited in tendem or unattended by the male and allowed other conspecific pairs also to lay eggs in the are without farming oviposition territory and eggs were laid endophytically in the floating leave of the *Salvinia* sp. plants and they recorded that females of *D. quadrimacurlata* oviposited in company with the male and eggs are inserted singly among the filaments of the algae on the water surface, unlike the typical endophytic oviposition reported in other zygoptera.

Kumar (1989) observed that *Tramea virgina* oviposits in large perennial ponds towards the decline of south-west monsoon. He observed modified egg laying behaviour as the female suddenly separated herself from male in rended and dip her abdomen in water to release eggs. Mathavan (1985) reported that on completion of copulation the female of *B. contaminata* separates from tandem and oviposits expiphytically on the blades of *Cynodon dectylon* (Perr) or other members of *Graminae* sp. plants or any floating surface under the protection of the male and he also reported females of *O. sabina* descends to the water surface and oviposits exophytically without protection of male.

Kumar (1989a) reported the oviposition of *N. t. tullia* occurred in isolated marshy tanks towards the outset of south-west monsoon during the month of September. Begum *et al.* (1990a) recorded oviposition of Rhodothemis rufa exophytically on the submerged leaves or twigs of the pond while female continuously guarded by the male.

Kumar (1985) reported the oviposition of *I. rapax* is exophytic and performed by female alone. The male abandon the female after copulation. The female about 2 m over water surface and suddenly dips and release the eggs in water.

Chowdhury and Karim (1994) found oviposition of *C. annulata* in upright tandam of the male and without the male gaurding the female Srivastava *et al.* (1994) observed that females of *P. decorum* oviposits endophytically on the *Potomogeton* plants in tandem at the water surface and by unattached with male in under water. The oviposition of this species occurred throughout the year except during late summer when aquatic vegetation undergone decomposition, during rains and in severe winter months.

Sangal *et al.* (1994) observed *O. s. sabina* and *N. t. intermedia* in both species oviposition is exophytic and male guards the female from intruding males during oviposition.

Parasites and Predators

The population dynamics of both larvae and adults of odonates are mostly affected due to the parasites and their predators. Kumar (1973) reported that predator such as spider waters bugs, beetles and fishes prey quite often upon different stages of odonate like eggs. Kumar (1977) reported *Lestes praemorasa praemorasa, Bavadera india, B. contaminata* and *Aethriamanta b. brevipennia* in garden spiders web at Dehradun.

Thomas and Gladstone (1989) stated that predators like birds Sitana. Calotes often feed on *D. trivialis* while they are generally taking rest. Srivastava *et al.* (1994) reported females of *P. decorum* victim of frog and fishes during submerged oviposition. Mitra (1996) recorded predation of tenerals of *Orthetrum sabina, Crocothemis s. servilla* and *Ischnura aurora aurora* by garden lizard *Calotes versicolor.*

Sarkar (1981) suggested that the complex epimerite of the gregarines may destroy the region of the midgut epithelium of adult Odonates at which it is auchored and they heavy infection leading to the complete blockage of the dragonfly midgut lumen, may disturb the process of digestion of the host.

Sarkar and Mazumdar (1983) reported the trophozoite and sporadin stages of Cephaline gregarines *Hoplorhynchus carusi* sp. in *Odonaticola brachvdiplaxi* sp. n recovered from the midguts of adult *P. decorum* and *Brachvadiplax farinos.*

Thomas and Gladstone (1989) observed ectoparasites like *Limnochares americana* in the posterior thoracic pleural region of *Diplacodes trivialis.*

Acknowledgement

We are deeply grateful to Professor P.S. Corbet (United Kindgom), Professor B. Kiauta (Holland), Professor S.P. Roy (Bhagalpur), Dr. Arun Kumar (Dehradun), Dr. V.K. Srivastava (Allahabad), Professor D.B. Tembhare (Nagpur) and Dr. M. Prasad (ZSI, Kolkata) for exchange of literature and valuable information to prepare this review article.

References

Begum, A., Biswas, B.R. and Elias, M. 1986. The ecology and seasonal fluctuations of mosquito larval in a lake in Dhaka city. *Bangladesh J. Zool.* 14 (1) : 41-48.

Begum, A., Bashar, M.A. and Biswas. V. 1990a. The mating behaviour and development of *Rhodothemis ruja* (Rambur) (Anisoptera : Libellulidae). *Indian Odonatol* 3: 1-9.

Begum. A., Bashar, M.A. and Biswas, V. 1990b. On the life history of *Neurothemis tullior tullia* (Drury) from Dhaka. Bangladesh. (Anisoptera : Libellulidae). *Indian Odonatol.* 3 : 11-20.

Bick, G.H. and Bick. J.C. 1965. Demography and behaviour of the damselfly *Argia apicalis* (Say). *Ecology*. 46 :461-472.

Biswas, V., Begum. A., Bashar. M.A. and Begum. S.A. 1994. Emergence pattern of some dragonfly (Odonata : Anisoptera) larval under the laboratory condition. *Adv. Oriental and Odonatol.* 23-30.

Chowdhury, S.H. and Karim. N. 1994. Observation on the reproductive behaviour of *Copera annulata* (Selys). *Adv. Oriental. Odonatal.* 67-76.

Chowdhury, S.H. and Mohiddin. M. 1994. Dragonfly phenology— a mechanism for optimal habitat utilisation. *Adv. Oriental. Odonatol.* 47-54.

Corbet, P.S. 1962. *A Biology of Dragonflies*. Witherby. London.

Corbet, P.S. 1980. Biology of Odonata. *Ann. Rev. Entomol.* 25 : 189-217.

Fraser, F.C. 1933. *The Fauna of British India, including Ceylon and Burma. Odonata.* Vol. 1. Taylor and Francis. London. 1-423.

Fraser, F.C. 1934. *The Fauna of British India, including Ceylon and Burma. Odonata.* Vol. 2. Taylor and Francis. London. 1-398.

Fraser, F.C. 1936. The Fauna of British India, including Ceylon and Burma. Odonata Vol. 3 Taylor and Francis. London 1-461.

Gangulee, P.K. 1979. Studies on the morphology, taxonomy, growth rate and seasonal abundance of dragonfly larvae (Anisoptera : Odonata) from Dhaka University campus and its adjoining areas. M. Sc. Thesis. University of Dhaka.

Kaushik, S., Sharma. S., Saxena, M.N. and Saxena, D.N. 1990. Habitat ecology of Anisopteran nymphs in certain waterbodies. *Acta. Hydrochim. Hydrobiol.* 18. 443-447.

Kumar, A. 1972. The phenology of dragonflies in the Dehra Dun Valley. *India Odonatol* 1(4) : 199-207.

Kumar, A. 1973. The life history of *Bradinopyga geminata* (Rambur) Odonata : Libellulidae). Gurkul Kangri Vishwavidyalaya *J. Sci.* 5(1-2) : 50-57.

Kumar, A. 1977. A note on Dragonflies caught in spider's webs. *Odonatologica*. 6(1) : 19-20.

Kumar, A. 1985a. Studies on the life history of Indian dragonflies *Ictinogomphus rapax* (Rambur) (Gomphidae : Odonata). *Ann. Entomol.* 3(1) : 29-38.

Kumar, A. 1985b. A review on the biology of Indian Dragonflies. *Proc. First Indian Symps. Odonatol.* 73-92.

Kumar, A. 1989a. Life history of *Tramea virginia* (Rambur) from Dehradun. India (Anisoptera : Libellulidae) *Indian Odonatol.* 2 : 5-14.

Kumar, A. 1989b. Pattern of emergence of Dragonflies (Insecta : Odonata) in relation to Monsoon in India. *Indian J. Phy. Nat. Sci.* 9(A) : 10-15.

Kumar, A. and Negi, B.K. 1989. The emergence and sex ratio of *Bradinopyga geminat* (Rambur) and *Crocothemis servilia servilia* (Drury) at Dehra Dun. India (Anisoptera : Libellulidae). *Fraseria.* 16 : 70-71.

Kumar, A. 1994. Population density, species diversity and evenness of dragonfly nymphs in a freshwater pond of Santal Pargana (Bihar). *J. Freshwat. Biol.* 6:231-235.

Kumar, A. 1995. Community structure and population dynamics of odonate larvae in certain wetlands of Santal Pargana, India. *Proc. Nat Acad. Sci. India* 65(B) : 401-410.

Kumar, A. 1995. Estimation of productivity of odonate larvae of a tribal pond of Santal Pargana. Bihar. *Environ and Ecol.* 13: 313-315.

Kumar, A. 1996. A comparative study on stomach-content and forage ratio of zygopteran and anisopteran larvae in a fish pond of Santal Pargana (Bihar) India. *Proc. Nat. Acad. Sci. India* 66 (B): 315-321.

Kumar, A. and Kumari. P. 1999. Population dynamics of the zygopteran larvae in a fish pond. *Fraseria.* 6(1/2) : 3-8.

Kumari, K.R.N. and Nair, N.B. 1985. Life history studies on *Pantala flavescens* Fab.(Odonata : Libelluilidae). *Proc. First Indian Symp. Odonatol.* 133-149.

Manu Thomas, M. and Gladstone, M. 1989. Biological notes on *Diplacodes trivialis. Frasocia.* 16 : 69.

Mathavan, S. 1985. Reproductive behaviour of *Brachythemis contaminata* and *Orthetrum sabina* (Odonata : Libellulidae). *Proc. First Indian Symp. Odonatol.* 93-106.

Mitra, A. 1996. A note on dragonfly predation by the garden lizard. *Calotes versicolor* (Daudin). *Ann. For.* 4(1) : 113-115.

Nasiruddin, M. and Begum, A. 1985. Description of the final instar nymph of *crocothemis servilia servilia* Drury (Anisoptera L Libellulidae) with notes on its emergence pattern. Chittagong Univ. Studies. II. *Sc.* 9(1) : 28-34.

Nirmala Kumari, K.R. and Balakrishna Nair, N. 1985. Life history studies on *Pantala flavescens* Fab. (Odonata : Libellulidae). *Proc. First Indian Symp. Odonatol.* 133-149.

Poyyamol, G. and Pandian, T. J. 1985. Studies on diel emergene pattern of some tropical dragonflies. *Proc. First Indian Symp. Odonotol.* 159-165.

Prasad, M. 1990 a. Field notes on the mating behaviour of *Crocothemis servilia servilia* (Drury) at a perennial pond in Calcutta. India (Anisoptera : Libellulidae). *Indian Odonatol.* 3 : 67-68.

Prasad, M. 1990b. Reproductive behaviour of *Ceriagrion coromandelianum* (Fabricius) and *Pseudagrion rubriceps* Selys. (Zygoptera : Coenagrionidae). *Ann. Entomol.* 8 (2) : 35-38.

Prasad, M. 1991. On some aspects of reproductive behaviour in *Brachythemis contaminta* (Fabricues) Anisoptera : Libellulidae). *Ann. Entomol.* 9(1) : 1-3.

Prasad, M. and Varshney, R.K. 1995. List of odonata of India in data on larval studies. *Ori. Insects.* 29 : 385-428.

Robert. P.A. 1958. *Leslibellules (Odonates),* Delachaux and Niestle. Neuchatel.

Rowe, R.J. 1978. *Ishnura aurora* (Brauer). A dragonfly with unusual mating behaviour (Zygoptera : Coenagrionidae). *Odonatol.* 7(4) : 375-382.

Roy, S.P. 1998. Energetics and trophic biology of larval odonates with special reference to their role in the management of aquatic ecosystem. *Fraseria.* 5: 47-56.

Sarkar, N.K. 1981. On the histopathological changes of the midgut of *Ceriagrion cerinorubelum* (Brauer) (Insecta : Odonata) due to infection of Acanthospora bengalensis. Sarkar and Haldar. *Abst. VIth. Int. Congr. Protozoel.* 326 pp.

Sarkar, N.K. and Mazumder, S.K. 1983. Two new species of cephaline gregarines frm Odonata of West Bengal. India (Protozoa. Apicomplexa. Eugregarinidar. Actinocephalidae). *Odonatol.* 12(2) : 173-178.

Samgal, S.K., Bhandari, P. and Saxena, A. 1994. Reproductive behaviour of *Orthetrun sabina sabina* (Drury) and *Neurothemis intermdedia intermedia* (Rambur) (Anisoptera : libellulidae). *Adv. Oriental. Odonatol.* 63-68.

Suri Babu, B. 1988. A discussion focusing on the seasonal ecology of *Pantala flavescens* in the Indian subcontinent. In : *Current Topics in Dragonfly Biology* (Ed. P.S. Corbet). 3 : 4.

Suri Babu, B. 1998a. Final instar larva of *Ischnura aurora aurora* (Brauer) (Zygoptera : Coenagriidae). *J. Bombay Nat. Hist. Soc.* 95 : 354-357.

Suri Babu, B. 1998b. Description of the larva of *Pseudafrion decorum* (Rambur. 1842) from central India (Zygoptera : Coenagrionidae). *Odonatologica.* 2 (4) : 473-477.

Suri Babu, B. 2000. Description of the larva of *Neutothemis intermedia* (Rambur) with notes on biology (Anisoptera : Libellulidae). *Odonatologica.* 29 (4) : 341-346.

Suri Babu, B. and Srivastava, B.K. 1990. Breeding biology of *Ceriagrion coromandelianum* (Fabricius) with special reference to seasonal regulation (Zygoptera : Coenagrionidae). *Indian Odonatol* 3 : 33-43.

Suri Babu, B., Srivastava, B.K. and Srivastava, V.K. 1995. Description of last instar larva of *Ictinogomphus angulosus* (Selys) with notes on biology (Anisoptera : Gumphidae). 2(1/2) : 9-11.

Suri Babu, B. Srivastava, B.K. and Srivastava, V.K. 1996. Morphology and biology of the final instar larva of *Trithemis pallidinervis* (Kirby). (Odonata : Libellulidae). *Bioved.* 7(2) : 107-110.

Suri Babu, B. and Srivastava, V.K. 2001. Annotations on the dragonfly fauna of Sagar. Madhya Pradesh. Central India (Odonata : Anisoptera). *Opuse. Zool. Flumin.* 193 : 1-7.

Srivastava, B.K. and Suri Babu, B. 1984. Some observations on oviposition of *Ischnura aurora* (Brauer) in Indian biotopes. (Zygoptera : Coenagrionidae). *Fraseria.* 6 : 24.

Srivastava, B.K. and Suri Babu, B. 1985a, reprodcutive behaviour of *ceriagrion coromandelianum* (Zygoptera: Coenagriidae). *Proc. First Indian Symp. Odonatol.* 209-216.

Srivastava, B.K. and Suri Babu, B. 1985b. On some aspects of reproductive behaviour in *Chloroneura* (Disparoneura) *Vuadrimaculata* (Rambur) (Zygoptera : Protoneuridae). *Odonatologica.* 14(3) : 219-226.

Srivastava, B.K. and Suri Babu, B. 1994. Identifying description of the final instar larva of *Enallagma parvum* Selys (Zygoptera Coenagriidae) from Sagar (M.P.) *Adv. Oriental. Odonatol.* 13-17.

Srivastava, V.K. and Srivastava, B.K. and Suri Babu, B. 1994. The behaviour of reproduction and oviposition in *Pseudagrion decorum* (Rambur) (Zygoptera : Peudagriinae) in Central India. Adv. Oriental. *Odonatol.* 77-84.

Tembhare, D.B. 1998. *Souvenir of Vth. South Asian Symposium of Odonatology.* Nagpur. pp. 11-12.

Thomas, M.M. and Gladstone, M. 1989. Biological notes on *Diplacodes trivialis. Fraseria.* 6: 69-74.

Tillyard, R.J. 1917. *The Biology of Dragonflies.* Cambridge Univ. Press, Cambridge. 396 pp.

Waage, J.K. 1979. Dual function of the damselfly penis : Sperm removal and transfer. *Science.* 203 : 916-918.

6

Conservation Assessment and Management Plan (CAMP) Process: A Tool to Provide Strategic Guidance for the Management of Threatened Odonata

• *B.A. Daniel, Sanjay Molur and Sally Walker*

Introduction

The biological resources are important to man's economic and social development. It is recognized that biodiversity is an assest and of tremendous value to mankind. But the threats to the species and the ecosystem are alarmingly increasing, resulting in the loss of species, which is a real concern today. Man has altered half of the habitable surface of the planet and so man is at the verge of mass extinction. Many life forms are disappearing from the ecoystem at an alarming rate. Lack of information interrupts our ability to understand the trend in Biodiversity loss. The present-day most important questions are how many species are there? Where do they occur? What is their ecological role? What are the species that are being threatened? What could be done to save the threatened species? A concrete answer is not available since our knowledge of the diversity of plants, animals, microbes and the ecosystem are limited.

India has been identified as one of the 24 mega-diversity areas in the world (Mittermeier *et al.*, 1998). the term biodiversity denotes the variety and variability of all life forms on earth and this can be considered at three levels – species diversity, ecological diversity, and genetic diversity. India is rich in biodiversity but India's biological wealth is threatened by over exploitation of natural resources resulting in devastation of habitats and ecosystems. Thus, there is an erosion or loss of biodiversity on an unprecedented scale everywhere. Species extinction, the disappearance of ecological associations, or the loss of genetic variants in an extant species can all be classed as biodiversity loss.

Odonates are part of the Biodiversity. The adults of dragonflies and damselflies are integral part of terrestrial ecosystm and the larvae are aquatic. This group of invertebrates are playing integral part and are directly or indirecly associated with the welfare of the animal as well as the human society. Fishes eat the odonate larvae and reptiles, insectivorous birds, and amphibians eat adults and larvae. Thus, they form as an integrated part of the food chain in the ecosystem. Odonates are natural biocontrol agents and particularly the larvae feed on the juveniles of certain vector of human and animal diseases. These are also used as biological indicators or aquatic pollution.

Why Odonates are to be Conserved?

Odonates are part of biodiversity so it should be conserved. They have its own cultural value, i.e., owing to the size, these diurnal flying insectrs has brought the attention of people so that they are included in the folklore of many countries and they are the subject of poetry and painting. Dragonflies are considered as potential bio-indicator. They are used to assess the water quality though they are less sensitive compared to other aquatic organisms. Some species are habitat specific and so can be used for mapping of the habitats which they represent. These insects are used as bio-control agents of certain vectors like mosquitoes. The advancement in the aeronautic science is mainly due to the knowledge obtained from the fight principles of the dragonflies. Dragonflies, therefore, have a potential health and economic value, which is not yet fully exploited. Most invertebrates are small and inconspicuous and their immense importance in nature is underestimated. Odonates being large and conspicuous draw attention to invertebrates and the neend for conservation.

Odonate Diversity and Present Status

So far about 5,000 species of Odonates has been reported from all over the world. India's odonate diversity is high which records 500 species and subspecies under 135 genera and 19 families (Tyagi, 1997). This forms abut 10% of the world odonate fauna. This rich diversity is fast disappearing due to destruction of their breeding and resting habitats. *Epiophlebia laidlawi* Tillyard was reported 60 years back from the Darjeeling Himalayas the only representative of anisozygoptera. This genus is represented only by two species. *E. laidlawi* is in India while the other in Japan (Moore, 1997). Thus, endemic species are more important to science, which need proper protection. Identifying the priority species that require immediate attention is the first step in conservation action. Species prioritisation of the dragonflies and damselflies of India is a prerequisite for odonata conservation. Government of India has listed *E. laidlawi* (Anisozygoptera) in Schedule I Part IV of the Wildlife (Protection) Act, 1991 and so this species is protected constitutionally and this is the only odonate species protected by Indian law. Apart from this, four species of dragonflies namely, *Cephalaeschna acutifrons* (Vulnerable), *Acanthaeshna victoria* (Vulnerable) (Family Aeshnidae); *Burmagomphus sivalikensis* (Critically endangered) (Gomphidae); and *Ephiophlebia laidlawi* (Vulnerable) (Epiophlebiidae) are listed in the IUCN Red Data Book, 1996 (IUCN, 1996). Odonate conservation cannot be treated separately since the national programmes such as formation of national parks take care of invertebrates also. But many of the odonate extent range area that does not come under any of the national programmes have to be focused. Dehra Dun valley in the State of Uttaranchal has more than 60 species of dragonflies and damselflies, some of which are endemic to that area e.g. *Burmagomphus sivalikensis* Laidlaw (Gomphidae). Severe habitat loss has been observed in Dehra Dun valley in the past resulting in the decline of this population (Tyagi, 1985). Such species which require conservation priority has to be identified. CAMP workshop help to prioritise fauna at Global/ National/Regional level. Based on that due conservation action plan can be planned for necessary species or subspecies.

Causes of Extinction

Humans have been a cause of species extinction, especially when they migrated into new habitats. Despite the long history of human impact, human activities are placing significantly more species at risk of extinction today than at any time in the past as a result of environmental changes affecting current population sizes, environmental carrying capacity, population density, the mean and variance in populations and the size, number and distances of suitable habitat patches and local populations. These are results of various human activities such as destruction or reduction of habitat due to forest clearance, damming of rivers, drainage and pollution resulting in distraction of fresh water habitat, use of pesticides and fertilizers, and introduction of exotic predator fishes etc.

Based on ecological theory, species that are most threatened with extinction are :

1. Large organisms,
2. Those highest on the food chain,
3. Those with small population sizes,
4. Those with small ranges or distributions,
5. Those tied in some aspect of their biology to habitats or ecosystems that are threatened,
6. Those that have evolved in isolation (e.g. Island fauna and flora),
7. Those with poor dispersal and colonization abilities,
8. Those with colonial nesting habits,
9. Migratory species,
10. Those dependent on unreliable resources, and
11. Those with little evolutionary experience of disturbances. (Heywood, 1995).

Among invertebrates most of the insects in Odonata are large and attractive. Certain odonates are short-range species that has restricted distribution and many are enedemic. Though not much population study is available on this group of invertebrates it could be assumed that due to multiple threat factor many invertebrates with small population size are in danger and odonata is not an exemption. The biology of this group is tied up with their habitats or ecosystems that are threatened. Many locally endemic odonates will be lost because of their susceptibility to habitat loss.

Extinction rate and conservation strategy (*With special reference to invertebrates)* : The natural extinction rate is inevitable and in the past, the slow alteration by man was barely distinguishable compared to natural extinction. But the present rate of global extinction is estimated at approximately 10-20,000 per annum, or between 1000 and 10,000 times faster than the natural rate (Wilson, 1988). Conservation biologists caution that unless current trends of species loss are reversed, more than one-quarter of the Earth's species may vanish in the next 50 years (Anon, 1992).

Invertebrates dominate most in the animal Kingdom. About two-thirds of the 1.4 million described species are invertebrate (Wilson, 1988) if these majority are insects. But in the 1996 Red Data List, very few of these invertebrate crustaceans, insects an molluscs are the highly assessed group. Among insects the orders with more than hundred threatened species are Odonata (4 reported from India), Hymenoptera and Lepidoptera (IUCN, 1996). Global assessment of these groups are possible may be because of its popularity. It is obvious that those invertebrates that have received minimal attention may have many threatened species.

Conservation Approach for Odonata

Our knowledge of the entire odonate diversity is patchy. Non-availability of information is one of the reason for not giving due importance towards its conservation. The available information about the reported dragonfly species is inadequate and it will be a time consuming efforts to acquire knowledge since the odonate specialists are few. One the other side habitat destruction is proceeding at an unprecedented rate. Conservation of dragonflies, species by species is not possible by the odonatologists since necessary information cannot be provided within a short period of time. Therefore, adopting a broad-brush approach, which must depends on what can be done on information, can only on serve most species. The risk of doing the wrong thing through lack of scientific knowledge is much less than the risk of delay (Moore, 1991). Ranking individual species conservation status globally using IUCN Red List criterion is in practice since the 1970s. Individual countries have also brought out Red Data Books on various taxonomic groups.

Conservation Breeding Specialist Group (CBSG) and Conservation Assessment and Management Plan (CAMP)

The Conservation Breeding Specialist Group (CBSG) is one of the 107 Specialist Groups of the IUCN, which is responsibe for coordinating the conservation needs and activities between *ex situ* facilities and *in situ* requirements. The CBSG developed a participatory methodology for assessing conservation status of species within a broad group of taxa. This method involved gathering field specialists on a particular taxon or a taxonomic group to provide data necessary for deriving status in the wild and population in captivity if available. This information was entered into a spreadsheet format and later converted into individual Taxon Data Sheets. A variety of factors related to population and its range is used to decide the status of the taxa under review. Other factors are considered to make a decision or for research recomendations. This workshop process is called Conservation Assessment and Management Plan or CAMP workshop. Thus, CAMP is one of the "tools" or processes developed by CBSG to bring about effective management and planning. Conservation Assessment and Management Plan process is designed to assess the status of, and degree of threat to, a group of taxa, or of the taxa of a particular region so that management actions can be developed for species in need and thus CAMP results provide strategic guidance for the management of threatened taxa.

The IUCN Red List criteria underwent several revisions over the five years between 1990 and 1995. The major differences between the old criteria and the new criteria are that the revised criteria are more objective, numerical and scientific as well as having greater applicability across taxon groups. The revised criteria are meant to be used for all organisms except micro-organisms. The new IUCN Red List Categries and Criteria have several specific aims :

- to provide a system that can be applied consistently by different people;
- to improve the objectivity by providing users with clear guidance on how to evaluate different factors which affect risk of extinction;
- to provide a system which will facilitate comparisons across widely different taxa;
- to give people using threatened species lists a better understanding of how individual species were classified.

IUCN Red List category has been applied for a wide spectrum of vertebrates, Indian mangrove species, reptiles, amphibians, mammals, fresh water fishes, medicinal plants of Southern India, non-timber forests products of Central India and soil invertebrates of Southern India.

Conservation Assessment and Management Plan (CAMP) : Process

CAMP workshops are organised to bring together a broad spectrum of expertise on wild and captive management of the taxa or group under review. The entire process is intensive and interactive taking place over a period of 3 or 5 day period, depending on the scope of the workshop and the number of species to be assessed. Ten to forty experts come together during the workshop to determine the status of threat for the taxa under review.

The CAMP process utilizes information from action plans that already have been formulated by the taxon based Specialist Groups as well as additional data from experts on the taxa, both published and unpublished. Participants develop the assessment of risks and formulate recommendation for action using a systematically designed Taxon Data Sheet. This sheet facilitates recording of detailed information about each taxon, including data on the status of population, their habitat in the wild and recommendations for intensive conservation action. An accompanying computerized CAMP data entry programme aides the collection of information, facilitates production of the report and allows information for all CAMP workshops to be accessed and queried by any interested parties.

The CAMP's attempt to be as quantitative or numerates as possible because IUCN/SSC Action Plans and biodiversity strategies, for biological and legislative reasons, ultimately must establish numerical objectives for population sizes and distribution if they are to be viable. In addition, numbers provide for more objectivity, less ambiguity, more comparability, better communication and hence cooperation (Byers *et al.*, 1999).

The Categories

The Categories (Fig. 6.1).

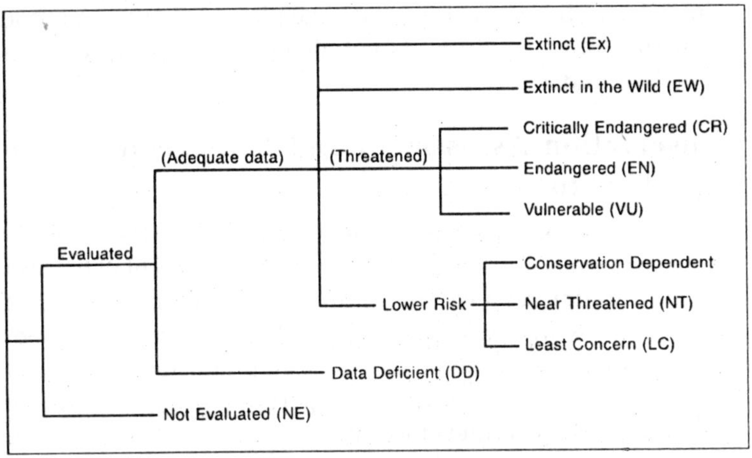

Fig. 6.1 : Relationships between the categories

A representation of the relationships between the categories is shown in Fig. 6.1.

Extinct (EX)

A taxon is Extinct when there is no reasonable doubt that the last individual has died.

Extinct in the Wild (EW)

A taxon is Extinct in the Wild when it is known only to survive in cultivation, in captivity or as a naturalised population (or populations) well outside the past range. A taxon is presumed extinct in the wild when exhaustive surveys in known and/or expected habitat, at appropriate times (diurnal, seasonal, annual), throughout its historic range have failed to record an individual. Surveys should be over a time frame appropriate to the taxon's life cycle and life form.

Critically Endangered (CR)

A taxon is Critically Endangered when it is facing an extremely high risk of extinction in the wild in the immediate future as described by any of the following criteria (A to E) :

A. Population reduction in the form of either of the following :

1. An observed, estimated, inferred or suspected reduction of at least 80% over the past 10 years or three generations whichever is longer, based on (and specifying) any of the following :
   (a) direct observation
   (b) an index of abundance appropriate for the taxon
   (c) a decline in area of occupancy, extent of occurrence and/or quality of habitat
   (d) actual or potential levels of exploitation
   (e) the effects of introduced taxa, hybridisation, pathogens, pollutants, competitors or parasites.
2. A reduction of at least 80%, projected or suspected to be met within the next ten years or three generations, which ever is linger, based on (and specifying) any of (b), (c), (d) or (e) above.

B. Extent of occurrence estimated to be less than 100 km^2, or area of occupancy estimate to be less than 10 km^2, and estimates indicating any two of the following:

1. Severely fragmented or known to exist at only a single location.

2. Continuing decline, observed, inferred or projected, in any of the following :
   (a) extent of occurrence
   (b) area of occupancy
   (c) area, extent and/or quality of habitat
   (d) number of locations or subpopulations
   (e) number of mature individuals.
3. Extreme fluctuations in any of the following :
   (a) extent of occurence
   (b) area of occupancy
   (c) number of locations or subpopulations
   (d) number of mature individuals.

C. Population estimated to number less than 250 mature individuals and either :

1. An estimated continuing decline of at least 25% within three years or one generation, whichever is longer, OR
2. A continuing decline, observed, projected, or inferred, in numbers of mature individuals and population structure in the form of either :
   (a) severely fragmented (i.e. no subpopulation estimated to contain more than 50 mature individuals).
   (b) all individuals are in a single subpopulations.

D. Population size estimated to number less than 50 mature individuals.

E. Quantitative analysis showing the probability of extinction in the wild is at least 50% within 10 years or three generations, whichever is the longer.

Endangered (EN)

A taxon is Endangered when it is not Critically Endangered but is facing a very high risk of extinction in the wild in the near future, as defined by any of the following criteria (A to E) :

A. Population reduction in the form of either of the following:

1. An observed, estimated, inferred or suspected reduction of at least 50% over the last 10 years or three generations whichever is longer, based on (and specifying) any of the following :
   (a) direct observation
   (b) an index of abundance appropriate for the taxon
   (c) a decline in area of occupancy, extent of occurrence and/or quality of habitat
   (d) actual or potential levels of exploitation
   (e) the effects of introduced taxa, hybridization, pathogens, pollutants, competitors or parasites.
2. A reduction of at least 50%, projected or suspected to be met within the next ten years or three generations, which ever is linger, based on (and specifying) any of (b), (c), (d) or (e) above.

B. Extent of occurrence estimated to be less than 5000 km^2, or area of occupancy estimated to be less than 500 km^2, and estimates indicating any two of the following :

1. Severely fragmented or known to exist at not more than five location.
2. Continuing decline, observed, inferred or projected, in any of the following :
   (a) extent of occurrence
   (b) area of occupancy
   (c) area, extent and/or quality of habitat
   (d) number of locations or subpopulations
   (e) number of mature individuals.
3. Extreme fluctuations in any of the following :
   (a) extent of occurrence
   (b) area of occupancy
   (c) number of locations or subpopulations
   (d) number of mature individuals.

C. Population estimated to number less than 2500 mature individuals and either :

1. An estimated continuing decline of at least 20% within five years or two generation, whichever is longer, OR
2. A continuing decline, observed, projected, or inferred, in numbers of mature individuals and population structure in the form of either :
   (a) severely fragmented (i.e. no subpopulation estimated to contain more than 250 mature individuals).
   (b) all individuals are in a single subpopulations.

D. Population estimated to number less than 250 mature individuals.

E. Quantitative analysis showing the probability of extinction in the wild is at least 20% within 20 years or five generations, whichever is the longer.

Vulnerable (VU)

A taxon is Vulnerable when it is not Critically Endangered or Endangered but is facing a high risk of extinction in the wild in the medium-term future, as defined by any of the following criteria (A-E) :

A. Population reduction in the form of either of the following :

1. An observed, estimated, inferred or suspected reduction of at least 20% over the last 10 years or three generations whichever is longer, based on (and specifying) any of the following:
   (a) direct observation
   (b) an index of abundance appropriate for the taxon
   (c) a decine in area of occupancy, extent of occurrence and/or quality of habitat
   (d) actual or potential levels of exploitation
   (e) the effects of introduced taxa, hybridization, pathogens, pollutants, competitors or parasites.

2. A reduction of at least 50% projected or suspected to be met within the next ten years or three generations, which ever is linger, based on (and specifying) any of (b), (c), (d) or (e) above.

B. Extent of occurrence estimated to be less than 20,000 km^2, or area of occupancy estimated to be less than 2000 km^2, and estimates indicating any two of the following :

1. Severely fragmented or known to exist at no more than ten locations.
2. Continuing decline observed, inferred or projected in any of the following :
   (a) extent of occurrence
   (b) area of occupancy
   (c) area, extent and/or quality of habitat
   (d) number of locations or subpopulations
   (e) number of mature individuals.
3. Extreme fluctuations in any of the following :
   (a) extent of occurrence
   (b) area of occupancy
   (c) number of locations or subpopulations
   (d) number of mature individuals.

C. Population estimated to number less than 10000 mature individuals and either :

1. An estimated continuing decline of at least 10% within 10 years or three generation, whichever is longer, OR
2. A continuing decline, observed, projected, or inferred, in numbers of mature individuals and population structure in the form of either :
   (a) severely fragmented (i.e. no subpopulation estimated to contain more than 1000 mature individuals).
   (b) all individuals are in a single subpopulations.

D. Population very small or restricted in the form of either of the following :

1. Population estimated to number less than 1000 mature individuals.
2. Population is characterized by an acute restriction in its area of occupancy (typically less than 100 km^2) or in the number of locations (typically less than 5). Such a taxon would thus be prone to the effects of human activities (or stochastic events whose impacts is increased by human activities) within a very short period of time in an unforeseeable future, and is thus capable of becoming Critically Endangered or even Extinct in a very short period.

E. Quantitative analysis showing the probability of extinction in the wild is at least 10% within 100 years.

Lower Risk (LC)

A taxon in Lower Risk when it has been evaluated, does not satisfy the criteria for any of the categories Critically Endangered, Endangered, or Vulnerable. Taxa included in the Lower Risk category can be separated into three subcategories :

1. **Conservation Dependent (cd)** Taxa which are the focus of a continuing taxon-specific or habitat-specific conservation programme targeted towards the taxon in question the cessation of which would result in the taxon qualifying for one of the threatened categories above within a period of five years.
2. **Near Threatened (nt)** Taxa which do not quaify for Conservation Dependant, but which are close to qualifying for Vulnerable.
3. **Least Concern (lc)** Taxa which do not qualify for Conservation Dependent or Near Threatened.

Data Deficient (DD)

A taxon is Data Deficient when there is inadequate information to make a direct, or indirect, assessment of its risk of extinction based on its distribution and/or population status. A taxon in this category may be well studied, and its biology well known, but appropriate data on abundance and/or distribution are lacking. Data Deficient is therefore not a category of threat or Lower Risk. Listing of taxa in this category indicates that more information is required and acknowledges the possibility that future research will show that threatened classification is appropriate. It is important to make positive use of whatever data are available. In many cases great care should be exercised in choosing between DD and a threatened status. If the range of a taxon is suspected to be relatively circumscribed, and a considerable period of time has elapsed since the last record of the taxon, threatened status may well be justified.

Not Evaluated (NE)

A taxon is Not Evaluated when it is has not yet been evaluated against the criteria.

Three kinds of assessment/recommendations are made for each taxon reviewed :

1. Assessing taxa to the New IUCN Red List Categories of Threat :

   The New IUCN Red List Categories are base both on population and distribution criteria. They provide a system that facilitates comparisons across widely different taxa. These criteria can be applied to any taxonomic unit at or below the species level, with the exception of micro-organisms.

2. Making recommendation for *in situ* and *ex situ* research and management activities :

   These recommendations aim to integrate research and practical management actions with known threats to contribute to the taxon's conservation.

3. Making recommendation for *ex situ* programmes :

   These recommendations are only established if they can contribute to the conservation of the taxon whether directly (as a genetic reservoir) or indirectly (as a research or education population). These recommendations form the foundation for development of regional strategic collection plans for the zoo and aquarium community and are now being used by and for botanical agencies as well.

The usefulness of CAMPs is twofold; they provide (1) a vehicle for involving the entire conservation community in biodiversity activities, and (2) a rapid assessment of conservation status of any plant or animal groups, e.g., automatic "prioritization".

CAMPs can help organisations in establishing their strategic plans and actions for conservation.

Uses of CAMP Results

The CAMP Process establishes priorities for global and regional species management and conservation. It provides a framework for intensive management both *in situ* and *ex situ.* CAMP documents can be used as guidelines by national and regional biodiversity agencies as well as regional *ex situ* programmes as they develop their own conservation action plans. The CAMP process has a long-term impact on global priority setting. It is intended that the CAMP process will ultimtely contribute to the wise use, would wide, of limited resources for species conservation. With the ever-increasing demand for CAMP Workshops, biodiversity mangers will, in the near future, have a set of comprehensive documents at their disposal, collaboratively and scientifically developed by the experts on the taxon or region. These documents will assist both *in situ* and *ex situ* managers with prioritization and decision making with regard to species conservation.

One possible recommendation to result from a CAMP workshop is for a Population and Habitat Viability Assessment (PHVA) Workshop to be conducted for a particular threatened species or subspecies. The assessment for each species involves an in depth analysis of information on the life history, population dynamics, ecology, and population history of the individual populations. Information in the demography, genetics and environmental factors pertinent to assessing the status of each population and its risk of extinction under current management scenarios and perceived threats is assembled in preparation for the PHVA. This information provides the basis for constructing simulation models of each population to evaluate the deterministic and stochastic effects and interactions of genetic, demographic, environmental and catastrophic factors on the population dynamics and extinction risks.

A CAMP should be followed up with a review process to assess changes in species status, results of the recommended management and new information. In 3-5 years, any number of changes might have taken place which would impact the status of a species and/or its habitat. Regular reviews or follow-up CAMPs can provide an effective monitoring strategy.

CAMP for Soil Invertebrates of Southern India

The Biodiversity Conservation Prioritisation Project (BCPP) India undertook a prioritisation exercise for species, sites and strategies for conservation. The Endangered Species Subgroup selected the Conservation Assessment and Management Plan Workshop Process and the IUCN Red List Criteria for assessing conservation status of species at a planning workshop held as part of the Project.

A Conservation Assessment and Management Plan workshop was conducted for selected Soil invertebrates of Sourthern India to assess their status in the wild. The workshop tool place from 24-28 February, 1997 in Chennai, hosted by the Zoological Survey of India, Sourthern Regional Station. Forty-six participants from 36 institutes with expertise ranging from field biology to forest management attended the workshop. A total of 95 taxa of soil invertebrates were assessed in the 5-day workshop. The selection of species for assessment within certain pre-decided groups was left to the discretion of the participants. Participants worked in 4 working groups for 5 days and asessed 95 taxa. Information for every taxa were entered on 'Taxon data sheets' in which details of each taxon such as distribution, population numbers, habitat structure, threats affecting the taxa, population decline and the qualityof data provided for the taxa are given here. This information was used to assess the status of the taxon and assign a category of threat according to the IUCN Red List categories. Taxon specific recommendations were also made after categorisation for use in conservation action planning.

Ninety-five taxa of soil invertebrates were assessed at the workshop of which 16 are Critically Endangered, 21 Endangered, 24 Vulnerable, 20 Lower Risk near threatened, 9 Lower Risk least concern, 4 Data Deficient and 1 as not Evaluated. The high percentage of threatened species (64%) in this list may not reflect the actual situation for all invertebrates in India because these 95 represent better studied invertebrates which also may have been selected for assessment due to concern for their conservation in the first instant.

The factors that are used in a categorisation of threat are population reduction, restricted distribution, population size, Number of mature individuals or population restriction and probability of extinction. The degree of threat depending on each or any of these five criteria determines the threat category. Of the threatened invertebrates, most are assessed so because of restricted distribution (91%) while only a few are due to population reduction (19%). Assessments were restricted to Southern India and nearly 50% of the taxa assessed were endemic to southern Indian. This being the case, most of the endemic species have restricted distribution, which makes them qualify for the criteria for threat.

Recommendations for future conservation action were proposed for every taxon of the 94 assessed. Of the threatened taxa, 67 were recommended for intensive surveys, 39 for monitoring, 34 for habitat management, 36 for life history studies, 14 for limiting factor research, 7 each for taxonomic studies and population and habitat viability assessments, 4 for limiting factor management and 7 for various taxon specific recommendations. Survey was considered very important because information on complete distribution of taxa is lacking. Monitoring and habitat management was considerd important because of constant human interference causing changes in population structure of invertebrates and also changes in quality of habitat in the wild.

Odonate Conservation

The IUCN/SSC Odonata Specialist Group has identified species prioritisation as one of the research priorities for dragonfly conservation. It is mentioned that odonatologists must first make judgements about which species are special and secondly which of the special species are threatened. It is no doubt that threatened special species should be given top priority. CAMP process will help to identify the priority species.

CAMP methodology has proved so effective for assessing status in the wild that it has been recognized by IUCN SSC Specialist Groups, governmental and non-governmental agencies, conservation action planners and policy makers all over the world. The CAMP workshop philosophy and methodology is emerging as an effecting means of conducting biodiversity inventory, identification and monitoring, thus satisfying agenda item 7 in the Convention on Biological Diversity.

Countries which are now party to the 1992 Convention on Biological Diversity are taking steps to implement various articles of the treaty. In Article 7 [Identification and Monitoring, subparagraph (a)] of CBD, identification of components of biological diversity important for conservation is recommended. It states, "*Species and communities which are : threatened; wild relatives of domesticated or cultivated species; of medicinal, agricultural or other economic value; or social; scientific or cultural importance; or importance for research into the conservation and sustainable use of biological diversity, such as indicator species; and described genomes and genes of social, scientific or economic importance*" (CBD, 1992). Thus out come of CAMP process will be a sincere effort to species conservation. A CAMP for Odonates of India has been planned.

References

Anonymus, 1992. *Conserving Biodiversity. A Research Agenda for Development Agencies.* National Academy Press, Washington, D.C. 127p.

Byers, O.S., Walker, M. Maunder, S. Ellis and U. Seal, 1999. *Conservation Assessment and Management Plan (CAMP) Process References Manual for Organizers.* IUCN/SSC Conservation Breeding Specialist Group: Apple Valley, MN.

Heywood, V.H., 1995. *Global Biodiversity Assessment*, UNEP, New York.

IUCN, (1996). *1996 IUCN Red List of Threatened Animals.* IUCN, Gland, Switzerland and Cambridge, UK. 448 p.

Moore, N.W., 1991. Observe extinction or conserve diversity/in Collins, N.M. And Thomas, J.A. (eds) 1991. *The conservation of insects and their habitats.* Academic Press. London. 15th Symposium of the Royal Entomological Society of London, 14-15 September 1989.

Moore, N.W., 1997. *Status survey and conservation Action Plan-Dragonflies.* IUCN, Gland, Switzerland.

Mittermeier, R.A., N. Myers, J.B. Thomsen, G.A.B. Da Fonseca and S. Olivieri, 1998. Biodiversity hotspots an major tropical wilderness areas : Approaches to setting conservation priorities. *Conservation Biology* 12(3): 516-520.

Tyagi, B.K., 1985. Dragonfly conservation in India : Present status and future prospets. *Rep. Odon. Specialist Group Int. Un. Conserv. Nat.* No. 6: 1-5.

Wilson, E.O., 1988. The Current state of biological diversity. pp. 3-18 in Biodiversity, E.O. Wilson and F.M. Peter, eds. 1988. *Biodiversity,* Washington, D.C.: National Academy Press.

7

Functional Morphology of Rectal Gills and Oxygen Consumption by Dragonfly Nymphs in Aquatic Ecosystem

• *Arvind Kumar*

Abstract

The present article elaborates the structure of respiratory organs and oxygen uptake by dragonfly larvae in freshwater ecosystems. Gills were found in branchial chamber which have filled the maximum posterior part of the abdominal cavity. The rectal gills were observed in six rows and recorded their tracheal supply from four to six main longitudinal tracheal trunks.

The oxygen consumption by the 3rd and 5th instar nymphs of dragonfly was also estimated and found that oxygen uptake was directly dependent upon the body size and age group of the larvae. The average oxygen uptake (mlO_2/larvae/hr.). was 0.0112 ± 0.0015 for 3rd instar larvae whereas it was 0.0160 ± 0.0020 for 5th instar larvae. It was also observed that the 5th instar larvae utilized about 0.0048 mlO_2/individual/hr. more oxygen than the 3rd instar larvae.

Key words : Rectal gills, Dragonfly, Oxygen-uptake, Respirometer.

Introduction

In tropical countries there exists a very interesting group of dragonfly insects which have bimodal gas exchange mechanism where adults respire by means of trachea and larval forms respire by dint of rectal gills. Respiration in aquatic dragonflies is of great interest from a comparative point of view in that it utilizes a tidal ventilation mechanism in an aquatic medium. In most animals where water is the respiratory medium the flow over the respiratory surfaces is generally undirectional and this probably requires less energy for the extraction of a given amount of oxygen. Tidal or reciprocating pumping mechanisms in water are only truly found here and in some sea-cucumbers. The respiratory surface in dragonfly larvae is also of interest since it is made up of tracheal gills so that the oxygen is transferred from an aquatic medium to a gaseous phase. The general nature of ventilation is well known but accounts of the detailed patterns are largely dependent upon direct observation. There are some excellent accounts, but inevitably they are based upon observation over a relatively short period and small number of individuals. Tonner (1936) divided the respiratory patterns into three main types. He and many other authors also observed the way in which the respiratory mechanisms for ventilating the rectal chamber had been made use of in the locomotion of these insects. Methods of more continuous recording previously used, mainly for the study of swimming were applied to the study of ventilation in this animal by recording pressures in the respiratory chambers using a condenser manometer (Hughes, 1958).

The oxygen uptake has long been considered as measure for the metabolic rates in the animals. The odonate larvae may be considered as a model for estimation of oxygen uptake because of their high abundance, ease of collection and sensitivity to water quality. Considerable study has been made on morphology, anatomy, histology of caudal and rectal gills of odonate larvae (Tillyard, 1917; Rich, 1918; Calvert, 1934; Hughes and Mill, 1966; Khan and Tembhare, 1984; Kumar, 1992), but the knowledge pertaining to estimation of oxygen uptake in odonate larvae is megre.

Materials and Methods

The dragonfly larvae were collected from a pond by the help of D-shaped aquatic insect net made up of nylon cloth. The mesh size was 40 to 90/cm^2. The common and dominant species are *Mesogomphus lineatus, Potamarcha obscura, Zyxomma petiolatum* and

Cordulegaster sp. belonging to the family Gomphidae of suborder. Anisoptera in the present investigation. The larvae were sorted into two groups, *i.e.*, 3rd instar (average weight - 4.792g) and 5th instar (average weight - 8.398g) in the size range of 1.3-1.8 m and 2.93-8.62 cm respectively. Both groups were maintained separately in two glass aquaria filled with natural pond water. The larvae were then picked up with the help of the small camel hair brush and introduced inside the small respirometer of 1.0 litre size. During each experiment, a group of twenty larvae were introduced into the respirometer. Water samples were taken at the interval of six hours. The measurement of respiration was done according to Lampret (1984) during the period of August to October, 2000. The oxygen content of inspired and expired water was determine by Winkler's volumetric method (Welch, 1948). Difference in the dissolved oxygen between the inspired and expired water was regarded as the oxygn consumed by twenty larvae within the period of 6 hours. Such experiments were repeated four time within 24 hours. Oxygen uptake per unit time was calculated with the help of following equation :

$$VO_2 = CIO_2 - CEO_2$$

where, VO_2 = Oxygen uptake by 20 individuals, for 6 hr. ($mlO_2/6hr$)

CIO_2 = Oxygen content of inspired water,

CEO_2 = Oxygen content of expired water,

The dissolved oxygen of ambient water of the respirometer used for oxygen uptake ranged from 5.933 to 6.840 mlO_2.

Results and Discussion

During the present investigation, following organs play an important role in respiration :

(a) Rectal gills : Gaseous exchange takes place across the tracheal gills which are round in the branchial chamber which fills most of the posterior part of the abdominal cavity. These gills are arranged in six rows and receive their tracheal supply principally from four of the six main longitudinal tracheal trunks (Fig. 7.a, 7.b and 7.c).

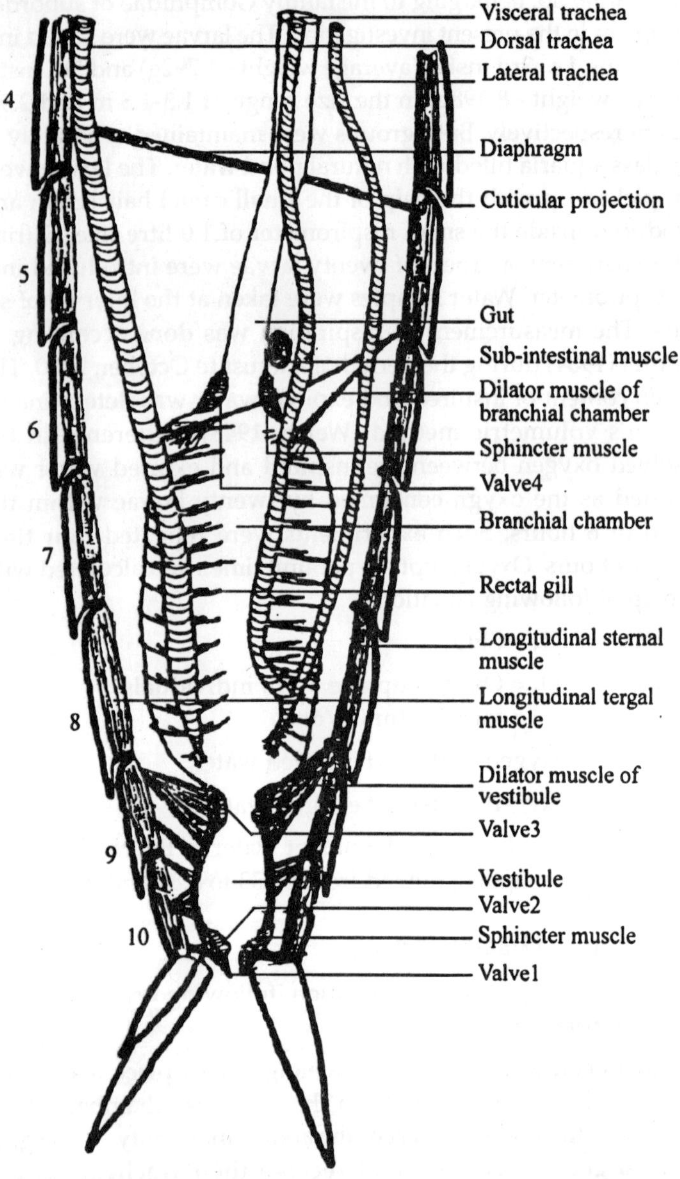

Fig. 7.a : Longitudinal section of posterior part of abdominal cavity showing the morphology of the respiratory organs in a dragonfly larva.

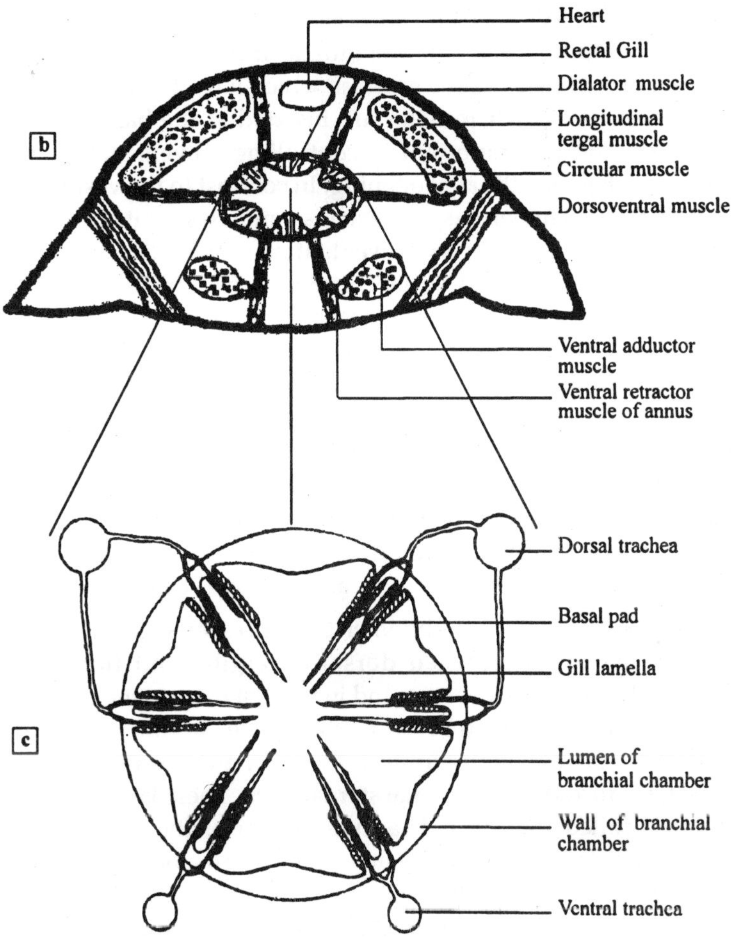

Fig. 7.b : Transverse section of the posterior part of abdominal cavity of dragonfly larva showing (b) Branchial chamber and (c) Gill lamellae and muscles.

The gills are contained within a cavity, the branchial chamber, which is sometimes considered to be part of the rectum, but Snodgrass (1954) homologizes it with the colon. This cavity can be closed by muscular valves at both ends. The anterior one (v_4) is the ileal valve and the posterior one (v_2) separates the branchial chamber from the vestibule. The vestibule, which is probably derived from the rectum, is very muscular and has both longitudinal and circular layers of muscle. It is also provided with six rows of extrinsic dilator muscles which have their origin on the exoskeleton. Other muscles are inserted at the bases of the paraprocts and are important in controlling movements of the valves which guard the anal opening. In contrast to the vestibule, the branchial chamber has few extrinsic muscles an its intrinsic musculature is relatively weak. Ventilation of the gills contained in this cavity is achieved by mechanisms which move water in and out of the anal opening. The mechanisms involve the structure of the exoskeleton, the hydrostatic function of the body cavity, the action of the extrinsic and instrinsic musculature of the branchial chamber and vestibule, and the muscles which control the different valves in the system. Of considerable importance are those muscles which change the volume of the whole abdominal cavity. During respiration these are principally the respiratory dorso-ventral muscles but in addition there are two transverse muscles. The nature and role of these structures are discussed below.

(b) Exoskeleton : The abdomen comprises ten segemnts each of which is strongly arched dorsally (tergite) and flattened ventrally. The sterna are convex and in the first nine segments each sternum is divided into a broad median plate (sternite) and two small lateral plates (laterosternites). The latter are movably hinged on the tergum and on the medial sternal plate. These lateral plates are often called pleurites. The main action of the respiratory muscles is to draw the pleurites upwards towards the tergites and thus distort the shape of the segments and reduce the volume of the abodominal cavity. It is clear that the elasticity of the exoskeleton can play an important role in the respiratory mechanism. Snodgrass (1954) noted that if the sternum of a single segment is pressed inwards, the lateral margins of the tergum are drawn closer to one another. On release of the pressure the more convex shape of the tergum acts as a spring which restores the sternum to its original position. The normal respiratory movements of the resting larva are comparable, and it is clear that these elastic forces play an important role in the inspiratory action of the abdomen. In addition to this type of movement the abdominal segements may be telescoped longitudinally, but this action plays little part in normal respiration although it is very important during swimming (Hughes, 1958). The abdomen terminates in five appendages, two lateral paraprocts and a median dorsal epiproct and a small cercus on each side. The parprocts and epiproct each have a valve (v_1) at the base which closes the vestibule or rectal chamber as a result of the contraction of a well-developed sphincter muscle.

(c) Muscles : The important muscles from the point of view of ventilation may be divided into those intrinsic to the branchial chamber and vestibule and the extrinsic muscles which either insert on them from the exoskeleton or are muscles of the abdominal wall itself. The intrinsic musculature is found mainly in the vestibule and comprises some fairly strong longitudinal muscles together with several circular strands. Also of this type are the strong muscles which close the anal valves and other sphincter muscles which operate the valves between the branchial chamber and vestibule, and between the branchial chamber and the next anterior region of the gut. The branchial chamber itself has few muscles, mainly bands of longitudinal fibres running between the rows of tracheal gills. There are also numerous fine circular fibres, but no dilator muscles. The main respiratory muscles, however, belong to the dorso-ventral group of muscle. In abdominal segments 4-9 these muscles are divided on each side into anterior and posterior parts and a respiratory dorso-ventral muscle. The latter are the principal expiratory muscles of the body wall and their contraction causes the pleurites to be lifted with a consequent arching of the tergum. The volume of the abdominal cavity tends to be reduced and the increase in the pressure within it will be transmitted to the contents of the respiratory chamber.

In addition to these muscles there is a pair transverse muscle. One of these is the sub-intestinal transverse muscle which lies at the anterior end of the sixth segment and is entirely ventral to the gut. It is attached laterally to the tergum on both sides. In addition there is a muscle which almost separates the abdominal cavity into two compartments. This, the so-called diaphragm or supra-intestinal muscle, has dorsal and ventral portions lying above and below the gut. There has been considerable debate regarding not only its function but also its morphological position since it was first mentioned by Amans (1881). He described it as an elliptical membrane stretched across the body cavity between the fourth and fifth segments in front of the Malpighian tubules. Several other authors have given a variety of descriptions of its position (Mill, 1965) and the degree to which it separates the abdominal cavity into two parts.

That the diaphragm does separate the abdominal cavity into two parts has been demonstrated by injecting methylene blue into the anterior region, when the dye remains restricted to the part of the body in front of the diaphragm. If the injection is into the hind segments it stays in this part except for dye pumped forwards by the heart. Correspondingly, X-ray photographs taken after radio-opaque material had been injected into the posterior region showed that the injected material was restricted to that part of the abdomen. The precise position of the diaphragm seems to be that described by Wallengren (1914), namely, it is attached ventrally to the anterior border of the fifth abdominal segment and it remains attached round the lateral borders of that segment for about two-thirds of its length. Dorsally it is not attached and allows the tracheal trunks to pass above it. During life this dorsal opening will largely be occluded and hence fluid will not be able to pass readily from one compartment to the other. Regarding the functions of the diaphragm some authors have maintained that it is important as an inspiratory muscle (*e.g.* Matula, 1911; Tonner, 1936), but in his more recent monograph Snodgrass (1954) claims that the elasticity of the exoskeleton would be sufficient to produce inspiration and that the diaphragm is not therefore required for this purpose. He suggests that it is more important during expulsion of the labial mask. However, as such a muscle is absent in the Zygoptera which also have a 'musk' but a different respiratory mechanism, it seems more probable that it is linked in some way to ventilation. A respiratory function is clearly indicated by neurophysiological experiments (Mill and Hughes, 1966) in which rhythmic bursts of electrical activity were recorded in this muscle at the same frequency as in the respiratory dorso-ventral muscles.

Many experiments show that activation of both the diaphragm and the sub-intestinal muscle is simultaneous and alternates with that of the respiratory dorso-ventral muscles. Tonner (1936) maintained that the two transverse muscles were inspiratory in action so long as they contracted simultaneously. Contraction of the diaphragm alone was thought to result only in expiration. Tonner does not show the anatomical arrangement of these muscles correctly and although his conclusion regarding their function is supported here, it seems that he laid insufficient stress on the elastic properties of the exoskeleton. It must be remembered that they were made with the abdomen dissected and hence the mechanical arrangement were disturbed. If, as now seems reasonable, the diaphragm is an inspiratory muscle it is difficult to be certain of the nature of its action. Apart from Tonner's suggestion perhaps it also functions by producing a fall in pressure within the posterior abdominal cavity which is transmited to the branchial chamber. Such an action would be analogous to the mammalian diaphragm but there is little evidence for such a function.

Oxygen Consumption

Data of the oxygen uptake rate of 3rd and 5th instar of dragonfly larvae have been presented in Figures 7.1 to 7.6. The oxygen uptake rate for the group of twenty 3rd instar dragonfly larvae ranged from 0.745 to 1.704 mlO_2 for 6 hours with mean oxygen consumption six hourly at the rate of 1.354±0.2220ml O_2. However, oxygen consumption in 3rd instar larvae (20 individuals) at hourly interval varies from 0.124 to 0.290mlO_2 with mean consumption of 0.225 + 0.025 mlO_2. The oxygen uptake rate for a group of twenty fifth instar larvae ranged from 1.537 to 2.235 mlO_2/6 hr with mean oxygen consumption at the rate of 1.810 ± 0.240 mlO_2/6 hr However, the oxygen consumption at hourly interval in these groups of twenty larvae varies from 0.256 to 0.372 mlO_2/6 hr with mean consumption of 0.301 ± 0.0038 mlO_2. Marked differences in oxygen uptake in 3rd and 5th instar larvae of the dragonfly have been recorded under such experimental condition.

In the present investigation, it was observed that the oxygen uptake in these larvae is directly dependent upon the body weight, size and age group of the larvae. It is evident from the Figures 7.1

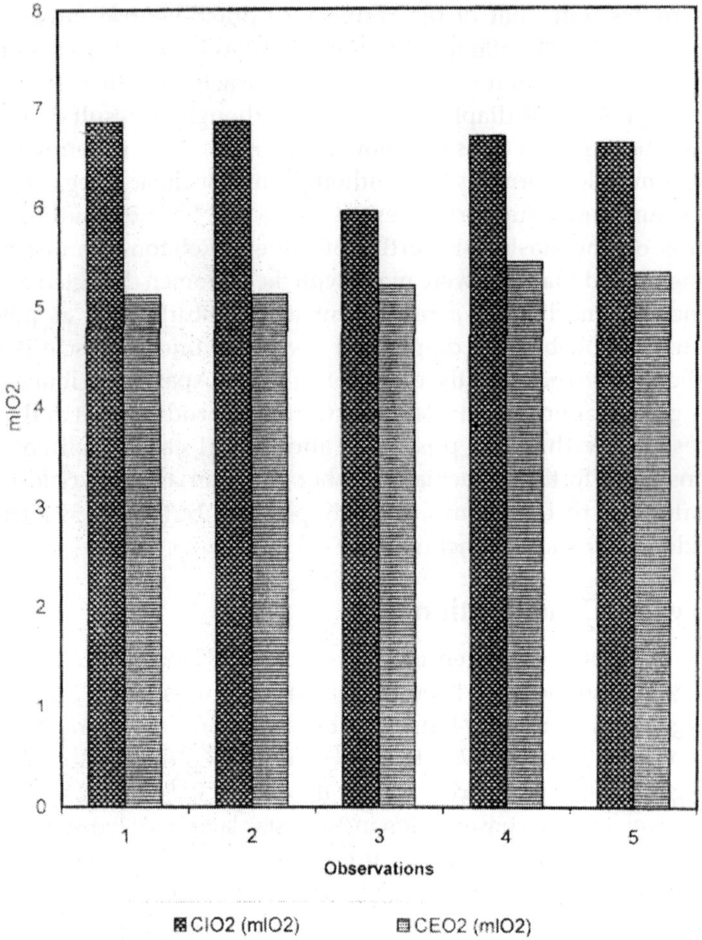

Fig. 7.1 : Value of oxygen uptake by a group of twenty 3rd instar dragonfly nymphs.

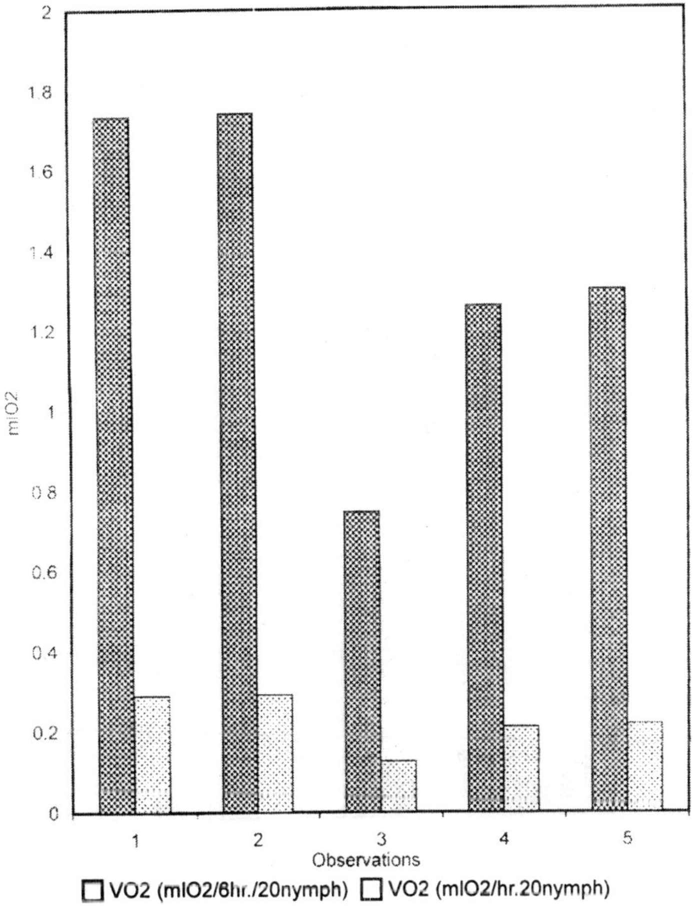

Fig. 7.2 : Value of oxygen uptake by a group of twenty 3rd instar dragonfly nymphs.

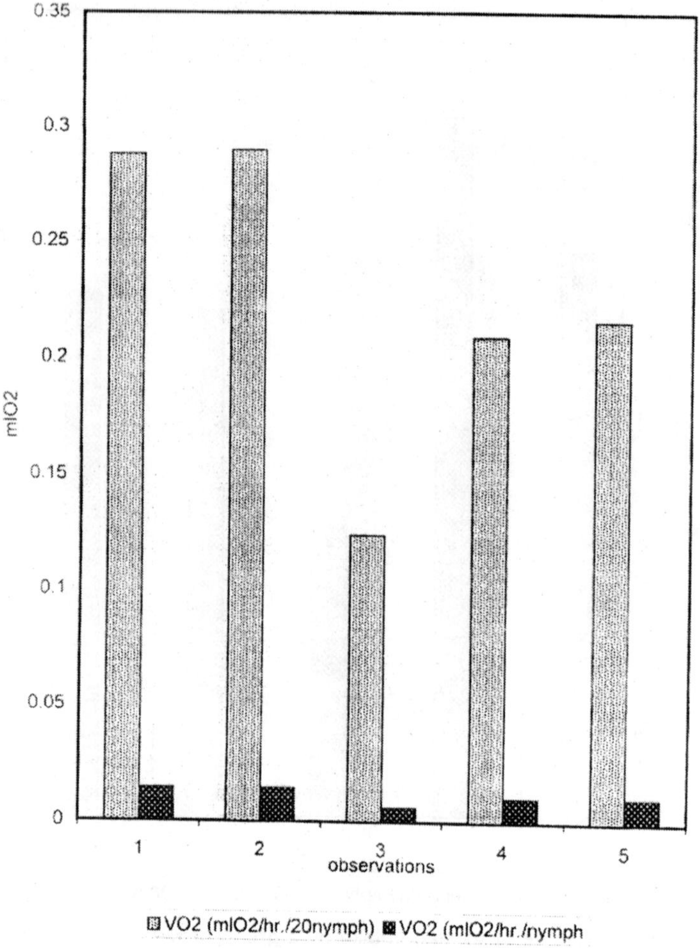

Fig. 7.3 : Value of oxygen uptake by a group of twenty 3rd instar dragonfly nymphs.

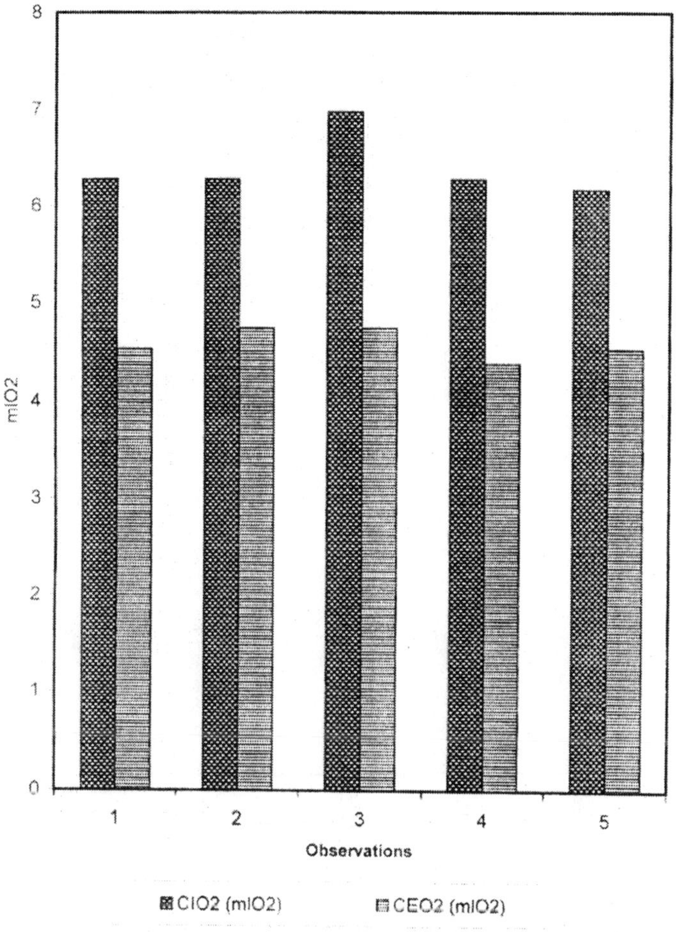

Fig. 7.4 : Value of oxygen uptake by a group of twenty 5th instar dragonfly nymphs.

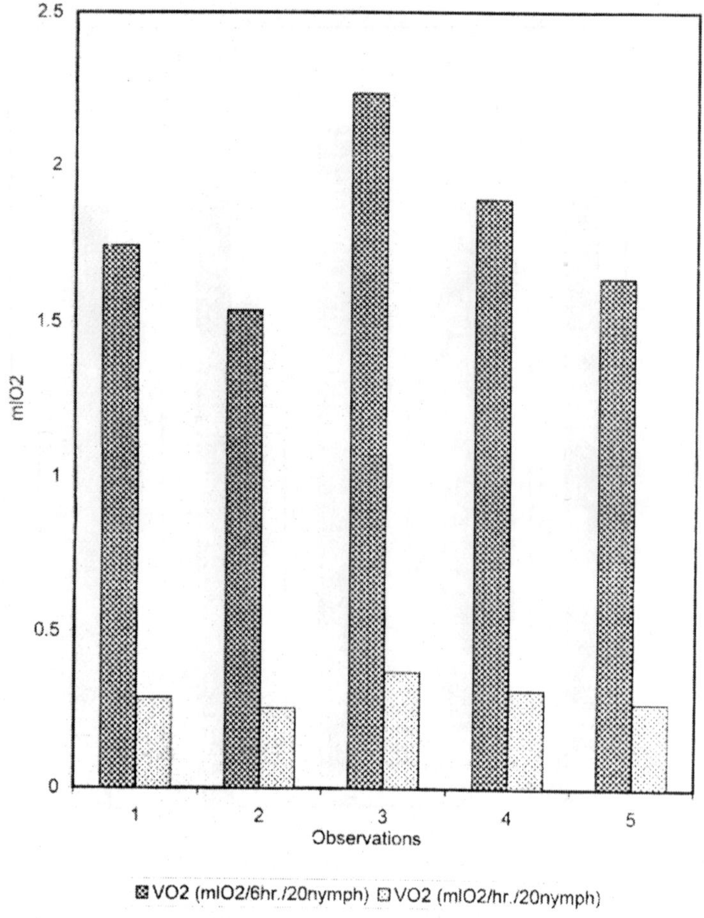

Fig. 7.5 : Value of oxygen uptake by a group of twenty 5th instar dragonfly nymphs.

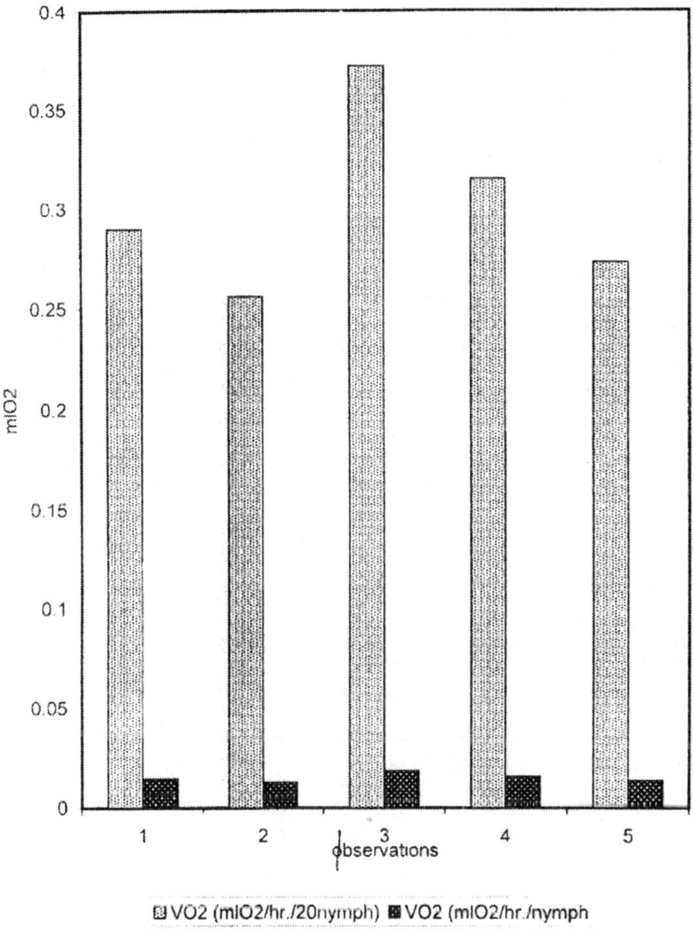

Fig. 7.6 : Value of oxygen uptake by a group of twenty 5th Instar dragonfly nymphs.

to 7.6 that the 5th instar larvae utilized more oxygen than the 3rd instar larvae. The average oxygen uptake (mlO_2/6hr/individual) is far greater in 5th instar larvae than the 3rd instar larvae.

The 5th instar larvae utilize more food and such they need more amount of oxygen for its aerobic oxidation, whereas, the 3rd instar larvae use less food in comparison to 5th instar larvae, hence require lesser amount of oxygen. It is established that metabolic rate increase exponentially with the increase of body size (Zeuthen, 1953). The exponent is always less than one, so the weight specific oxygen uptake decreases with increase in body weight (Hemingsen, 1960). Grodzinski *et al.* (1975) and Lampret (1984) also supported this view, Peters (1983) and Schmidt-Nielsen (1984) emphasized the importance of body size in the studies and stated that in general the smaller animals have higher weight specific metabolic rate (mlO_2/g/hr) than the larger animals. Hughes (1984) scaled the respiratory surface area in relation to oxygen up-take in vertebrates and concluded the decrease in weight specific VO_2 is always in accordance with the conservation of energy and decrease in respiratory surface area.

Therefore, the oxygen uptake in dragonfly larvae is dependent upon body size, weight, age, availability of food in the environment and also with the surface area of respiratory organs.

References

Amans, P. 1881. Recherches anatomiques et physiologique sur la larve de 1' Aeschna grandis. *Rev. Sci. nat. (Montpellier) ser.* 3 : 63-74.

Calvert, P.P. 1934. Report on Odonata. Internal organs of larvae and the respiration and rectal tracheation of Zygopteran larvae in general. *Ent. News.* 26 : 385-395.

Grodzinski, W.; Klekowski, R.Z and Duncan, A. 1975. Methods for ecological bioenergeties, International Biological Programme Committee. Osney Mead, Oxford.

Hemingsen, A. 1960. Energy metabolism as related to body size, respiratory surfaces and its evolution. Rept. Steno. Meml. Hosp. Nord. Insulin lab. 9 : 1-110.

Hughes, R. D. and Mill, P.J. 1966. Patterns of ventilation in dragonfly larvae. *J. Exp. Biol.* 44 : 317-334.

Hughes, G.M. 1958. The co-ordination of insect movements. III. Swimming in Dytiscus, Hydrophilus and a dragonfly nymphs. *J. Exp. Biol.* 35 : 567-583.

Hughes, G.M. 1984. Scaling of respiratory areas in relation to oxygen consumption of vertebrates. *Experientia.* 40 : 519-524.

Khan, M.W. and Tembhare, D.B. 1984. Histological histochemical and experimental studies on the rectum of the nymphs of the dragonfly. *Pantala flavescens* (Fabr.). Proc. Ist. Indian Symp. of Odonatology. School of Boil Sc. Madurai Kam. Univ.

Kumar, A. 1992. Some aspects of eco-physiology of larval odonates of wetlands of Santhal Parganas. Ph. D. Thesis, Bhagalpur University.

Lampert, W. 1984. The measurement of respiration. In "A manual methods for assessment of secondary productivity in freshwaters". Bl. Sc. Pub. London. pp. 413-468.

Matula, J. 1911. Untersuchungen tiber die Funktionen der Zentralnevensystems bei Insekten. Pfligers. Arch. ges. *Physiol.* 138 : 388-456.

Mill, P.J. 1965. An anatomical study of the abdominal nervous and muscular systems of dragonfly (Aesnidae) nymphs. *Proc. Zool. Soc. Lond.* 145 : 57-73.

Mill, P.J. and Hughes, G.M. 1966. The nervous control of ventilation in dragonfly larvae. *J. Exp. Biol.* (in the press).

Peters, R.H. 1983. *The ecological implications of body size.* Cambridge University Press, Cambridge.

Rich, S.G. 1918. Gill chamber of dragonfly nymphs. J. Morph. 31 : 317-349.

Schmidt-Nielsen, K. 1984. Scaling : *Why is animal size so important?* Cambridge Universtiy Press, Cambridge.

Snodgrass, R.E. 1954. *The dragonfly larva.* Smithson. Misc. Coll. 123 : 1-38.

Tillyard, R.J. 1917. *The biology of Dragonflies.* Cambridge University Press, Cambridge.

Tonner, F. 1936. Mechanik und Koordination der Atem-und Schwimmbewegung bei Libellenlarven. *Z. wiss. Zool.* 147 : 433-454.

Wallengren, H. 1914. Physiologisch-biologische Studien uber die Atmung bei den Arthopoden. II. Die Mechanik der Atembewegungen bei Aeschnalarven. *Acta. Univ. lund. N.F. Afd.* 2, 10: 1-24.

Welch, P.S. 1948. *Limnological Methods.* McGraw Hill Book Co. I. N.C., New York, London.

Zeuthen, E. 1953. Oxygen uptake as related to body size in organisms. *Q. Rev. Biol.* 28 : 1-12.

8

ANALYTICAL STUDIES ON THE FOOD AND FEEDING HABITS OF TROPHICALLY DISTINCT AQUATIC ODONATE LARVAE OF UDHUWA LAKE IN SANTAL PARGANA, JHARKHAND, INDIA

• *Chandan Bohra*

Abstract

This investigation showed a special food preference of odonate larvae which was conducted in a lake during the year of 2000. This study revealed the fact that odonates are primarily carnivorous and feed on cladocera, copepoda, rotifera, rhizopoda, aquatic insects, fish fries and other animal life present in the suroundings. According to observations, the food preference of zygoptera larvae is: Rhizopoda>Cladocera>Aquatic insects>Rotifers> Copepoda> Algae while the food preferences of Anisopteran larvae is utterly different being - Cladocera>Rotifera>Aquatic insect>Copepoda > Algae>Rhizopoda.

Key words : Forage Ratio, Odonate larvae, Udhuwa lake, Feeding habit.

Introduction

Freshwater insects are the dominant members of the littoral as well benthic fauna, occupying different trophic levels of the food chain. They are the chief converters of plant materials into animals protoplasm and play a significant role in the energy-flow pathway of the systems in which they inhabit. Most evolutionary lineages in insects seem to be evolved from scavenging to other feeding habits (Daly *et.al.,* 1978). The ephemerid and odonates are the earliest known living insects which have had scavenging mode of feeding. In the littoral fauna insects such as mayfly nymphs (Ephemeroptera) and chironomid larvae (Diptera) are the primary converters of plant materials into animal protoplasm as well as odonate larvae who occupy top position at the predatory food chain. They formed an important component of the zoobenthos of the natural eutrophic systems and considered to be an important part of diet of cultured fishes. They form an important link in energy transfer in aquatic ecosystems.

Odonates (dragonflies and damselflies) are usually the most consipicuous insect group near any body of inland waters and commonly called as "devil's darning needle", "snake feeder", snake doctor" and "horse stinger". The common name "mosquito hawk" refers to their value in destroying mosquitoes, gnats and other pests (Usinger *et.al.*, 1956). The nymphal odonates are predaceous like the adult and spending their entire nymphal period in the inland water. They usually form a high proportion of biomass in fresh waters. Thus, they occupy an important position at the top of the food webs and energy-flow pathway of the fresh water ecosystem. As dominant members of the benthic fauna, they are considered promising organisms for use in organic pollution biomonotoring because of their ease of collection, large number of species and sensitivity to the water quality.

Considerable work has been done on the food and feeding habits of odonate nymphs (Fischar, 1971; Pritchard, 1964 and Hassan, 1976), but literature pertaining to the gut content analysis of those nymphs under field conditions and their foraging ratio (FR) is rather scanty. Therefore, in the present investigation the gut content analysis of nymphal stages of Odonata has been carried out with a view to assess their role in freshwater ecosystem and their food preference in Udhuwa lake in Santal Pargana.

Study Area

Santal Pargana division lies between 23° 40′ and 25° 18′ North Latitude and between 86° 28′ and 87° 57′ East Longitude and stituated north-eastern portion of the Jharkhand State. This division comprises six districts, *viz.* Dumka, Deoghar, Godda, Pakur, Jamtatra and Sahibganj. There are two major lakes in

Rajmahal area in Sahibganj district of Jharkhand. Government of Bihar has notified these lakes as "Udhuwal Lake Bird Sanctuary" in the year 1991. It is situated about 42 km from Sahibganj and about 10 kms South-east of Rajmahal on the bank of Ganges. The main waterbodies of the sanctuary are name as "Pataura jheel" with 155 ha and Brahma-Jamalpur (Berhale jheel) with 410 ha of area. Pataura jheel is perennial and its average depth is about 2 m. Udhuwal Nala connects the jheel with the river Ganga. This jheel is 2 km distant from Udhuwa Bazar.

Brahma-Jamalpur or Berhale jheel or Barka Jalla is nearer to the Jamnagar village. This village is at a distance of 3 km from Udhuwa Bazar on way to Rajmahal. It is shallow water body which is dried up in summers. This jheel is about 2 km from Jamnagar.

Pataura is an extensive waterbody in comparison to Berhale. Both of these jheel together constitute Udhuwa lake. Udhuwa Nala with the back water of river Ganga is the water supply line for the Udhuwa Lake.

Materials and Methods

The fresh water larvae were collected from the lake by an larvae collecting net made by nylon cloth (mesh size 40-80/cm^2) in different seasons, *viz.*, summer, monsoon, winter and spring of the year 2000. The larvac of each haul were sorted out, killed in 70 per cent alcohol and transferred to 3 per cent formalin for preservation. The gut content analysis of odonate larvae was made. The specimens were dissected out under water or 3 per cent formalin. The intact foregut was removed and dissected further on a glass slide. The techniques as suggested by Mecom and Cummins (164) and Shapas and Hilsenhoff (1970) were used with some modifications to characterise feeding habits. The food items were counted using method of Brown (1961) Approximate per cent volumes of food items (Rotifera, Copepoda, Cladocera, algae, diatoms, detritus and other matters) were recorded for each season. Plankton of foraging medium were collected by a plankton net made of bolting silk no. 25 (mesh size 0.03-0.04 mm) and counted by Lacky microtransact counting method (Lacky, 1938).

The forage ratio (FR) is obtained as per Usinger (1956)

$$FR = \frac{n/N}{n'/N'}$$

where, n = the number of any organism in the guts;

N = the total number of organisms in the guts;

n′ = the number of the same organisms in the environment inhabited by the insect, and

N′ = the total number of food organisms in the environment.

A (FR) of 1.0 indicates that an organism is being taken at randon according to its relative abundance in the environment. A (FR) of> 1 indicates that an organism is either being selected in preference to other organisms, or that it is more available than others; A (FR)<1 indicates that an organisms is either less preferable or less available.

Results

Seasonal variation in the percentage composition of the major dietary component of dragon fly nymphs (Anisoptera) and damsel fly (Zygoptera) have been summarised in the Figures 8.1-8.8). The dominant species of sub order Anisoptera were *Mesogomphus lineatus* (Selys), *Potamarcha* sp. and *Zyxomma* sp. and of sub order Zygoptera were *Ischnura* sp. and *Rhodischnura* sp. The gut content analysis of these nymphs revealed that they are primarily carnivorous in habit, feeding on microcrustacea (Cladocera and Copepoda), rotifers, rhizopods, aquatic larvae and other animal tissues present in the lake. Only a small percentage of plant materials could be detected in the gut. They are mainly algae belonging to Chlorophyta, Bacillariophyta and Myxophyta.

The seasonal variation in the feeding habits show that both dragon fly nymphs and damsel fly nymphs prefer to take Rotifera, Cladocera, Copepoda, Rhizopoda and aquatic larvae (mayfly nymphs, chironomus larvae, mosquito larvae and pupae). Cartain ingested materials of animal origin could not be identified due to semi digestion of food items in the foregut. In damsel fly nymphs the Rotifera, Cladocera, aquatic larvae and Rhizopoda form the maximum percentage of the food items. But in dragonfly nymphs

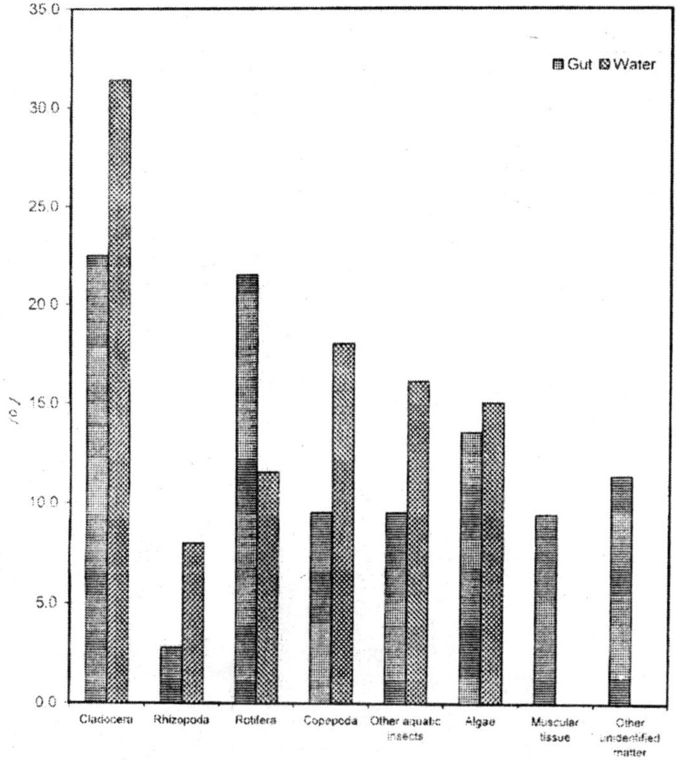

Fig. 8.1 : Percentage composition of dietary items in the gut of dragonfly nymph and in surroundings (Summer).

rhizopods were recorded in very small quantity and other food items, *viz.*, rotifers, Cladocera, Copeoda, aquatic larvae and other animal tissues were in maximum percentage in the foregut. The members of the algal community belonging to the Chlorophyta, Bacillariophyta and Myxophyta were also recorded in small percentage in different seasons of the year.

The food items of the nymphs were identified (Needham and Needham, 1962) as follows :

Rotifera : *Brachionus* sp., *Brachionus calciflorus, Keretella* sp.

Cladocera : *Daphania carinata; Moina dubia; Simocephalus elizabephae; Mysis* sp.; *Cladorus* sp.; *Diaphanosoma* sp.

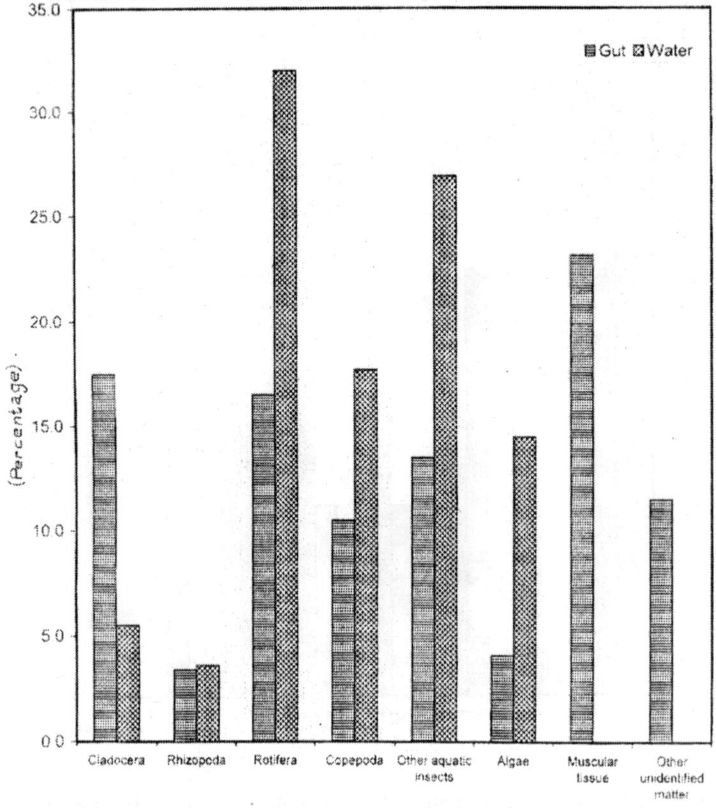

Fig. 8.2 : Percentage composition of dietary items in the gut of dragonfly nymph and in surroundings (Monsoon).

Rhizopoda : *Arcella* sp., *Centrophyxis* sp.

Aquatic larvae : May fly nymphs; Chironomus larvae; Mosquito larvae and pupae.

Fisheries : *Lobeo, Catla catla.*

Chlorophyta : *Scenedesmus* sp., *Pediastrum* sp., *Chlorella* sp., *Pendorina* sp., *Hydrodictyon* sp., *Oedogonium* sp.

Bacillariophyta : *Amphora* sp., *Navicula* sp. *Gomphonema* sp., *Fragillaria* sp., *Cymbella* sp.

Myxophyta : *Anabaena* sp.

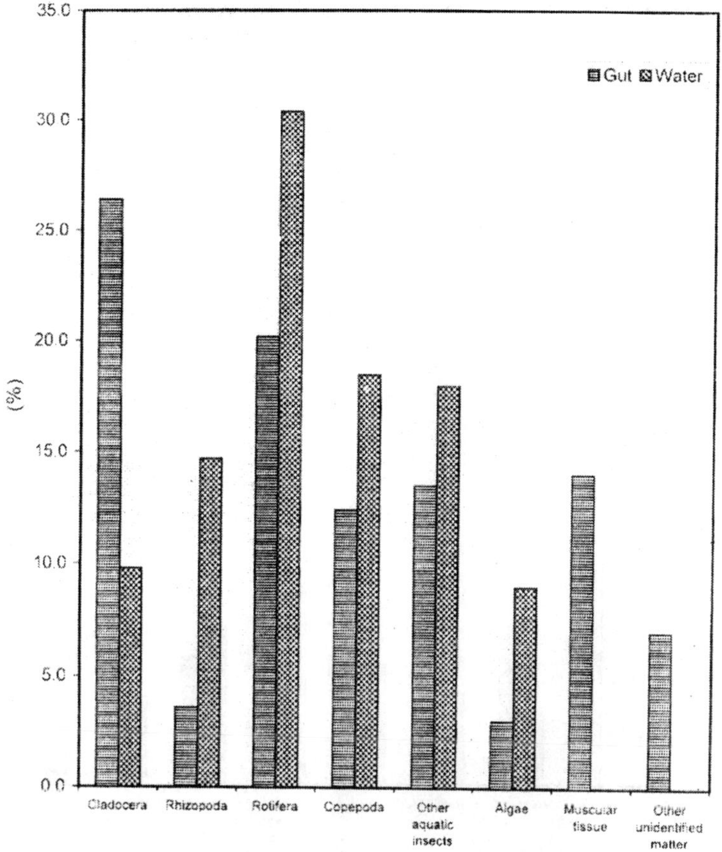

Fig. 8.3 : Percentage composition of dietary items in the gut of dragonfly nymph and in surroundings (Winter).

The forage ratio of dragonfly nymphs and damsel fly nymphs have been presented in the (Table 8.1 and 8.2) respectively. The forage ratio (FR) of dragonfly nymphs for Cladocera varied from 0.7 to 3.18% with the average of 2.12%; for Rhizopoda 0.27% to 0.96% with the average of 0.46%; for Rotifera 0.51% to 1.86% with the average of 0.94%; for Copepoda 0.52% to 0.67% with the average of 0.60%; for other aquatic larvae 0.50% to 0.75% with the average of 0.58% and for algâe 0.28% to 1.2% with the average of 0.65% in 2000.

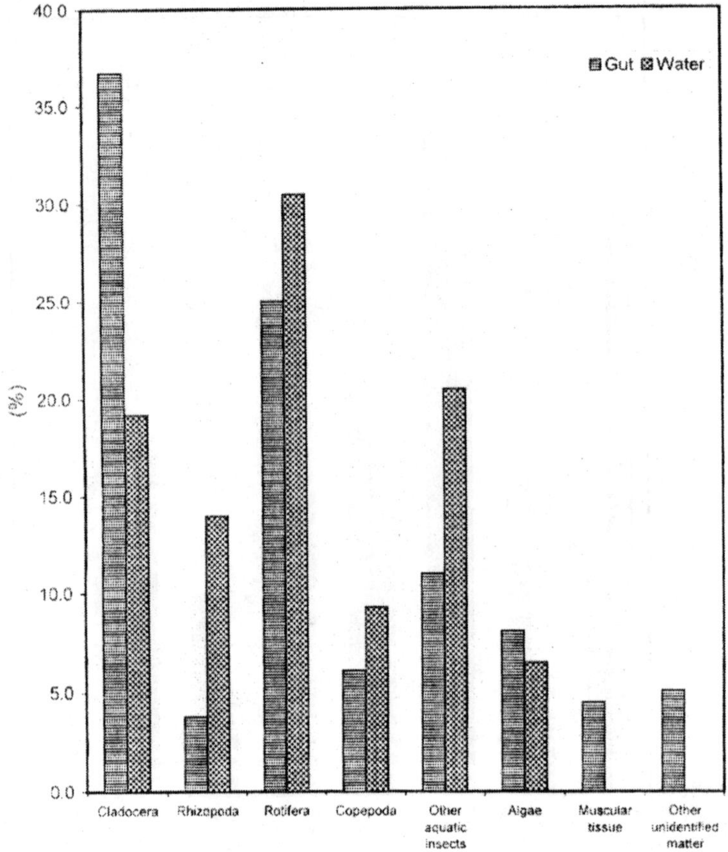

Fig. 8.4 : Percentage composition of dietary items in the gut of dragonfly nymph and in surroundings (Spring).

Table 8.1 : Forage ratio (FR) of Dragonfly nymphs

Season	Cladocera	Rhizopoda	Rotifera	Copepoda	Other aquatic insects	Algae
Summer	0.71	0.35	1.86	0.52	0.58	0.90
Monsoon	3.18	0.96	0.51	0.59	0.50	0.28
Winter	2.69	0.27	0.66	0.67	0.75	0.33
Spring	4.91	0.27	0.75	0.65	0.51	1.12
Average	**2.12**	**0.46**	**0.94**	**0.60**	**0.58**	**0.65**

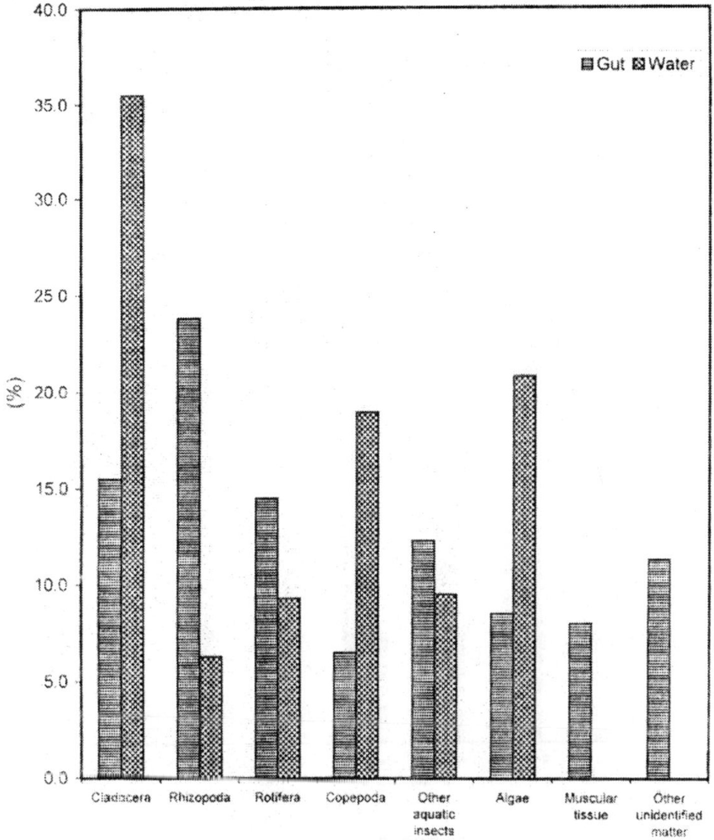

Fig. 8.5 : Percentage composition of dietary items in the gut of damselfly nymph and in surroundings (Summer).

Table 8.2 : Forage Ratio (FR) of Damsel fly nymph.

Season	Cladocera	Rhizopoda	Rotifera	Copepoda	Other aquatic insects	Algae
Summer	0.43	3.77	1.56	0.39	1.29	0.70
Monsoon	3.29	4.78	0.33	0.30	0.74	0.34
Winter	2.13	0.76	0.5	0.60	1.60	0.21
Spring	1.79	0.48	0.66	1.20	0.83	0.33
Average	**4.91**	**1.79**	**0.77**	**0.62**	**1.11**	**0.39**

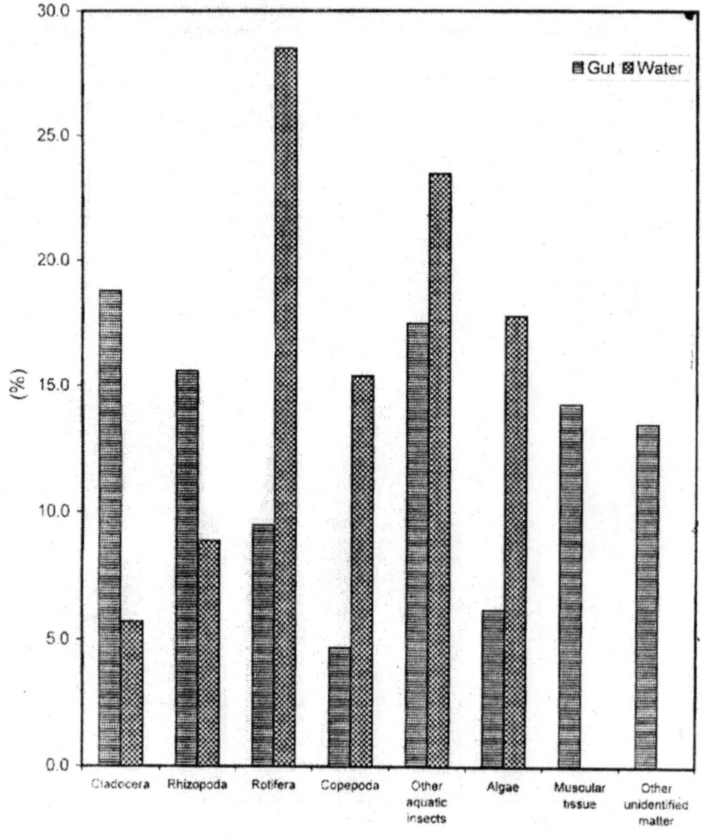

Fig. 8.6 : Percentage composition of dietary items in the gut of damselfly nymph and in surroundings (Monsoon).

The forage ratio (FR) of the nymphs of damselfly for Cladocera varied from 0.43 to 3.29% with the average of 1.91% for Rhizopoda 0.48% to 3.77% with the average of 1.79%; for Rotifera 0.33% to 1.56% with the average of 0.77%; for Copepoda 0.30% to 1.20% with the average of 0.62%; for other aquatic larvae 0.74% to 1.60% with the average of 1.11% and for algae 0.21% to 0.70% with the average of 0.39% during the period of observation.

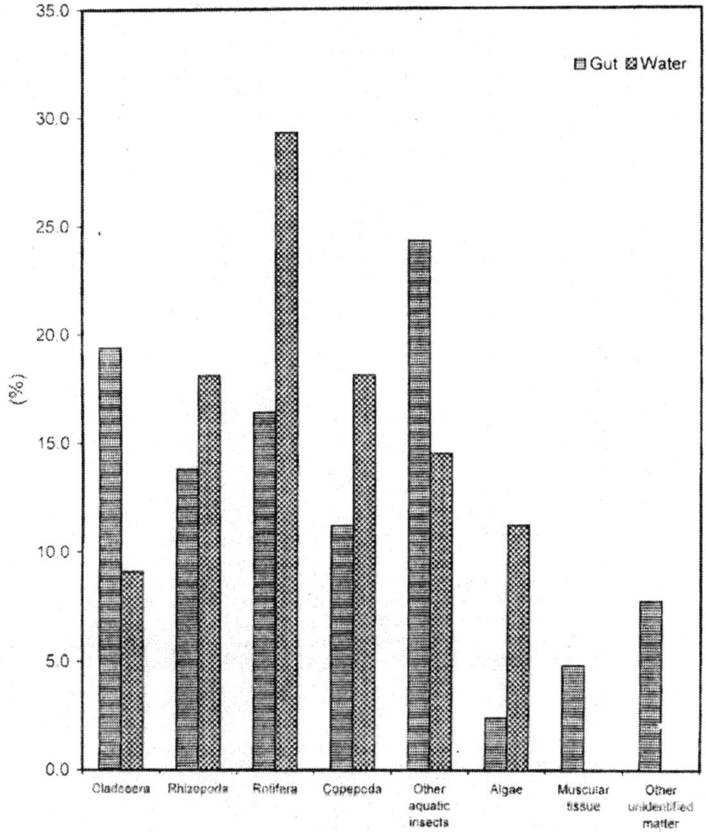

Fig. 8.7 : Percentage composition of dietary items in the gut of damselfly nymph and in surroundings (Winter).

Discussion

The gut contents analysis of Odonata nymphs show that the animal diet comprises the maximum percentage of dietary components whereas non-animal diet forms insignificant percentage. This view was supposted by Benke (1972, 1976) while he had made estimations of dragonfly production and prey turn over in South Corolina, U.S.A. He had expressed his views regarding feeding habits of Odonata as "stomach analysis for

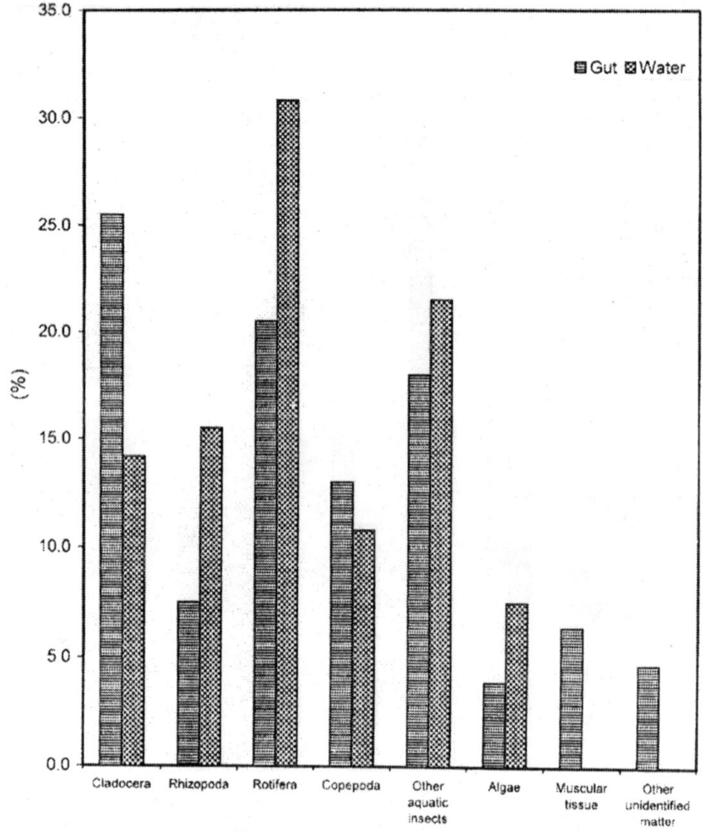

Fig. 8.8 : Percentage composition of dietary items in the gut of damselfly nymph and in surroundings (Spring).

dominant dragon flies did not reveal significant non-animal contributions, since odonate nymphs generally attack only moving prey, ingestion of non-animal items is negligible and probably insignificant."

In the present analysis Rotifera, Cladocera, Copepoda, Rhizopoda, other aquatic larvae and ramains of other aquatic animals were recorded in the guts of these nymphs. A small percentage of algal community belonging to the Chlorophyta, Bacillariophyta and Myxophyta were also recorded in different seasons of the year. However, Microcrustacea, Rhizopoda, Rotifera and aquatic larvae predominate in the gut contents. According to Lawton (1970) in *Pyrrhosoma nymphula* microcrustacea always comprised well over 50% of prey consumed but formed<15% of this diet by weight.

The gut contents of dragon fly nymphs the percentage composition of Cladocera, Rhizopoda, Copepoda and other aquatic larve is high in comparison to the environmental waters, except in summer for Caldocera and spring for other aquatic larvae. The dragonfly nymphs have some preference for Caldocera as their forage ratio (FR) for it was always found more than 1.0 with the average of 2.12% except in the summer season (0.71%). The dragonfly nymphs have also some preference for Rotifera as their forage ratio (FR) was found 0.94%.

In the guts or damsel fly (Zygoptera) nymphs percentage composition of Cladocera and Rhizopoda was found higher than environmental water except in the summer for Cladocera and in winter and spring for Rhizopoda. The forage ratio (FR) for Caldocera was found more than 1.0 except in summer (0.43) with the average of 1.91%. The forage ratio (FR) for Rhizopoda was found more than 1.0 except in winter and spring seasons. The forage ratio (FR) for other aquatic larvae was also found more than 1.0 except in monsoon and summer seasons. Thus, the damsel fly nymphs have some preference for Caldocera, Rhizopoda and other aquatic larvae.

However, diet of these nymphs may vary from season to season depending on the abundance of particular prey types. Field observation (Hasan, 1974) on the abundant prey of libellulid larvae (Anisoptera) in Ibadan, Nigeria indicated that the abundance of each type of prey fluctuated seasonally and also varied from lake to lake. The availability of food also affects the rate of larval development of Odonata. In a semi-natural enclosure vegetation provided hiding placs for prey (Hasan, 1974). In such conditions these nymphs have very hard time to find their specific food prey. It has been demonstrated by Hasan (1974) that this probably accounts for much longer larval duration (58 days) of *Palpolcura lucia lucia* (Anisoptera) in nature than in the laboratory (39.8 days). He concluded that the former figure is probably lose to what happens in eutrophic waters. Therefore, due to unavailability of proper food during the active period of development or body growth retards the emergence of adults in such water bodies. Thus, it may be concluded that the nymphal Odonata populations are largely regulated in numbers by the supply of food available to them in their habitat.

The switchover to an algal diet by Odonata nymphs did not appear to be related to a decline in numbers of potential animal prey, rather, it was probably a response to an increased availability of an alternative, apparently suitable food source which was only present in those seasons. Winterbourn (1971, 1974) suggested that there are transition from an algal to a carnivorous diet and from carnivorous to algal diet in the last instar larvae of Trichoptera and Plecoptera respectively. However, no other study was available in which a high proportion of Odonata nymphal population has been shown to switchover from predation to algal feeding.

The food and feeding analysis of odonate larvae revealed that many food items were undoubtedly missed due to differential digestion rates of the larvae. The food items digested most rapidly are seldom observed in the gut contents. Such items probably include certain soft bodied animal diet.

Thus, the feeding habits of odonate larvae is complicated by their generalised feeding capabilities. It was analysed that the food may never become the factor limiting the larval population in this aquatic system, since they can use a diverse types of food materials for nutrition. This ability to use a range of materials probably encourages exploitation of the aquatic environment to full extent.

Acknowledgement

The author is grateful to Professor (Dr.) Arvind Kumar, Post Graduate Department of Zoology, S.K. University, Dumka-814 101 for his guidance and thankful to Dr. T.P. Dubey, Principal, B.S.K. College, Barharwa for his continuous encouragement.

9

ODONATA DIVERSITY IN WESTERN HIMALAYA, INDIA

• *M. Prasad*

Introduction

Biodiversity is a contraction of biological diversity. Therefore, biological diversity refers to variety within the living organisms of the World. Biodiversity may be defined in terms of genes, species and ecosystems. India is very rich in biodiversity due to its unparallel geographic location, diversified climatic conditions and enormous ecodiversity etc. There are several biogeographical zones in India and Western Himalaya is one of the biogeographical zones of India.

Odonata (damselflies and dragonflies) are amphibiotic insects, their eggs and larvae are aquatic. Adults are areal in habits. Its aquatic stages are completed in waterbodies, whether it may be permanent or temporary. Adults are mostly restricated themself around the aquatic habitats. Adults are predacious, while larvae are cornivorous. Adults have a peculiar method of feeding. They capture their prey while on wing with the help of mouth parts and forwardly posed fore legs and consume their prey while on wing. Larvae are also known as naiads or nymph are aquatic in nature and differ morphologically with adults. Metamorphosos is incomplete, *i.e.*, Hemimetabolous, so there is no pupal stage.

Status of the Taxon

Global Status of Odonata

Earliest Protodonata, *Meganeura monyi* was recorded in the Carboniferous era. Protozygopterans were appeared in the lower Permian period. True Zygoptera were discovered during Upper Permian period. In the Mezozoic period we have come across the Anisozygoptera. In the Suborder Zygoptera fore and hind-wings are almost of same shape and size, with a definite discoidal cell formed below the arculus. Zygopterans of the families Coenagrionidae, Platycnemididae and Lestidae are found at the sides of freshwater waterbodies which may be permanent or temporary. Members of families Platystictidae, Protoneuridae, Chlorolestidae (synlestidae), Calopterygidae, Chlorocyphidae and Euphaeidae are observed breeding in the hill streams and sometimes seen near the waterbodies present in the plains.

Suborder Anisozygoptera which has a single superfamily Heterophlebioidea comprised of all families, all of which except family Epiophlebiidae have become extinct all over the world. This family is represented by a single species in India. Suborder Anisoptera which includes relatively robust built dragonflies as compared to suborder Zygoptera. Wings are usually held horizontally or descended when at rest. Dragonflies of the families Gomphidae, Aeshnidae, cordulegasteridae and Corduliidae are found near hilly tracts and breed in the hill streams, waterfalls, etc. Family Libellulidae is commonly found mostly in the plains near waterbodies, but sometimes members of family Gomphidae and Libellulidae were observed near hill streams.

Kirby (1890) has listed about 1800 species of Odonata under 319 genera and has expresssed his opinion that number of Odonata species in the whole World might be about four times of the number. Recently taxonomy of Odonata has also been revised on the World-wide basis, resulting in new synonymies, etc. Davies and Tobin (1984 and 1985) and Tsuda (1991) have recorded about 6,000 species/subspecies of Odonata belonging to 630 genera under 28 families from all over the world. Out of these only 499 species and subspecies are known from the India.

Indian Status of Odonata

Earliest consolidated list of Odonata in India is made by Fraser (1933, 1934 and 1936) in three Volumes of "Fauna of British India including Burma and Ceylon". Since then many workers in India and abroad have studied the Indian Odonata and described a series of new taxa from different part of India. Tsuda (1991) in his world list of Odonata has mentioned that approximately 906 species and subspecies of Odonata are known so far from continental Asia. Kumar and Prasad (1981) while dealing with the Odonata of Western Himalaya have published a detailed account of 162 species/subspecies from Western Himalaya. They have also discussed for the first time the zeogeography of Odonata of Western Himalaya. Recently Prasad and Varshney (1995) have pubished a revised updated check-list of 499 species and subspecies of Indian Odonata under 139 genera in 7 superfamilies, 17 families and 32 subfamilies.

Western Himalayan Status of Odonata

Geographical features and climate of Western Himalaya

In the present study term "Western Himalaya" is used for the Himalayan ranges lying in Jammu and Kashmir (31°-36' N. lat., 72°-80' E. long.), Himachal Pradesh (31°-30′ N. lat., 77° E. long.) and Kumaon and Garhwal ranges in Uttar Pradesh (28°-30' N. lat., 73°-77' E. long.) (Dey, 1974). Himalayan ranges are mostly of intricated and diverse character in the places highly deformed. In Jammu and Kashmir it is in the form of flat alluvial valley.

Kashmir lies on the southern side of Himalaya and is bounded by Pir Punjal range and with its happy valley. Himachal Pradesh comprises of mountaineous terrain and with a large number of rivers and streams, *i.e.,* Sutlej, Beas, Ravi and Chenab etc. In Uttar Pradesh, Kumaon and Garhwal ranges of Himalayas includes the numerous peaks like, Nanda Devi, Chaukhamba, Trisul and Kamst etc. Most of the main streams of River Ganga and Jamuna arises from the ranges of Himalayas.

In winter the mean temperature at about 2200 mtrs. is 44°F. (07°C) and in summer it is about 65°F. (18°C.) approximately. In the valley during May and June mean temperature is about 32°C.

to 37°C. Average annual mean rainfall is about 2,000 mm in the Western Himalaya.

Brief Descriptions of Odonata Biotopes in Western Himalaya

In Western Himalaya aquatic habitats may be divided into two major groups :

A. Permanent (Perennial) waterbodies

B. Temporary (seasonal) waterbodies

A. Permanent (Perennial) waterbodies

A number of permanent waterbodies are found in Western Himalaya. These may be in form of rivers, torrential hill streams in hilly terrain, slow running marshy streams present in flatter areas of the valley, slow running streams inside the deep forests, freshwater lakes and perennial ponds present in the lower altitudes of the Western Himalaya.

In Western Himalaya hill streams are usually isolated by steep sides of the hills in their upper reaches and gradient is high. In these hill streams water supply are permanent and its flow is very swift. The streams also have a number of rocky pools which are found due to accumulation of water in between the boulders. In the streams submerged aquatic vegetation.

At the lower altitudes or in flatter valley the nature of streams changes considerably. The gradient becomes less and the adjacent area is open. These streams are without steep hills. The climate becomes warmer and the temperature of water varies appreciably during the day time. The bed of streams are stable consisting of small boulders, stones, pebbles, gravel and sand. These streams are about 5-8 mtrs wide with a depth of water about 30-50 cm except in the monsoon period when it floods heavily. In these streams submerged and water vegetation are much more.

In the valley which is comparatively plains, there are a number of smaller slow running streams. These streams are generally tributaries of large rivers or streams which passes through the grassy areas. These streams form a net work of distributaries and rejoin in the main streams. Some of these distributaries or springs dry up during summer. The water flow in these streams or springs are slow and have a plenty of aquatic vegetations. Its bottom are muddy with a lot of debris and decaying leaves.

In Western Himalaya there are a number of lakes and ponds within forested areas. These lakes are deep with bottom full of mud with a lot of debris and decaying leaves. In the foot hills a lower range of Western Himalaya there are a number of perennial ponds. They are small in size and have a lot of aquatic vegetation and turbid water.

B. Temporary (seasonal) waterbodies

During period whole range of Western Himalaya get dotted with seasonal ponds. These ponds exist only for 3 to 4 months during June-July to September. These ponds formed as a result of the filling of monsoon rains water in the hollow depression.

Faunal Diversity of Odonata in Western Himalaya

The Himalaya is extended from Kashmir in the west and Arunachal Pradesh in the east. Himalayan mountain chain is divided into the Western Himalaya from Kashmir to Uttar Pradesh (Kumaon and Garhwal Hills), Central Himalaya in Nepal and the eastern Himalaya from Darjeeling (West Bengal) to Arunachal Pradesh. Odonata fauna of Western Himalayas has been studied most extensively by Calvert (1898), Fraser (1933, 1934 and 1936), Ander (1944), Bhasin (1953), Mani *et. al.*, (1955), Singh and Baijal (1954, Singh (1955 and 1963), Singh *et. al.*, (1955), Asahina (1960, 1962, 1978 and 1981), Schmidt (1961), Sahni (1964a, 1964b, 1965a, 1965b, 1970, 1972a and 1972b), St. Quentin (1970), Sangal and Kumar 1970a, and 1970b), Kumar (1970, 1971a, 1971b, 1971c, 1972a, 1972b, 1972c, 1973a, 1973b, 1973c, 1973d, 1976, 1978, 1979a, 1979b, 1980, 1982a, 1982b, 1983, 1984a, 1984b, 1984c, 1984d, 1985, 1988a 1988b and 1989), Kumar and Juneja (1976), Kumar and Khanna (1983), Kumar and Prasad (1976, 1977a, 1977b, 1977c, 1977d, 1977e, 1977f, 1978 and 1981), Kumar and Negi (1983 and 1989), Kumar and Mitra (1998a and 1998b), Kiauta

(1972 and 1975), Mitra (1973, 1975, 1995 and 1998), Atkinson (1974), Bhargava and Prasad (1974), Prasad (1974, 1976a, 1976b, 1977, 1978, 1988, 1997 and 1998), Prasad and Bhargava (1975), Prasad and Singh (1976 and 1994), Prasad and Kumar (1977a, 1977b, 1980 and 1981), Prasad and Varshney (1995), Singh and Prasad (1974, 1975, 1976a, 1976b, 1977 and 1980), Wojtusiak (1974), Chowdhary and Das (1975), Tyagi (1982, 1983, 1997, and 1998), Tyagi and Tyagi (1989), Tyagi *et. al.*, (1986), Chandra (1983), Cafri *et. al.*, (1983), Lieftinck (1984), Davies and Tobin (1984 and 1985), Sangal and Tyagi (1985), Hamalainen (1989a and 1989b,) Tsuda (1991), Mitra (1995 and 1999), and Gupta and Mitra (1998) etc., have recorded 155 species and subspecies of Odonata belonging to 76 genera under 14 families from Western Himalayas (Table 9.1). In this article controversial species, which are described for the first time from Western Himalaya are not included. Out of these 64 species/subspecies of Odonata from Jammu and Kashmir, 90 species/subspecies from Himachal Pradesh and 120 species/ subspecies from Uttar Pradesh are reported from these States so far. Detailed familywise distribution of Odonata in the three states are provided in Table 9.2. Among these, Zygopterans (damselflies) (Figures 9.1, 9.3, 9.6, 9.8, 9.10, 9.11, 9.12 and 9.13) families are represented by 9 families, viz., Coenagrionidae, Platycnemididae, Platystictidae, Protoneuridae, Lestidae, Synlestidae, Calopterygidae, Chlorocyphidae, and Euphaeidae. These above families together represent approximately 1/2.6 of 59:155 known Odonates from Western Himalayas. Coenagrionidae is the single family with representation of 25 species/subspecies under 9 genera, *i.e.*, 1/2.3 (25:59) of known Western Himalayan component of damselflies.

Anisopteran (dragonflies) (Figures 9.2, 9.4, 9.5, 9.7, 9.9, 9.14, 9.15, 9.16 and 9.17) component of Western Himalayas Odonates comprises of 96 species/subspecies under 49 genera and 5 families. This is 1/1.6 of the known odonates from Western Himalayas. Reported 5 families of anisopterans are : Gomphidae, Aeshnidae, Cordulegasteridae, Corduliidae and Libellulidae. Last family, *i.e.*, Libellulidae is having largest species component, among not only anisopterans but also amongsts the whole constituents-including zygopterans. This is represented by 53 species/subspecies. This is thus a little more than 1/1.8 of 53:96 anisopterans and a little more than 1/2.9 (53/155) of all known odonates from Western Himalayas. Western Himalayas odonates *vis-a-vis*, India's whole component of this group of insect is 155/499, *i.e.*, approximately 1/3.2.

Table 9.1 : Distributional range of Odonata species/Subspecies recorded from Western Himalayas

Sl. No.	Name of the species/subspecies	Western Himalaya		
		J&K	H.P.	U.P.
	Order Odonata Suborder Zygoptera Superfamily Coenagrionoidea Family Coenagrionidae Subfamily Psudagrioninae Genus *Ceriagrion* Selys			
1.	*Ceriagrion cerinorubellum* (Brauer)	-	+	+
2.	*Ceriagrion coromandelianum* (Fabricius)	+	+	+
3.	*Ceriagrion fallax cerinomelas* Lieftinck	-	+	-
	Genus *Pseudagrion* Selys			
4.	*Pseudagrion decorum* (Rambur)	-	+	+
5.	*Pseudagrion laidlawi* Fraser	-	-	+
6.	*Pseudagrion rubriceps rubriceps* Selys	+	+	+
7.	*Psudagrion spencei* Fraser	-	-	+
	Subfamily Coenagrioninae Genus *Cercion* Navas			
8.	*Cercion calamorum* (Ris)	-	+	+
	Genus *Himalagrion* Fraser			
9.	*Himalagrion exclamationis* Fraser	-	-	+
	Subfamily Ischnurinae Genus *Aciagrion* Selys			
10.	*Aciagrion pallidum Selys*	-	-	+
	Genus *Enallagma* Charp.			
11.	*Enallagma cyathigerum cyathigerum* Charp.	+	+	-
12.	*Enallagma parvum* Selys	-	-	+
13.	*Enallagma? risi* Schmidt	+	-	-
	Genus *Ischnura* Charp.			
14.	*Ischnura aurora aurora* (Brauer)	+	+	+
15.	*Ischnura focipata* Morton	+	+	+
16.	*Ischnura inarmata* Calvert	+	-	-
17.	*Ischnura pumilio* (Charp.)	+	-	-
18.	*Ischnura rufostigma annandalei* Laidlaw	+	-	-
19.	*Ischnura rufostigma rufostigma* Selys	+	+	+
20.	*Ischnura rufostigma mildredae* Fraser	+	-	-

Contd...

Table 9.1 – *contd...*

Sl. No.	Name of the species/subspecies	Western Himalaya		
		J&K	H.P.	U.P.
21.	*Ischnura senegalensis* (Rambur)	-	+	+
	Genus *Rhodischnura* Laidlaw			
22.	*Rhodischnura nursei* (Morton)	-	-	+
	Subfamily Agriocnemidinae			
	Genus Agriocnemis Selys			
23.	*Agriocnemis clauseni* Fraser	-	-	+
24.	*Agriocnemis corbeti* Kumar and Prasad	-	-	+
25.	*Agriocnemis pygmaea* (Rambur)	+	+	+
	Family Platycnemididae			
	Subfamily Calicnemidinae			
	Genus *Calicnemia* Strand.			
26.	*Calicnemia eximia* (Selys)	+	+	+
27.	*Calicnemia mortoni* (Laidlaw)	-	-	+
28.	*Calicnemia pulverulans* (Selys)	-	-	+
	Genus *Coeliccia* Kirby			
29.	*Coeliccia didyma didyma* (Selys)	-	+	-
30.	*Coeliccia renifera* (Selys)	+	+	+
	Subfamily Platycnemidinae			
	Genus *Copera* Kirby			
31.	*Copera ciliata* (Selys)	-	+	+
32.	*Copera marginipes* (Rambur)	+	+	+
33.	*Copera vittata* (Selys)?	-	+	+
	Genus *Platycnemis* Charp.			
34.	*Platycnemis dealbata* Selys	+	-	-
	Family Platystictidae			
	Subfamily Platystictinae			
	Genus *Drepanosticta* Laidlaw			
35.	*Drepanosticta carmichaeli* (Laidlaw)	+	-	+
	Genus *Protosticta* Selys			
36.	*Protosticta davenporti* (Fraser)	-	-	+
	Family Protoneuridae			
	Subfamily Disparoneurinae			
	Genus *Elattoneura* Cowley			
37.	*Elattoneura coomansi campioni* (Fraser)	-	-	+
	Genus *Prodasineura* Cowley			
38.	*Prodasineura automnalis* (Fraser)	-	-	+
	Superfamily Lestoidea			
	Family Lestidae			
	Subfamily Lestinae			
	Genus *Lestes* Leach			
39.	*Lestes barbarus* (Fabricius)	+	-	-
40.	*Lestes praemorsus praemorsus* (Selys)	+	-	+

Contd...

Table 9.1 – *contd...*

Sl. No.	Name of the species/subspecies	Western Himalaya		
		J&K	H.P.	U.P.
41.	*Lestes thoracicus* Laidlaw	-	-	+
42.	*Lestes viridulus* Rambur	-	+	+
	Subfamily Sympectmatinae			
	Genus *Indolestes* Fraser			
43.	*Indolestes cyaneus* (Selys)	+	+	+
	Genus *Sympecma* Burmeister			
44.	*Sympecma annulata annulata* (Selys)	+	-	-
45.	*Sympecma annulata kashmirensis* Ander	+	-	-
	Family Synlestidae			
	Subfamily Megalestinae			
	Genus *Megalestes* Selys			
46.	*Megalestes major* Selys	+	+	+
	Superfamily Calopterygoidea			
	Family Calopterygidae			
	Subfamily Calopteryginae			
	Genus *Neurobasis* Selys			
47.	*Neurobasis chinensis cheninsis* (Linn.)	-	+	+
	Family Chlorocyphidae			
	Genus *Rhinocypha* Rambur			
	Subgenus *Aristocypha* Laidlaw			
48.	*Rhinocypha* (*Aristocypha*) *immaculata* (Selys)	-	-	+
49.	*Rhinocypha* (*Aristocypha*) *quadrimaculata* Selys	+	+	+
50.	*Rhinocypha* (*Aristocypha*) *spuria* (Selys)	-	-	+
51.	*Rhinocypha* (*Aristocypha*) *trifasciata* (Selys)	-	+	+
	Subgenus *Heliocypha* Fraser			
52.	*Rhinocypha* (Heliocypha) *biforata biforata* Selys	-	-	+
	Subgenus *Paracypha* Fraser			
53.	*Rhinocypha* (*Paracypha*) *unimaculata* Selys	+	+	+
	Genus *Libellago* Selys			
54.	*Libellago lineata lineata* (Burm.)	-	+	+
	Family Euphaeidae			
	Genus *Anisopleura* Selys			
55.	*Anisopleura comes Hagen*	-	+	+
56.	*Anisopleura lestoides* Selys	-	-	+
57.	*Anisopleura subplatystyla* Fraser	-	+	-
	Genus *Bayadera* Selys			
58.	*Bayadera indica* (Selys)	+	+	+
	Genus *Epallaga* Charpeinter			
59.	*Epallage fatima* (Charpeinter)	+	-	-

Contd...

Table 9.1 – *contd...*

Sl. No.	Name of the species/subspecies	Western Himalaya		
		J&K	H.P.	U.P.
	Suborder Anisoptera			
	Superfamily Aeshnoidea			
	Family Gemphidae			
	Subfamily Gomphinae			
	Genus *Anisogomphus* Selys			
60.	*Anisogomphus bivittatus* (Selys)	-	+	+
61.	*Anisogomphus caudalis* Fraser	-	-	+
62.	*Anisogomphus occipitalis* (Selys)	-	-	+
	Genus *Anormogomphus* Selys			
63.	*Anormogomphus heteropterus* Selys	-	+	+
64.	*Anormogomphus kiritschenkoi* Bartenef.	-	+	+
	Genus *Burmagomphus* Williamson			
65.	*Burmagomphus hasimaricus* Fraser	-	-	+
66.	*Burmagomphus sivalikensis* Laidlaw	-	+	+
67.	*Burmagomphus vermicularis* Martin	-	+	-
	Genus *Macrogomphus* Selys			
68.	*Macrogomphus montanus* Selys	-	-	+
	Genus *Platygomphus* Selys			
69.	*Platygomphus dolabratus* Selys	-	-	-
	Genus *Davidius* Selys			
70.	*Davidius aberrans aberrans* (Selys)	-	-	+
71.	*Davidius kumaonensis* Fraser	-	-	+
	Genus *Nepogomphus* Fraser			
72.	*Nepogomphus modestus* (Selys)	-	+	+
	Subfamily Onychogomphinae			
	Genus *Onychogomphus* Selys			
73.	*Onychogomphus biforceps* (Selys)	-	-	+
74.	*Onychogomphys bistrigatus* (Hagen in Selys)	-	-	+
75.	*Onychogomphus duaricus* Fraser	-	-	+
76.	*Onychogomphus risi* (Fraser)	-	-	+
77.	*Onychogomphus schmidti* Fraser	-	-	+
	Genus *Ophiogomphus* Selys			
78.	*Ophiogomphus reductus* Calvert	+	-	-
	Genus *Paragomphus* Cowley			
79.	*Paragomphus lineatus* (Selys)	+	+	+
	Subfamily Lindeniinae			
	Genus *Ictinogomphus* Cowley			
80.	*Ictinogomphus rapax* (Rambur)		+	+
	Family Aeshnidae			
	Subfamily Aeshninae			
	Genus *Aeshna* Fabricius			
81.	*Aeshna juncea mongolica* Bartenef	+	-	-
82.	*Aeshna mixta mixta* Latreille			

Contd...

Table 9.1 – *contd...*

Sl. No.	Name of the species/subspecies	Western Himalaya		
		J&K	H.P.	U.P.
	Genus *Anax* Leach	+	-	-
83.	*Anax guttatus* (Burm.)	-	+	+
84.	*Anax immaulifrons* Rambur	+	+	+
85.	*Anax imperator imperator* Leach	+	-	+
86.	*Anax nigrofasciatus nigrolineatus* Fraser	-	-	+
87.	*Anax parthenope parthenope* (Selys)	+	+	+
	Genus *Hemianax* Selys			
88.	*Hemianax ephippiger* (Burm.)	-	+	+
	Genus *Gynacantha* Rambur			
89.	*Gynacantha khasiaca* MacLachlan	-	-	+
	Subfamily Brachytroninae			
	Genus *Cephalaeschna* Selys	-	+	-
90.	*Cephalaeschna orbifrons* Selys			
	Genus *Gynacanthaeschna* Fraser			
91.	*Gynacanthaeschna sikkima* (Karsch)	-	-	+
	Genus *Polycanthagyna* Fraser			
92.	*Polycanthagyna erythromelas* (MacLachlan)	-	+	-
93.	*Polycanthagyna ornithocephala* (MacLachlan)	-	+	-
	Superfamily Cordulegasteroidea			
	Family Cordulegasteridae			
	Subfamily Chlorogomphinae			
	Genus *Chlorogomphus* Selys			
94.	*Chlorogomphus atkinsoni* (Selys)	-	-	+
95.	*Chlorogomphus olympicus* Fraser	-	+	-
	Subfamily Crodulegasterinae			
	Genus *Anotogaster* Selys			
96.	*Anotogaster basalis basalis* Selys	-	+	+
	Genus *Cordulegaster* Leach			
97.	*Cordulegaster brevistigma brevistgma* (Selys)	+	+	+
98.	*Cordulegaster brevistigma folia* Fraser	-	-	+
99.	*Cordulegaster parvistigma* (Selys)	-	+	-
	Genus *Neallogaster* Cowley			
100.	*Neallogaster ornata* Asahina	-	+	-
	Superfamily Libelluloidea			
	Family Corduliidae			
	Subfamily Macromiinae			
	Genus *Epopthalmia* Burm.			
101.	*Epopthalmia vittata vittata* Burm.	-	-	+
	Genus *Macromia* Rambur			
102.	*Macromia moorei* Selys	-	+	+
	Family Libellulidae			
	Subfamily Tetrathemistinae			
	Genus *Hylaeothemis* Ris			
103.	*Hylaeothemis gardeneri* Fraser	-	-	+

Contd...

Table 9.1 – *contd...*

Sl. No.	Name of the species/subspecies	Western Himalaya		
		J&K	H.P.	U.P.
	Genus *Tetrathemis* Brauer			
104.	*Tetrathemis platyptera* Selys	-	-	+
	Subfamiy Brachydiplactinae			
	Genus *Brachydiplax* Brauer			
105.	*Brachydiplax sobrina* (Rambur)	-	+	+
	Subfamily Libellulinae			
	Genus *Cratilla* Kirby			
106.	*Cratilla lineata calverti* Foerster	-	+	+
	Genus *Lathrecista* Kirby			
107.	*Lathrecista asiatica asiatica* (Fabr.)	-	+	-
	Genus *Libellula* Linn.			
108.	*Libellula quadrimaculata grigorievi* Schmidt	+	-	-
	Genus *Orthetrum* Newman			
109.	*Orthetrum anceps* (Schn.)	-	+	+
110.	*Orthetrum brunneum brunneum* (Fons.)	+	+	+
111.	*Orthetrum cancellatum cancellatum* (Linn.)	+	+	-
112.	*Orthetrum chrysis* (Selys)	-	+	-
113.	*Orthetrum glaucum* (Brauer)	-	+	+
114.	*Orthetrum japonicum internum* MacL.	+	+	+
115.	*Orthetrum luzonicum* (Brauer)	+	+	+
116.	*Orthetrum martensi* Asahina	+	+	-
117.	*Orthetrum pruinosum neglectum* (Ramb.)	+	+	+
118.	*Ortehtrum ramburi* (Selys)	+	+	-
119.	*orthetrum sabina sabina* (Drury)	+	+	+
120.	*Ortehtrum taeniolatum* (Schn.)	-	+	+
121.	*Orthetrum triangulare triangulare* (Selys)	+	+	+
	Genus *Potamarcha* Karsch			
122.	*Potamarcha congener* (Ramb.)	-	+	+
	Subfamily Sympetrinae			
	Genus *Acisoma* Rambur			
123.	*Acisoma panorpoides panorpoides* Rambur	+	+	+
	Genus *Brachythemis* Brauer			
124.	*Brachythemis contaminata* (Frbr.)	+	+	+
	Genus *Bradinopyga* Brauder			
125.	*Bradinopyga geminata* (Rambur)	-	-	+
	Genus *Crocothemis* Brauer			
126.	*Crocothemis servilia servilia* (Drury)	+	+	+
127.	*Crocothemis erythreae erythreae* (Brulle)	-	+	+
	Genus *Diplacodes* Kirby			
128.	*Diplacodes lefebvrei* (Rambur)	+	-	+
129.	*Diplacodes nebulosa* (Fabr.)	-	+	+
130.	*Diplacodes trivialis* (Rambur)	+	+	+
	Genus *Neurothemis* Brauer			

Contd...

Table 9.1 – *contd...*

Sl. No.	Name of the species/subspecies	Western Himalaya		
		J&K	H.P.	U.P.
131.	*Neurothemis fluctuans* (Fabr.)	+	-	-
132.	*Neurothemis fulvia* (Drury)	-	+	+
133.	*Neurothemis intermedia intermedia* (Ramb.)	-	+	+
134.	*Nurothemis tullia tullia* (Drury)	-	+	+
	Genus *Sempetrum* Newman			
135.	*Sympetrum commixtum* (Selys)	+	+	+
136.	*Sympetrum fonscolombi* (Selys)	+	-	-
137.	*Sympetrum haematoneura* Fraser	+	-	+
138.	*Sympetrum hypomelas* (Selys)	-	+	+
139.	*Sympetrum meridionale* (Selys)	+	-	-
140.	*Sympetrum* (?) *Valgatum flavun* Bartenef	+	-	-
	Subfamily Trithemistinae			
	Genus *Trithemis* Brauer			
141.	*Trithemis aurora* (Burm.)	+	+	+
142.	*Trithemis festiva* (Rambur)	+	+	+
143.	*Trithemis kirbyi kirbyi* Selys	+	+	+
144.	*Trithemis pallidinervis* (Kirby)	-	+	+
	Subfamily Palpopleurinae			
	Genus *Palpopleura* Rambur			
145.	*Palpopleura sexmaculata sexmaculata* (Fabr.)	+	+	+
	Subfamily Trameinae			
	Genus *Rhyothemis* Hagen			
146.	*Rhyothemis triangularis* Kirby	+	+	-
147.	*Rhothemis variegata variegata* (Linn.)	-	+	+
	Genus *Pantala* Hagen			
148.	*Pantala flavenscens* (Fabr.)	+	+	+
	Genus *Tramea* Hagen			
149.	*Tramea basilaris burmeisteri* Kirby	-	+	+
150.	*Tramea virginia* (Rambur)	+	+	+
	Genus *Tholymis* Hagen			
151.	*Tholymis tillarga* (Fabr.)	-	+	+
	Genus *Zyxomma* Rambur			
152.	*Zyxomma petiolatum* Rambur	-	+	+
	Subfamily Urothemistinae			
	Genus *Selysiothemis* Ris			
153.	*Selysiothemis nigra* (Vander Linden)	+	-	-
	Genus *Urothemis* Brauer			
154.	*Urothemis signata signata* (Rambur)	-	-	+
	Subfamily Zygonychinae			
	Genus *Zygonyx* Hagen			
155.	*Zygonyx torrida isis* Fraser	-	-	+

Table 9.2 : Family-wise distribution of Odonata species/subspecies in different States of Western Himalaya

Name of the family	Number of species/ subspecies in		
	J&K	H.P.	U.P.
Coenagrionidae	12	12	19
Platycnemididae	4	6	7
Platystictidae	1	-	2
Protoneuridae	-	-	2
Lestidae	5	2	4
Synlestidae	1	1	1
Calopterygidae	-	1	1
Chlorocyphidae	2	4	7
Euphaeidae	2	3	3
Gomphidae	2	8	19
Aeshnidae	5	7	8
Cordulegasteridae	1	5	4
Corduliidae	-	1	2
Libellulidae	29	40	41
Total	64	90	120

Common Species/Subspecies of Odonata in Western Himalaya

35 species/subspecies of Odonata (14 species/subspecies, Zygoptera; 21 species/subspecies, Anisoptera) are common to all the three States are : *Ceiagrion coromandelianum* (Fabricius), *Pseudagrion rubriceps rubriceps* Selys, *Ischnura aurora aurora* (Brauer), *Ischnura forcipata* Morton, *Ischnura urfostigma rufostigma* Selys, *Agriocnemis pygmaea* (Rambur), *Calicnemia eximia* (Laidlaw), *Coeliccia renifera* (Selys), *Copera marginipes* (Rambur), *Indolestes cyaneus* (Selys), *Megalestes major* Selys, *Rhinocypha* (*Aristocypha*) *quadrimaculata* Selys, *Rhinocypha* (*Paracypha*) *unimaculata* Selys, *Bayadera indica* (Selys), *Paragomphus lineatus* (Selys), *Anax immaculifrons* Rambur, *Anax parthenope parthenope* (Selys), *Cordulegaster brevistigma brevistigma* (Selys), *Orthetrum brunneum brunneum* (Fons.), *Orthetrum japonicum internum* MacLachlan, *Orthetrum luzonicum* (Brauer), *Orthetrum pruinosum neglectum* (Rambur), *Orthetrum sabina sabina* (Drury), *Orthetum triangulare trangulare* (Selys), *Acisoma panorpoides panorpoides* Rambur, *Brachythemis contaminata* (Fabricius), *Crocothemis servilia servilia*

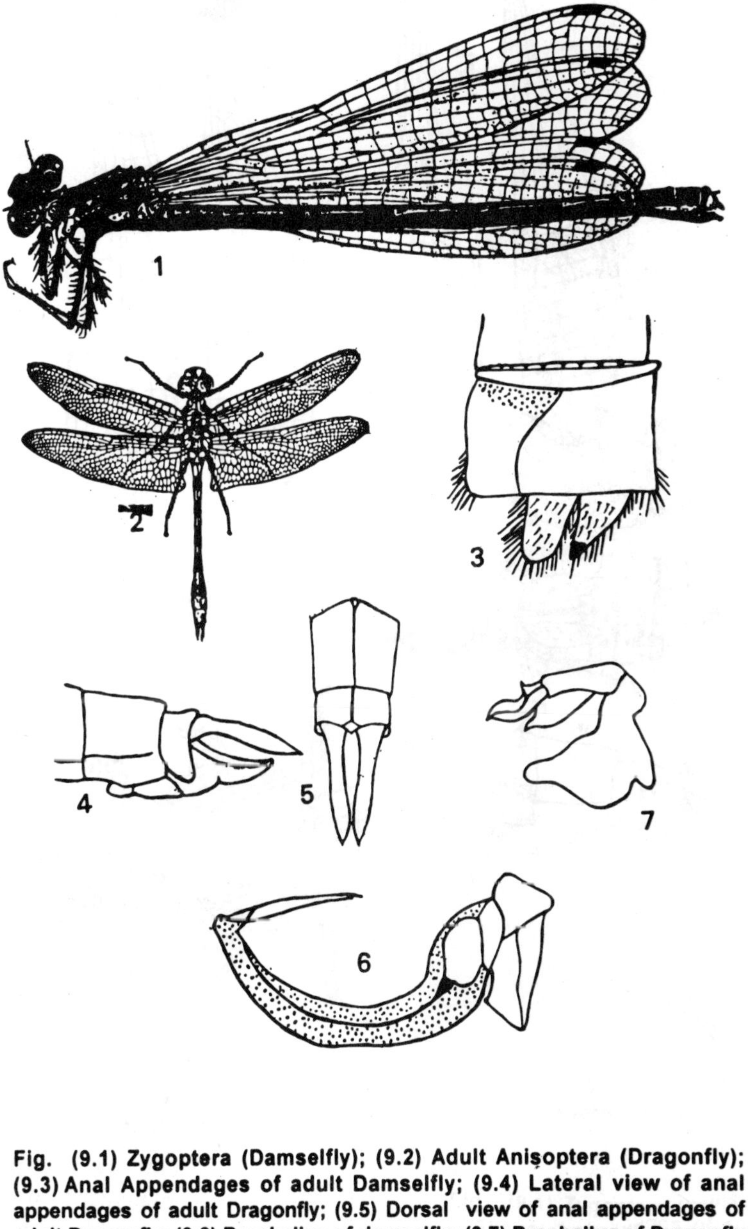

Fig. (9.1) Zygoptera (Damselfly); (9.2) Adult Anisoptera (Dragonfly); (9.3) Anal Appendages of adult Damselfly; (9.4) Lateral view of anal appendages of adult Dragonfly; (9.5) Dorsal view of anal appendages of adult Dragonfly; (9.6) Prophallus of damselfly; (9.7) Prophallus of Dragonfly

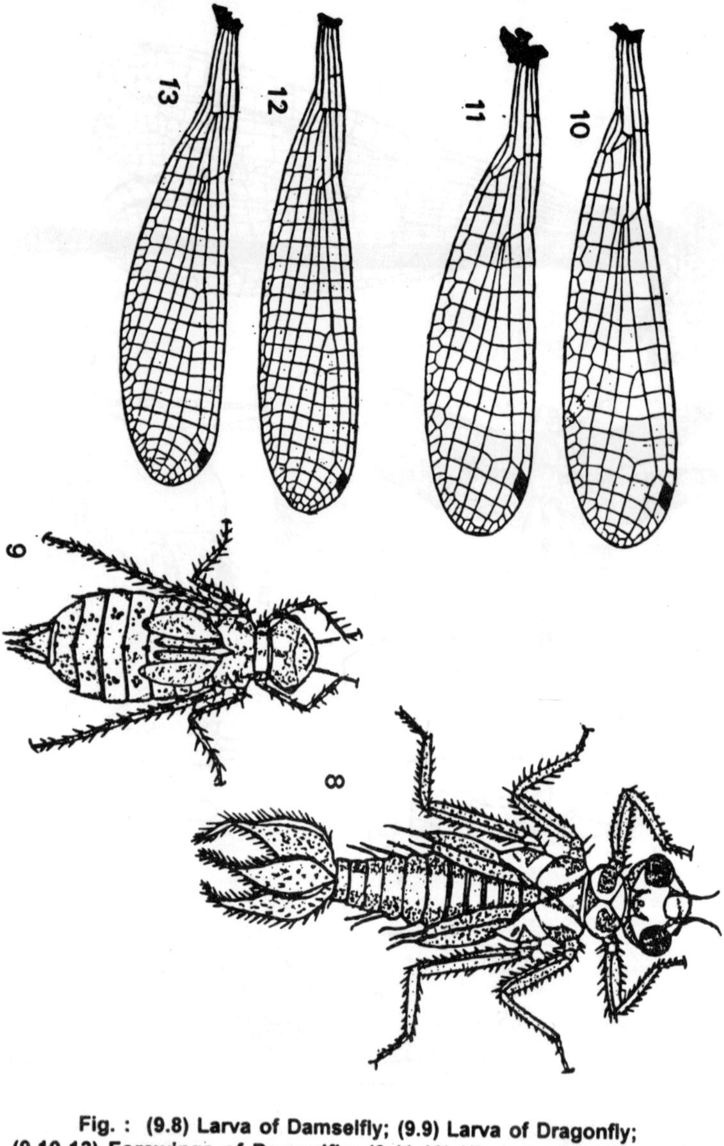

Fig. : (9.8) Larva of Damselfly; (9.9) Larva of Dragonfly; (9.10-12) Forewings of Damselfly; (9.11-13) Hindwings of Damselfly

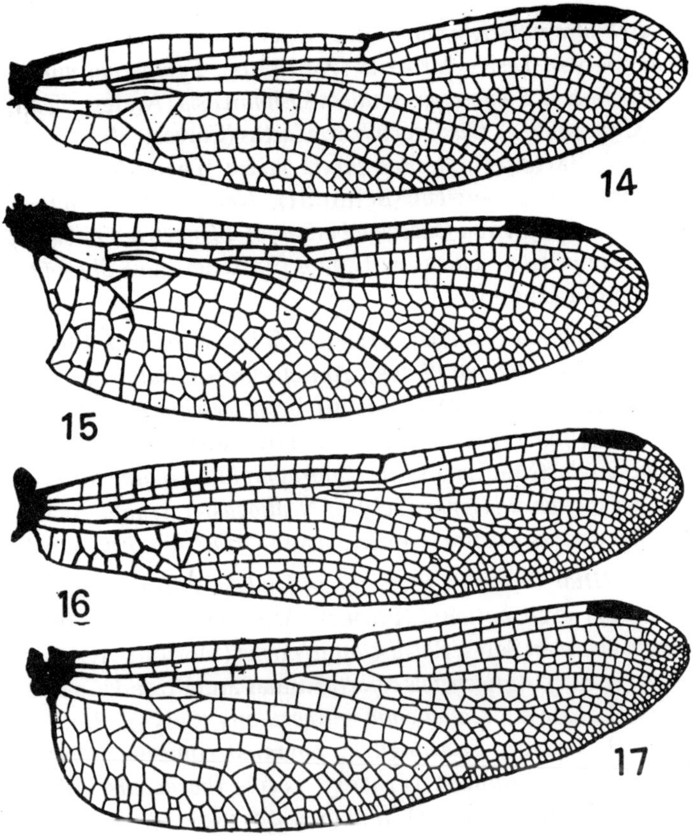

Fig. : (9.14) and (9.16) Forewings of Dragonfly : (9.15) and (9.17) Hindwings of Dragonfly.

(Drury), *Diplacodes trivialis* (Rambur), *Sympetrum commixtum* (Selys), *Trithemis aurora* (Burm.), *Trithemis festiva* (Rambur), *Trithemis kirbyi kirbyi* (Selys), *Palpolleur sexmaculata sexmaculata* (Fabricius), *Pantala flavescens* (Fabricius) and *Tramea virginia* (Rambur).

Odonata species/subspecies which confined their distribution within the states are as follows :

(A) *Jammu and Kashmir : Enallagma? risi* (Schmidt), *Ischnura inarmata* (Calvert), *Ischnura rufostigma annandalei* Laidlaw, (Cafri,

1983), *Ischnura rufostigma mildredae* Fraser (Cafri, 1983), *Platycnemis dealbat* (Selys), *Lestes barbarus* (Fabricius), *Sympecma annulata annulata* (Selys), *Sympecma annulata kashmirensis* (Ander), *Epallago fatima* (Charpeinter), *Ophiogomphus reductus* (Calvert), *Aeshna juncea mongolica* (Bartenef), *Aeshna mixta mixta* (Latreille), *Libellula quadrimaculata griogrievi* (Schmidt), *Neurothemis flactuance* (Fabricius), *Sympetrum fonscolombi* (Selys), *Sympetrum meridionale* (Selys), *Sympetrum? vulgatum flavum* (Bartenef) and *Slysiothemis nigra* (Vander Linden).

(B) *Himachal Pradesh : Ceriagrion fallax cerinomelas* Lieftinck, *Coeliccia didyma didyma* (Selys), *Anisopleura subplatystyla*, Fraser, *Burmagomphus vermicularis* (Martin), *Cephalaeschna orbifrons* Selys, *Polycanthagyna erythromelas* (MacLachlan), *Polycanthagyna ornithocephala* (MacLachlan), *Chlorogmphus olympicus* Fraser, *Cordulegaster parvistigma* (Selys), *Neallogaster ornata* Asahina, *Lathrecista asiatica asiatica* (Fabricius) and *Orthetrum chrysis* (Selys).

(C) *Uttar Pradesh : Pseudagrion laidlawi* Fraser, *Pseudagrion Himalagrion exclamationis* (Fraser), *Aciagrion pallidum* (Selys), *Enallagma parvum* Selys, *Rhodischnura nursei* (Morton), *Agriocnemis clauseni* (Fraser), *Agriocnemis corbeti* Kumar and Prasad, *Calicnemia mortoni* (Laidlaw), *Calicnemia pulverulans* (Selys), *Protosticta davenporti* Fraser, *Ellatoneura coomansi compioni* (Fraser), *Prodasineura autumnalis* (Fraser), *Lestes thoracicus* Laidlaw, *Rhinocypha* (*Aristocypha*) *immaculata* (Selys), *Rhinocypha* (*Aristocypha*) *spuria* (Selys), *Rhinocypha* (*Heliocypha*) *biforata biforata* (Selys), *Anisopleua lestoides* (Selys), *Anisogomhus caudalis* (Fraser), *Anisogomphus occipitalis* (Selys), *Burmagomphus hasimaricus* (Fraser), *Macrogomphus montanus* (Selys), *Platygomphus dolabratus* Selys, *Davidius kumaonensis* (Fraser), *Davidius aberrans aberrans* (Selys), *Onychogomphus biforceps* (Selys), *Onychogomphu bistrigatus* (Hagen in Selys), *Onychogomphus dyaricus* (Fraser), *Onychogomphus risi* (Fraser), *Onychogomphus schimidti* (Fraser), *Anax nigrofasciatus nigrolineatus* (Fraser), *Gynacantha khasiaca* (MacLachlan), *Gynacanthaeschna sikkima* (Karsch), *Chlorogomphus atkinsoni* (Selys), *Cordulegaster brevistigma folia* (Fraser), *Epopthalmia vittata vittata* Burm. *Hylaeothemis gardeneri* (Fraser), *Tetrathemis platyptera* (Selys), *Bradinopyga geminata* (Rambur), *Urothemis signata signata* (Rambur) and *Zygonyx torrida isis* (Fraser).

Out of 155 species/subspecies of Odonata 35 species/ subspecies are commonly occurring in all the three states. Among these, *Ceiagrion coromandelianum* (Fabricius), *Pseudagrion rubriceps rubriceps* Selys, *Ischnura aurora aurora* (Brauer), *Agriocnemis pygmaea* (Rambur), *Orthetrum sabina sabina* (Drury), *Crocothemis servilia servilia* (Drury) and *Diplacodes trivialis* (Rambur) and found almost throughout the year near permanent/temporary waterbodies. Most of these Odonata are multivoltine and completed their three generations in a year. Out of 155 species/subspecies of Odonata 19 species/subspecies of Odonata confined their distribution only in Jammu and Kashmir, 12 in Himachal Pradesh and 41 in Uttar Pradesh. State-wise distribution of these, Odonata are provided separately in the text. Only 5 species, excluding controversial species, which are described as new to science from Western Himalayas are endemic. Most of these species are not reported second time since their description. These species are known only from their type localities.

Endemicity

It includes the species which are found only in Western Himalayas. Only one genus, *i.e., Himalagrion* Fraser is endamic to Western Himalaya. According to Emerson (1955) "Endemic genera are found only in a single major zoogeographical region". A total of 5 species of Odonata, *i.e.,* about 3.2 per cent of total Western Himalayan Odonata species which are endemic to different ecosystem. Species which are endemic to Western Himalayas are : *Ischnura inarmata* Calvert, *Agriocnemis corbeti* Kumar and Prasad, *Davidius kumaonsis* Fraser, *Cordulegaster parvistigma* (Selys) , and *Hylaeothemis gardeneri* Fraser. In this article 7 other Odonata species *Viz., Coenagrion kashmirus* Chowdhary and Das, *Himalagrion pithoragarhicum* Sahni, *Calicnemia doonensis dooensis* Sangal and Tyagi, *Acrogomphus mohani* Sahni, *Anaciaschna kashmirensis* Singh and Baijal, *Nannaphya katrainensis* Singh and *Orthetrum guptai* Baijal described from Western Himalaya are not included due to their controversial position. All these species are known only from their type localities. Thus, they are presumed to have evolved locally (Roonwal and Verma, 1977).

Larval Studies (Figs. 9.8 and 9.9)

Odonata are hemimetabolous insects and their immature stages are called larva or nymph or naiad. Morphologically both adults and larva are quite different and have no resemblance with each other. Laidlaw (1917 and 1920) and Fraser (1919) have initiated their studies on larvae of Odonata in India. Subsequently Kumar (1970-1998), Kumar and Prasad (1977c, 1977d, 1977f and 1978) Sangal and Kumar (1970a and 1970b) have published descriptions of the last instar larvae and notes on biology of many species/subspecies from the Dehra Dun Valley, Uttar Pradesh. In Table 9.3 a list of 47 species and subspecies of Odonata of Western Himalaya especially from Uttar Pradesh of which larval descriptions have been published is given. An asterisk (*) indicates those species/subspecies in which the life history has been studied and described from Western Himalaya. Out of 155 species and subspecies descriptions of 47 species/subspecies (30.3%) of the larva is known from Western Himalaya.

Table 9.3 : Odonata species/subspecies of which larvae have been studied in Western Himalaya

No.	Name of species/subspecies	State where studied in Western Himalaya
	Zygoptera	
1.	*Ceriagrion coromandelianum* (Fabr.)	Uttar Pradesh
2.	*Pseudagrion laidlawi* (Fraser)	-do-
3.	**Pseudagrion rubriceps rubriceps* (Selys)	-do-
4.	*Ishnura aurora aurora* (Brauer)	-do-
5.	*Ischnura senegalensis* (Rambur)	-do-
6.	*Rhodischnura nursei* (Morton)	-do-
7.	*Agriocnemis corbeti* Kumar and Prasad	-do-
8.	*Agriocnemis pygmaea* (Rambur)	-do-
9.	*Calicnemis miles?* (Laidlaw)	-do-
10.	*Copera marginipes* (Rambur)	-do-
11.	*Elatoneura campioni* (Fraser)	-do-
12.	**Lestes praemorsa praemorsa* (Selys)	-do-
13.	*Neurobasis chinensis chinensis* (Linn.)	-do-
14.	*Rhinocypha (Aristocypha) quadrimaculata* (Selys)	-do-
15.	*Rhinocypha (Aristocypha) trifasciata* (Selys)	-do-
16.	*Rhinocypha (Heliocypha) biforata biforata* (Fraser)	-do-
17.	*Rhinocypha (Paracypha) unimaculata* (Selys)	-do-

Contd...

Table 9.3 – *contd...*

No.	Name of species/subspecies	State where studied in Western Himalaya
18.	*Anisopleura lestoides* (Selys)	Uttar Pradesh
19.	*Bayadera indica* (Selys) (Ainsoptera)	-do-
20.	*Anisogomphus occipitalis* (Selys)	-do-
21.	*Burmagomphus sivalikensis* (Laidlaw)	-do-
22.	*Paragomphus lineatus* (Selys)	-do-
23.	**Ictinogomphus rapax* (Rambur)	-do-
24.	*Anax guttatus* (Burm.)	-do-
25.	**Anax immaculifrons* (Rambur)	-do-
26.	*Anax nigrofasciatus nigrolineatus* (Fraser)	-do-
27.	*Anax parthenope parthenope* (Selys)	-do-
28.	*Macromia moorei* (Selys)	-do-
29.	**Orthetrum brunneum brunneum* (Fonscolombe)	-do-
30.	**Orthetrum pruinosum neglectum* (Rambur)	-do-
31.	*Orthetrum sabina sabina* (Drury)	-do-
32.	*Orthetrum taeniolatum* (Schneider)	-do-
33.	*Potamarcha congener* (Rambur)	-do-
34.	**Acisoma panorpoides panorpoides* (Rambur)	-do-
35.	**Brachythemis contaminata* (Fabricius)	-do-
36.	**Bradinopyga geminata* (Rambur)	-do-
37.	**Crocothemis servilia servilia* (Drury)	-do-
38.	**Diplacodes trivialis* (Rambur)	-do-
39.	**Neurothemis tullia tullia* (Drury)	-do-
40.	*Sympetrum commixtum* (Selys)	-do-
41.	*Trithemis aurora* (Burmeister)	-do-
42.	**Trithemis festiva* (Rambur)	-do-
43.	**Pantala flavescens* (Fabr.)	-do-
44.	*Tramea basilaris burmeister* Kirby	-do-
45.	*Tramea virginia* (Rambur)	-do-
46.	*Tholymis tillarga* (Fabr.)	-do-
47.	*Zyxomma petiolatum* (Rambur)	-do-

A total of 155 species and subspecies of Odonata which are found in various ecosystem of Western Himalaya are divided into highmontane, fast running rivers and streams form, slow running rivers and streams form, forms available near the streams passes through the forests, forms present within the sides of aquatic vegetations, forms present near permanent and temporary waterbodies and forms found at cemented waterbodies. Odonata which are found at different ecosystem in Western Himalaya are arranged systematically and listed below :

1. *High montane fast running rivers and streams form*

Some species of Odonata which are confined this habitat are : *Rhinocypha* (*Aristocyha*) *quadrimaculata* (Selys), *Rhinocypha* (*Paracypha*) *unimacutala* (Selys), *Rhinocypha* (*Aristocypha*) *trifasciata* (Selys), *Anisopleura lestoides* (Selys), *Anisopleura comes* (Hagen), *Bayaderaindica* (Selys), *Neurobasis chinensis chinensis* (Linn.) *Anormogomphus heteropterus* (Selys), *Anisogomphus occipitalis* (Selys), *Anisogomphus bivittatus* (Selys), *Paragomphus lineatus* (Selys), *Onychogomphus biforceps* (Selys), *Onychogomphus risi* (Fraser), *Nepogomphus modestus* (Selys), *Anax immaculifrons* (Rambur), *Anotogaster basalis basalis* (Selys), *Macromia moorei* (Selys), *Orthetrum anceps* (Schneider), *Orthetrum glaucum* (Brauer), *Orthetrum japonicum internum* (MacLachlan), *Orthetrum pruinosum neglectum* (Rambur), *Orthetrum taeniolatum* (Schneider), *Orthetrum triangulare triangulare* (Selys), *Crocothemis servilia servilia* (Drury), *Sympetrum commixtum* (Selys), *Sympetrum hypomelas* (Selys), *Trithemis aurora* (Burmeister), *Trithemis festiva* (Rambur) and *Pantala flavescens* (Frbr.).

2. *Slow running rivers and streams forms of hills and montane areas*

Odonata species and subspecies which were observed and found on the slow runnng rivers and streams of hills and mountain of Western Himalaya are : *Drepanosticta carmichaeli* (Laidlaw), *Prodasineura autumnalis* (Fraser), *Calicnemia eximia* (Selys), *Calicnemia pulverulans* (Selys), *Ischnura senegalensis* (Rambur), *Ischnura aurora aurora* (Brauer), *Megalestes major* (Selys), *Libellago lineata lineata* (Burmeister), *Rhinocypha* (*Aristocypha*) *immaculata* (Selys), *Rhinocypha* (*Aristocypha*) *quadrimaculata* (Selys), *Rhinocypha* (*Aristocypha*) *trifasciata* (Selys), *Rhinocypha* (*Heliocypha*) *biforata biforata* (Selys), *Rhinocypha* (*Paracypha*) *unimaculata* (Selys), *Neurobasis chinensis chinensis* (Linn.), *Anisopleura lestoides* (Selys), *Bayadera indica* (Selys), *Anormogomphus heteropterus* (Selys), *Anormogomphus kiritschenkoi* Bartenef, *Anisogomphus bivittatus* (Selys), *Anisogomphus caudalis* (Fraser), *Anisogomphus occipitalis* (Selys), *Burmagomphus sivalikensis* (Laidlaw), *Macrogomphus montanus* Selys, *Nepogomphus modestus* (Selys), *Onychogomphus biforceps* (Selys), *Onychogomphys bistrigatus* (Hagen in Selys),

Onychogomphus risi (Fraser,) *Paragomphus lineatus* (Selys), *Anax guttatus* (Burmeister), *Anax immaculifrons* (Rambur), *anax imperator imperator* (Leach), *Hemianax ephippiger* (Burmeister), *Gynacantha khasiaca* (MacLachlan), *Anotogaster basalis basalis* (Selys), *Epopthamis vittata vittata* (Burmeister), *Macromia moorei* (Selys), *Orthetrum anceps* (Schn.), *Orthetrum brunneum brunneum* (Fonscolombe), *Orthetrum glaucum* (Brauer), *Orthetrum japonicum internum* (MacLachlan), *Orthetrum pruinosum neglectum* (Rambur), *Orthetrum sabina sabina* (Drury), *Orthetrum taeniolatum* (Schn.), *Orthetrum triangulare triangulare* (Selys), *Palpoleura sexmaculata sexmaculata* (Fabricius), *Crocothemis servilia servilia* (Drury), *Sympetrum commixtum* (Selys), *Sympetrum hypomelas* (Selys), *Trithemis aurora* (Brumeister), *Trithemis festiva* (Rambur), *Trithemis kirbyi kirbyi* (Selys), *Trithemis pallidinervis* (Kirby), *Pantala falvescens* (Fabr.) and *Zygomyn torrida isis* (Fraser) etc.

3. *Odonata found near streams passes through forests*

Odonata found near or within the streams passes through the forested areas are given below. Beds of these streams are full of stone and pebbels etc. Sides of these streams are full of aquatic vegetations. *Calicnemia eximia* (Selys), *Calicnemia pulverulans* (Selys), *Coeliccia renifera* (Selys), *Copera marginipes* (Rambur), *Drepanosticta carmichali* (Laidlaw), *Prodasineura autumnalis* (Fraser) *Aciagrion pallidiun* (Selys), *Libellago lineata lineata* (Burmeister), *Rhinocypha (Aristocypha) immaculata* (Selys), *Rhinocypha (Aristocypha) quadrimaculata* (Selys), *Rhinocypha (Aristocypha) trifasciata* (Selys), *Rhinocypha (Heliocypha) biforata biforata* (Selys), *Rhinocypha (Paracypha) unimaculata* (Selys), *Neurobasis chinensis chinensis* (Linn.), *Orthetrum triangulare triangulare* (Selys), *Orthetrum pruinosum neglectum* (Rambur), and *Palpopleura sexmaculata sexmaculata* (Fabricius).

4. *Forms found within and sides of aquatic vegetations*

Very small number of Odonata species/subspecies are found within these habitats are following : *Calicnemia eximia* (Selys), *Calicnemia pulverulans* (Selys), *Coeliccia renifera* (Selys) and *Drepanosticta carmichaeli* (Laidlaw) etc.

5. *Permanent and temporary waterbodies forms*

In the foot-hills and lower range of Western Himalaya a large number of permanent perenial and temporary waterbodies, *viz.*, ponds and lakes etc., are found. However, these ponds and lakes have rich assemblege of Odonata, which are as follows : *Ceriagrion coromandelianum* (Fabr.), *Ceriagrion cerinorubellum* (Brauer), *Pseudagrion decorum* (Rambur), *Pseudagrion laidlaw* (Fraser), *Pseudagrion rubriceps rubriceps* (Selys), *Pseudagrion spencei* (Fraser), *Cercion calamorum* (Ris), *Enallagma parvum* (Selys), *Ischnura aurora aurora* (Brauer), *Ischnura forcipata* Morton, *Ischnura rufostigma rufostigma* (Selys), *Ischnura senegalensis* (Rambur), *Rhodischnura nursei* (Morton), *Agriocnemis clausoni* (Fraser), *Agriocnemis corbetti* (Kumar and Prasad), *Agriocnemis Pygmaea* (Rambur), *Copera marginipes* (Rambur), *Lestes praemorsus pramorsus* (Selys), *Lestes thoracicus* (Laidlaw), *Lestes viridulus* (Rambur), *Paragomphus lineatus* (Selys), *Ictinogomphus rapax* (Rambur), *Anax guttatus* (Burmeister), *Hemianax ephippiger* (Burmeister), *Brachydiplax sobrina* (Rambur), *Potamarcha congener* (Rambur), *Orthetrum glaucum* (Brauer), *Orthetrum luzonicum* (Brauer), *Orthetrum pruinosum neglectum* (Rambur), *Orthetrum sabina sabina* (Drury), *Acisoma panorpoides panorpoides* (Rambur), *Brachythemis contaminata* (Fabricius), *Crocothemis servilia servilia* (Drury), *Diplacodes trivialis* (Rambur), *Diplacodes nebulosa* (Fabricius), *Neurothemis fulvia* (Drury), *Neurothemis intermedia intermedia* (Rambur), *Neurothemis tullia tullia* (Drury), *Trithemis palidinervis* (Kirby), *Rhyothemis variegata variegata* (Linn.) *Pantala flavescens* (Fabricius), *Tramea basilaris burmeisteri* (Kirby), *Tramea virginia* (Rambur), *Tholymis tillarga* (Fabricius), *Zyxomma petiolatum* Rambur and *Urothemis sinata signata* (Rambur).

6. *Odonata found at the cemented waterbodies*

Bradinopyga geminata (Rambur) is found at the cemented waterbodies which may be in the garden or near the boulders present within the sides of streams. Sometimes this species is also observed on the cemented walls of the houses.

Threata and Conservation

Human activities have mainly affected the grasslands and scrub jungle, etc., of India. A large areas of the land have been converted into agricultural lands or huge industrial complexes. therefore, due to the shrinkage of the grasslands and freshwater ecosystem, the Odonata population as well as diverstiy of species are reducing very fast. The concept of Odonata conservation in India or even in insects in general is very recent. It is a milestone in our efforts to conserve and protect several other species of Odonata which are vanishing fast today due to destruction of their breeding and perching habitats. *Tyagi* (1985) has suggested that (i) water level of dams, marshes and ponds at a constant level should be maintained; (ii) natural bushes and trees stand adjacent to both standing and running waterbodies should be maintained and (iii) siltation of freshwater bodies should be avoided. To make it more effective, a list of endangered and rare species alongwith their pictorial illustrations and distributional map should be prepared.

Summary

This article deals with the Odonata diversity of 155 species and subspecies of Odonata in Western Himalaya, India (Jammu and Kashmir Himalaya, Himachal Pradesh and Uttar Pradesh Himalayas). Brief descriptions of geographical features, Odonata biotopes, a list of 47 species/subspecies of Odonata of Western Himalaya of which larval descriptions and life history of 14 species are known from Western Himalaya are provided.

Acknowledgements

The author is thankful to Dr. J.R.B. Alfred, the Director, Zoological Survey of India, Calcutta, Drs. G.K. Srivastava, Addl. Director and S.K. Mitra, Joint Director of this Department for ecouragement and facilities to carry out this work.

References

Ander, K. 1944. Einige Odonaten aus Kaschmir und ostturkestan. *Forh. K. fusiogr. Sallsk.*, 14 : 168-178.

Asahina, S. 1960. Notes on the relationship between Himalayan and Japanese insect-fauna. *J. Bengal nat. Hist. Soc.*, 31 : 69-75.

Asahina, S. 1962. *Anax nigrofasciatus* Oguma and *Anax nigrolineatus* Fraser (Odonata : Aeshnidae). *Jap. J. Zool.*, 13 : (2) : 249-255.

Asahina, S. 1978. A new and some known species of Odonata from Kashmir. *Senekenbergiana biol.*, 59 (1/2) : 115-120.

Asahina, S. 1981. A revision of the Himalayan dragonflies of the genus *Cephalaeschna* and its allies (Odonata, Aeshnidae). *Bull. natn. Sci. Mus.* (A), 7(2):57-77.

Atkinson, E.T. 1974. Fauna of the Himalayas containing species of Kumaon, Garhwal, Nepal and Tibet. *Cosmo Publication*, Delhi (Reprint).

Bhargava, R.N. and Prasad, M. 1974. *Orthetrum glaucum* (Brauer) preying upon *Palpopleura sexmaculata sexmaculata* (Fabricius) (Odonata: Libellulidae). *J. Bombay nat. Hist. Soc.*, 71 (1) : 164.

Bhasin, G.D. 1953. A systematic catalogue of main identified collections at Forest Research Institute, Dehra Dun. Pt. 12. Order Odonata. *Indian Forest Leafl.* (N.S.)., 121 (3) : 63-78.

Cafri, S. Del. Centina, P. and Terzani, F. 1983. Odonati reccolti nel Jammu-Kashmir, Himachal Pradesh, Uttar Pradesh (India). *G. it. Ent.*, 1 : 261-264.

Calvert, P.P. 1898. Odonata (Dragonflies) from the Ocean and from Kashmir collected by Dr. W.L. Abbott. *Pro. Acad. nat. Sci. Philad.* 1:141-154.

Chandra, M. 1983. Additions to the Odonata fauna of district Solan, Himachal Pradesh. *Fraseria*, No. 5 : 17-18.

Chowdhary, S.K. and Das, S.M. 1975. Notes on the life history of *Coenagrion kashmirus* n. sp. (Odonata: Zygoptera : Coenagriidae). *3rd All. India Corngr. Zool.*, : 60-61. (Abstr.).

Dey, A.K. 1974. Physiographic provinces of India and their resources other than human. *Everyman's Science*, **9** (2) : 55-62.

Davies, D.A.L. and Tobin, P., 1984. The dragonflies of the World : A systematic list of the extant species of Odonata, Vol. I. *Soc. Internat. Odonatol. Rapid. Comm.* (*Suppl.*), No. 3 : 1-127.

Davies, D.A.L. and Tobin, P. 1985. The dragonflies of the World : A systematic list of the extant species of Odonata, Vol., II, *Soc. Internat. Odonatol. Rapid. Comm.* (*Suppl.*), No. 5 : 1-151.

Emerson, A.E. 1955. Geographical origins and dispersal of termite genera. *Fieldiana* (*Zool*). 37 : 465-521.

Fraser, F.C. 1919. Description of new Indian Odonata larvae and exuviae. *Rec. Indian Mus.*, 16 : 459-467.

Fraser, F.C. 1933. The Fauna of British India, including Ceylon and Burma, Odonata, Vol. I. *Taylor and Francis Ltd.*, London. : 1-423.

Fraser, F.C. 1934. The Fauna of British India, including Ceylon and Burma, Odonata, Vol. II. *Taylor and Francis Ltd.*, London, : 1-398.

Fraser F.C. 1936. The Fauna of British India, including Ceylon and Burma, Odonata, Vol. III. *Taylor and Francis Ltd.*, London : 1-461.

Gupta, I.J. and Mitra, T.R. 1998. First record of the Odonata from Pong Wetland, Himachal Pradesh. *VI South Asian Symp. Odonatol.*, Nagpur. : 7.

Hämalainen, M. 1989a. Synonymic notes on some species of the genus *Mortonagrion* (Zygoptera : Coenagrionidae). *Indian Odonatol.*, 2 : 1-4.

Hämalainen, M. 1989b. Odonata from Dehra Dun Valley (Uttar Pradesh, India), with notes on synonymy of some West Himalayan species. *Odonatologica*, **18** (1) : 13-20.

Huxley, J. 1953. Evolution in action, *A Pelican Book.* :167.

Kiauta, B. 1972. Scientific results of the Yugoslow, 1969, Himalayan Expedition. Odonata. *Biol. Vest.*, **20** : 109-119.

Kiauta, B. 1975. Cytotaxonomy of dragonflies with special reference to Nepalese Fauna II. Lectures delevered at the Tribhuvan University, Kathmandu, *Nepal Research Centre*, Kathmandu. : X+77.

Kumar, A. 1970. Bionomics of *Orthetrum pruinosum neglectum* (Rambur) (Odonata : Libellluliidae). *Bull. Ent.*, 11 : 85-93.

Kumar, A. 1971a. The larval stages of *Orthetrum brnneum brunneum* (Fonscolombe) with a description of the last instar larva of *Orthetrum taeniolatum* (Schneider) (Odonata : Libellulidae). *J. nat. Hist.*, 5 : 121-132.

Kumar, A. 1971b. Studies on the phenology of the dragonflies of the Dehra Dun Valley (India). *Ist. Europ. Symp. Odonatol.*, Gent. : 29-30 (Abrstr.).

Kumar, A. 1971c. Taxonomic studies on the last instar larvae of Odonata from the Dehra Dun Valley, India. *Ph.D. Thesis, University of Meerut.*

Kumar, A. 1972a. Studies on the life history of *Trithemis festiva* (Rambur, 1842) (Odonata : Libellulidae). *Odonatologica*, 1 (2) : 103-112.

Kumar, A. 1972b. The lfie history of *Lestes praemorsa praemorsa* Selys (Odonata : Lestidae). *Treubia*, 28 : 3-20.

Kumar, A. 1972c. The phenology of dragonflies in the Dehra Dun Valley, India. *Odonatologica*, 1(4) : 199-207.

Kumar, A. 1973a. Descriptions of the larvae of *Anax nigrofasciatus nigrolineatus* Fraser, 1935 and *A. parthenope parthenope* (Selys, 1839) from India, with a key to the known larvae of the Indian representatives of the genus *Anax* Leach, 1815 (Anisoptera : Aeshnidae). *Odonattologica*, 2 (2) : 83-90.

Kumar, A. 1973b. Descriptions of the last instar larvae of Odonata from the Dehra Dun Valley (India), with notes on biology. I. Suborder Zygoptera. *Oriental Ins.*, 7(1) : 83-118.

Kumar, A. 1973c. Descriptions of the last instar larvae of Odonata from the Dehra Dun Valley (India), with notes on biology. II. Suborder Anisoptera. *Oriental Ins.*, 7 (2) : 291-331.

Kumar, A. 1973d. The life history of *Bradinopyga geminata* (Rambur) (Odonata : Libellulidae). *G.K.V.J. Sci. Res.*, 5 (1-2) : 50-57.

Kumar, A. 1976. Biology of Indian dragonflies with special reference to seasonal regulation and larval development. *Bull. Ent.*, 17 : 37-47.

Kumar, A. 1978. Some field notes on the Odonata around a freshwater lake in Western Himalaya (Renuka, Himachal Pradesh). *J. Bombay nat. Hist. Soc.*, **74** (3) : 506-510.

Kumar, A. 1979a. On the occurrence of multivoltine generations in some Indian dragonflies. Sci. and Cult., **45** (3) : 126-127.

Kumar, A. 1979b. Studies on the life history of Indian dragonflies, *Pseudagrion rubriceps* (Selys) (Coenagrionidae : Odonata). *Rec. Zool. Surv. India*, 75 : 371-381.

Kumar, A. 1980. Studies on the life history of Indian dragonflies, *Ceriagrion coromandelianum* (Fabr.) (Coenagriidae : Odonata) *Rec. zool. Surv. India*, 76 : 249-258.

Kumar, A. 1982a. An annotated list of Odonata from the Pithoragarh district, Western Himalaya, Uttar Pradesh, India. *Notul. Odonatol.*, 1 (9) : 145-147.

Kumar, A. 1982b. An annotated list of Odonata of Himachal Pradesh. *Indian J. Phy. Nat. Sci.*, 2 : 55-59.

Kumar, A. 1983. An annotated list of Odonata from the Jammu Region (Western HImalaya, Jammu and Kashmir, India). *Fraseria,* No. 4 : 15-16.

Kumar, A. 1984a. On the first record of dragonflies (Insecta : Odonata) from Ladakh (Jammu and Kashmir). India. *Fraseria,* No. 6 : 25-26.

Kumar, A. 1984b. On the life history of *Pantala flavescens* (Fabricius) (Libellulidae Odonata). *Ann. Entomol.*, **2** (1) : 43-50.

Kumar, A. 1984c. Hitherto unknown female of *Burmagomphus sivalikensis* Laidlaw (Odonata : Gomphidae). *Ann.* **2** (2) : 49-52.

Kumar, A. 1984d. Studies on the life history of Indian dragonfly *Anax immaculifrons* (Rambur) (Aeshnidae : Odonata). *Entomon,* 9(2) : 127-133.

Kumar, A. 1984e. Studies on the life history of Indian dragonflies, *Diplacodes trivialis* (Rambur) (Libellulidae : Odonata). *Rec. zool. Surv. India,* 81 (3 and 4) : 13-22.

Kumar, A. 1984f. Studies on the life history of Indian dragonflies, *Acisoma panorpoides panorpoides* Rambur, 1842 (Libellulidae : Odonata). *Rec. zool. Surv. India,* 81 (3 and 4) : 205-213.

Kumar, A. 1985. Studies on the life history of Indian dragonflies, *Ictinogomphus rapax* (Rambur) (Gomphidae:Odonata). *Ann. Entomol.* 3 (1) : 29-38.

Kumar, A. 1988a. On the andromorphic female of *Neurothemis tullia tullia* (Drury) (Anisoptera : Libellulidae). *Notul. Odonatol.*, 3 (1) : 14-15.

Kumar, A 1988b. Studies on the life history of *Neurothemis tullia tullia* (Drury) from Dehra Dun, India (Odonata : Libellulidae). *Indian Odonatol.*, 1 : 35-44.

Kumar, A. 1989. Life history of *Tramea virginia* (Rambur) from Dehra Dun, India (Anisoptera : Libellulidae). *Indian Odonatol.*, 2 : 5-14.

Kumar, A. and Juneja, D.P. 1976. The Odonata of Renuka Lake (Western Himalaya : Himachal Pradesh). *Newsl. zool. Surv. India*, 2 (3) : 95-96.

Kumar, A. and Khanna, V. 1983. A review of the taxonomy and ecology of Odonata larvae from India. *Oriental Ins.*, 17 : 127-157.

Kumar, A. and Mitra, A. 1998a. Odonata diversity at Sahastradhara (Surpher springs), Dehra Dun, India with notes on their habitat ecology. *VI South Asia Symp. Odonatcl.*, Nagpur, : 11.

Kumar, A. and Mitra, A. 1998b. Odonata diversity at Sahastradhara (Sulphur springs), Dehra Dun, India, with notes on their habitat ecology. *Fraseria* (NS), **5** (1/2) : 37-45.

Kumar, A. and Negi, B. K. 1983. The eggs of *Ictinogomphus rapax* (Rambur) (Odonata : Gomphidae). *Fraseria*, No 5:18.

Kumar, A. and Negi, B.K. 1989. The emergence and sex ratio of *Bradinopyga geminata* (Rambur) and *Crocothemis servilia servilia* (Drury) at Dehra Dun, India (Anisoptera : Libellulidae). *Fraseria, No.* 16 : 70-71.

Kumar, A. and Prasad. M. 1976. On the occurrence of *Orthetrum garwalicum* Singh and Baijal, 1954, (Odonata: Libellulidae) in Kinnaur (Western Himalaya : Himachal Pradesh). *Newsl. zool. Surv. India*, 2(3) : 94.

Kumar, A. and Prasad, M. 1977a. A note on dragonflies caught in spider's webs. *Odonatologica*, **6**(1) : 19-20.

Kumar, A. and Prasad, 1977b. Reproductive behaviour in *Neurobasis chinensis chinensis* (Linn.) (Zygoptera : Calopterygidae). *Odonatologica*, **6** (3) : 163-171.

Kumar, A. and Prasad, M 1977c. On the larva of *Tramea virvinia* (Rambur, 1842) from India, with notes on the larvae of Indian representative of genus *Tramea* Hagen, 1861 (Libellulidae : Odonata). *J. Bombay nat. Hist. Soc.*, **74**(1) : 199-202.

Kumar, A. and Prasad. M. 1977d. On the larvae of genus *Rhinocypha* (Odonata:Chlorocyphidae) from Garhwal Hills. *Oriental Ins.*, **11** (4) : 547-554.

Kumar, A. and Prasad, M. 1977e. Odonata of ponds, tanks and paddy fields at and around the Dehra Dun Valley (Western Himalaya). *Newsl. zool. Surv. India*, **3** (5) : 270-273.

Kumar, A. and Prasad, M. 1977f. Last instar larvae of two Odonata species from Western Himalaya. *Entomon*, **2** (2) : 225-230.

Kumar, A. and Prasad, M. 1978. On new species of *Agriocnemis* Selys, 1869 (Coenagriidae : Odonata) with description of its larva from Dehra Dun Valley, India. *J. Bombay nat. Hist. Soc.*, **75**(1) : 174-179.

Kumar, A. and Prasad, M. 1981. Field ecology, zoogeography and taxonomy of the Odonata of Western Himalaya, India. *Rec. zool. Surv. India, Occ. pap.*, No. 20: 1-118.

Laidlaw, F.F. 1917. A list of the dragonflies from the India empire, with special reference to the collection of the Indian Museum. *Rec. Indian Mus.*, 13 : 321-348.

Laidlaw, F.F. 1920. Notes on some interesting larvae of dragonflies (Odonata) in the collection of the India Museum. *Rec. Indian Mus.* **19** : 185-187.

Lieftinck, M.A. 1984. Further notes on the specific characters of *Calicnemia* Strand, with key to the males and remarks on some larval forms (Zygoptera : Platycnemididae) *Odonatologica*, **13** (3) : 351-375.

Mani, M.S., Singh, S., Gupta, V.K. and Baijal, H.N. 1955. Entomological Survey of the Himalayas. Pt. IX. First annotated check-list of insects from the North-West (Punjab) Himalayas. *Agra Univ. J. Res.* (Sci.), **4**(2) : 471-512.

Mitra, A. 1995. A note on the seasonal ecology of *Pantala flavescens* (Fabricius) (Anisoptera:Libellulidae) in Dehra Dun Valley, India. *Fraseria* (NS), **2**(1/2) : 13-15.

Mitra, A. 1999. Two new Odonata record from the Western Himalaya, India (Anisoptera : Macrodiplactidae, Libellulidae). *Notul. Odonatol.*, **5**(3) : 39.

Mitra, T.R. 1973. *Sympetrum tandicola* Singh, 1955, a synonym of *Pantala flavescens* (Fabricius) (Odonata : Libellulidae). *Ent. Res. J. Var.*, 85(1) : 30-31

Mitra, T.R. 1875. A review of India species of *Agriocnemis Selys* (Insecta : Odonata : Zygoptera : Coenagrionidae), with a note on *Agriocnemis nainitalensis* Sahni. *Dr. B.S. Chauhan Comm.* Vol. : 403-409.

Mitra, T.R. 1995. Taxonomic status of four Odonata species described by D.N. Sahni. from Western Himalaya. *J. Bengal Nat. Hist. Soc.*, New Sr. **14** (1) : 71-72.

Prasad, M. 1974. The Odonata of Garhwal Hills. *Bull. Ent.*, 15 : 41-55.

Prasad, M. 1976a. On the occurrence of *Urothemis signata signata* (Rambur) (Odonata: Macrodiplactidae) from Eastern U.P., India. *Newsl. zool. Surv. India.*, **2** (4) : 142.

Prasad, M. 1976b. Odonata of district Kangra (H.P.), *Rec. zool. Surv. India.* **71** : 95-119.

Prasad, M. 1977. On the collection of Odonata (Insecta) from U.P. Tarai district Nainital and Lakhimpur Kheri. *Newsl. zool. Surv. India,* 3(3) : 112-114.

Prasad, M. 1978. Studies on the male accessory genital structure of *Neurothemis fulvia* (Drury), *N. intermedia intermedia* (Rambur) and *N. tullia tullia* (Drury) from Western Himalaya (Odonata : Libellulidae : Sympetrinae). *Bull. zool. Surv. India,* **1**(1) : 87-90.

Prasad, M. 1988. Introduction to the external morphology of the Odnate male accessory genitalia, with descriptions of sixty three cases in North-West Indian species. *Indian Odonatol.*, **1** : 45-88.

Prasad, M. 1997. Bharitya biyadh patangon me jaiw vividhta (In Hindi). *Bharat ki Prani Bibidhta,* : 119-122.

Prasad, M. 1988. Faunal diversity in India : Odonata. *Zool. Surv. India, Calcutta.,* : 171-178.

Prasad, M. and Bhargava, R.N. 1975. Observations on the coitus of *Orthetrum pruinosum neglectum* (Rambur) (Insecta : Odonata). *Cheetal,* **17**(1) : 53.

Prasad, M. and Kumar, A. 1977a. Extension of distribution of some dragonflies (Odonata) in Western Himalaya (U.P.) *Newsl. zool. Surv. India.,* 3(4) : 170-174.

Prasad, M. and Kumar, A. 1977b. On the occurrence of *Cynacantha khasiaca MacLachlan* (Odonata : Aeshnidae) from Kumaon Hills (Western Himalaya). *Newsl. zool. Surv. India,* **3**(6) : 340.

Prasad, M. and Kumar, A. 1980. Pashchmi Himalaya ke byadh patangon ka nirikhan. (In Hindi). *Prani Jagat,* **1** : 69-81.

Prasad, M. and Kumar, A. 1981. Studies on the intraspecific variations in *Trithemis festiva* (Rambur) (Odonata : Libellulidae). *J. Bombay nat. Hist. Soc.,* **77** (2) : 238-246.

Prasad, M. and Singh, A. 1976. Odonata of Doon Valley. II. Zygoptera. *Rec. zool. Surv. India, 70* : 121-131.

Prasad, M. and Singh, A. 1994. Fauna of Conservation Areas 5. Rajaji National Park : Odonata., *Zool. Surv. India,* : 195-215.

Prasad. M. and Varshney, R.K. 1995. A check-list of the Odonata of India including data on larval studies. *Oriental Ins.,* **29** : 385-428.

Ram, R. and Srivastava, V.D. 1984. *Ortehtrum mathew* Singh and Baijal, 1954, a synonym of *Pantala flavescens* (Fabricius) (Odonata : Libelulidae). *Bull. zool. Surv. India,* **5** : 181.

Roonwal, M. L. and Verma, S.C. 1977. Re-Survey of the termite fauna of Rajasthan, India and its zoogeography. *Rec. zool. Surv. India,* **72** : 425-480.

Sahni, D.N. 1964a. Survey of the Insect fauna of Nainital, Odonata (Anisoptera). *Agra Univ. J. Res.* (Sci.), **13** (2) : 89-98.

Sahni, D.N. 1964b. Some new species of Odonata from Kumaon Hills (India). *Agra Univ. J. Res.* (Sci.), **13** (3) : 79-86.

Sahni, D.N.1965a. Studies on Odonata (Zygoptera) of Nainital. *Indian J. Ent.,* **27** (2) : 205-216.

Sahni, D.N. 1965b. Studies on Odonata (Anisoptera) of Nainital. *Indian J. Ent.,* **27** (2) : 277-289.

Sahni, D.N. 1970. Studies on the Odonata of Kumaon Pt. II. Suborder Anisoptera, Family Libellulidae (Contd.), Gomphidae and Aeshnidae. *Bull. Ent.,* **11**(2) (1972) : 121-130.

Sahni, D.N. 1972a. Studies on the Odonata of Kumaon Part I. Suborder Anisoptera, Family Libellulidae. *Bull. Ent.,* **11** (1970) : 37-54.

Sahni, D.N. 1972b. Studies on the Odonata of Kumaon Part III. Suborder Zygoptera. *Bull. Ent.* **12** (2) : 60-84.

Sangal, S.K. and Kumar, A. 1970a. Studies on the taxonomy of the larvae of Doon Valey Odonata I, *Crocothemis servilia servilia* (Drury) and 2. *Bradinopyga geminata* (Rambur) (Family Libellulidae). *J. nat. Hist.,* 4 : 33-38.

Sangal, S.K. and Kumar, A. 1970b. Studies on the taxonomy of larvae of Doon Valley Odonata II, 3. *Anax guttatus* (Burmeister); 4. *Anax immaculifrons* Rambur (Family Aeshndae). *J. nat. Hist.* 4 : 305-313.

Sangal, S.K. and Tyagi, B.K. 1985. *Calicnemia miniata doonensis,* a new subspecies from the Dehra Dun Valley (Dehra Dun, India) (Zygoptera : Platycnemididae). *Proc. First Indian Symp. Odonatol.,* : 193-197.

Schemidt, E. 1961. Ergeboisse der deutchen Afganishtan Expedition 1956 der Landessamlungen Fur Nuturkunde Karlsruche sowic der Expedition J. Klapperich, Bonn 1952-53 und. Dr. K. Lindberg, Lund (Schweden) 1957-60. Libellen (Odonata). *Beitr. Natruk. Forsch. Sudw. Dtl.,* **19** : 399-435.

Singh, A. and Prasad, M. 1974. New records of Odonata from North-West India.*J. Bombay nat. Hist. Soc.,* **70** (2) : 403-405.

Singh, A. and Prasad, M. 1975. New record of *Rhinocypha bifasciata* Selys (Odonata : Zygoptera : Chlorocyphidae) from North West India. *Newsl. zool. Survey India* **1** (1) : 3.

Singh, A. and Prasad, M. 1976a. On the identity of *Rhinocypha trifasciata* and *R. bifasciata* (Odonata : Zygoptera : Chlorocyphidae). *Oriental Ins.*, **10** (4) : 553-556.

Singh, A. and Prasad, M. 1976b. Odonata of Doon Valley-I. Anisoptera. *Rec. zool. surv. India*, **70** : 21-38.

Singh, A. and Prasad, M. 1977. Odonata (Insecta) of Corbett National Park (Uttar Pradesh : India). *J. Bombay nat. Hist. Soc.*, **73** (2) : 419-421.

Singh, A. and Prasad, M. 1980. A new species of *Tholymis* Hagen (Odonata : Libellulidae : Libellulinae) from Uttar Pradesh, India. *Zool. Beitr.*, Berlin, **26** : 65-68.

Singh, S. 1955. Entomological Survey of the Himalaya. Pt. V. On two new species of Odonata. *Agra Univ. J. Res. (Sci.)*, 4 (1) : 171-174.

Singh, S. 1963. Entomological Survey of the Himalayas. Pt. XXIV. Fourth and final annotated check-list of the insects from the North-West (Punjab) Himalayas. *Agra Univ. J. Res. (Sci)*, **12** (1) : 363-393.

Singh, S. and Baijal. H.N. 1954. Entomological Survey of Himalayas. Part II. On a collection of Odonata. *Agra Univ. J. Res.* (Sci.), 3 (2) : 385-400.

Singh, S. Baijal, H.N. Gupta, V.K. and Mathew, K. 1955. Entomological Survey of the Himalayas. Pt. XIV. Notes on some insects collected by the second Entomological Expedition of the North-West Himalaya (1955), with description of three new species of Odonata. *Agra Univ. J. Res.* (Sci.), 4 (Suppl.). : 741-766.

Srivastata, V.D. and Prasad, M. 1991. Animal resources of India : Odonata, Insecta. *Zoological Survey of India*, : 253-258.

ST. Quentin, D. 1970. Odonata aus Nepal. *Khumbu Himal.*, 3 (3) : 389-411.

Tsuda, S. 1991. A distributional list of World Odonata.: 1-362 (Osaka).

Tyagi, B.K.1982. Notes on some dragonflies from the Dehra Dun Valley (Uttar Pradesh : India). *Fraseria, No.* 3 : 9-10.

Tyagi, B.K. 1983. A preliminary list of the little known dragonflies of North-Western India. *Fraseria*, No. 5 : 18-19.

Tyagi, B.K. 1985. Dragonfly conservation in India : Present status and future prospects. *Rep. Odon. specialist, Group Int. Un. Conserv.*, No. 6 : 1-5.

Tyagi, B.K. 1997. Dragonfly wealth of India-Extant species-A preliminary systematic list. *Zoos. Print*, : 3-10.

Tyagi, B.K. 1988. *Calicnemis miniata doonensis* Sangal and Tyagi, 1984. *Fraseria*, No. 14 : 65.

Tyagi, B.K. and Tyagi, A. 1989. The story of *Calicnemia* species from Dehra Dun Valley. *Fraseria*, No. 16: 68-69.

Tyagi, B.K., Tyagi, A. and Sangal, S.K. 1986. An uptodate catalogue and bibliography of dragonfly fauna of the Dehra Dun Valley (India). *Occ. Pub. S.I.O. Nat. Off., India,* No.2 : 1-14.

Wojtusiak, J.A. 1974. A dragonfly migration in the high Hindukush (Afghanistan) with a note on high altitude records of *Aeshna juncea mongolica* Bartenef and *Pantala flavescens* (Fabr.) (Anisoptera : Aeshnidae : Libellulidae). *Odonatologica,* 3 : 137-142.

10

DESCRIPTION OF TERRITORIALITY AND REPRODUCTION OF *AGRIOCNEMIS PYGMAEA* (RAMBUR, 1842) (ZYGOPTERA : COENAGRIONIDAE)

• *B. Suri Babu*

Abstract

The territoriality and reproductive behaviour of *Agriocnemis pygmaea* (Ramb.) has been studied in detail in a perrennial pond "Dalpat Sagar", Jagdalpur Bastar District, Madhya Pradesh, India. The territoriality is strongly demonstrated by some of the males. Males copulate both Isochromatic and heterochromatic forms of females with equal consideration. Precopulatory courtship display is present and brief 4 to 7 Sec ($\bar{x}$: 4.2, N = 9). Intramale sperm translocation has occurred only after the seizure of the female and lasted for 12-18 Sec ($\bar{x}$: 15.35, N = 8). The copulatory wheel was formed during the perched condition; and stage I lasted for 20 to 42 min ($\bar{x}$: 28.05 N =14). While stage II lasted for 2 to 6 min ($\bar{x}$: 4.75, N = 14). The oviposition is endophytic and performed both in tandem and female alone. Behavioural comparisons at various levels have been drawn with the other members of the family Coenagrionidae.

Key words : *Agriocnemis pygmaea,* rendezvous, territoriality, aggressive behaviour, sperm transfer, copulatory wheel, oviposition.

Introduction

The territoriality and reprodcution behaviour of Indian Odonates have received very littel attention of odonatologists. A survey of literature revealed that Kumar and Prasad (1977) have described reproductive behaviour of *Neurobasis C. chinensis.*

Srivastava and Suri Babu (1984, 1985a, b) and Srivastava *et. al.* (1994) have studied the reproductive behaviour of *Ischnura aurora, Ceriagrion coromandelianum, Chloroneura quadrimanulata* and *Peudagrion decorum* respectively. However, detailed observations on the reprodcutive behavioural patterns on alien Zygopteran species are available (Corbet 1981; Cordero, 1989; Fincke, 1982; Meskin, 1986; 1988 and 1993 and Utzeri *et. al.* 1983). A survey of literature revealed the lack of information on the reproductive behaviour of the Genus *Agriocnemis* Selys, 1877. So, the present work was therefore undertaken to provide some information on the territoriality and reproduction of Agriocnemis pygmaea. This species is common in Gagdalpur (M.P.) India and are commonly distributed throughout India, Australia, Sri Lanka, China, Formosa, Java, Malaysia, Philippines, Singapore.

The species in the present study are smaller in size usually with abdomen 18.5-20.5 mm with hindwing 10.5-14.00 m. long. Males with postocular spots rounded and pale green in colour, thorax black on dorsum with apple green antehumeral stripus. Abdomen blackish in colour except segments 8, 9 and 10 which are brick-red coloured.

Females are dimorphic in colour, Isochromatic forms are commner than heterochromatic forms. Isochromatic females are similar to male except the thorax and the postocular spots are blue coloured. The heterochromatic forms with postocular spots violet in colour. Body dark pink with thorax black on dorsum and ante humeral stripes light violet. Abdominal segments 8, 9 and 10 are black on dorsum.

Biotope and Associated Fauna

The biotope comprised a perennial pond in Jagdalpur town in Bastar district of Madhya Pradesh, India at 19° 05′N and 82° 01′ E at about 566 m above mean sea level. The rain fed pond is surrounded by a number of houses and a metal road on southern, western and eastern sides while the main tank bund sprawling east-west on northern side adjoining. The pond covers an area of 40 hectares with catchment area of 300 hectares. The pond starts filling by the onset of south-west monsoon and becomes full towards the tagend of monsoon. During the months of August, September and October abundant odonata adult fauna of different species appear over the pond. The pond gets polluted due to bathing, washing and wallowing of cattle and other domestic uses of surrounding population. The common grasses and vegetation like *Scripus littoralies, Cyperus indicus, Lantana camera, Euphorbia* sp. and *Ipomea aquatica* growing on the bank of the pond. *Hydrilla verticillata, Potamogeton natans, P. pectinatus, Vallisneria spiralis* are common rooted submerged plants. The main floating vegetation like *Eichhornia crassipus, Nymphaea stellata* and *Trapa bispinosa* are also growing in the pond. The fallen and rotting leaves from the tress over the bank and algal filaments mainly comprised of *Spirogyra* sp. covered the surface of the shallow water. The water is found slightly alkaline (pH range 7.5 to 9). The ambient and water temperature measured in shade by celsius thermometer range between 25° to 28° C and 21 to 23° C respectively. The sumrise at 6.15 hrs and sunset at 18.00 hrs when investigations were done.

In addition to *A. pygmaea* the other species were observed to be reproductively active at the pond during the study period, are *Ceriagrion coromandelianum* (Fabr.) *Pseudagrion decorum* (Ramb.), *Cercion malayanum* (Sel.), *Enallagma parvum* (Sel.), *Ischnura senegalensis* (Ramb.), *Ictinogomphus rapax* (Ramb.), *Anax guttatus* (Burm.), *Anax. p. partthenope* (Sel.), *Brathythemis contaminata* (Fabr.), *Crocothemis s. servilia* (Drur.), *Neurothemis i. intermedia* (Ramb.) and *Trithemis pallidinervis* (Kirb.).

Materials and Methods

The observations were made from 15th September to 30th October, 1977. the tenerals, mature males and females were captured with an insect net and marked laterally on the thorax and forewings with permanent ink and released again in the field. Some tenerals which have emerged from the larva collected from the pond were also marked and released. A total of 245 teneral males, 118 teneral females 530 matured males and 410 matured females were marked. The marking of damselflies seemed to be not affecting their behaviour. All observations were made recording in the morning on the arrival of damselflies and continued upto late evening until all the damselflies vacated the site, using a field binocular. All the activities and behaviour were noted daily in a note book, exept on rainy days. At times damselflies in tandem and incopula were captured for verification of their markings. The duration of various events given herein were recorded with the help of a stopwatch. The length of territorial occupations and maturation period were abtained from the recorded data of the dates marked, and recaptured of each individuals. In addition to the study period fortnightly visits were made for one year to know the flight period.

Observation

Rendezvous

The male and females were found in the littoral zone of the pond, where they perform reproductive activity and oviposition by females. The males used to arrive at the breeding place earlier by 9.00 to 10.00 hrs and each male occupied a base perch which is about 15 to 30 cm above the water sruface with its wings folded, abdomes held horizontally and with the face towards the back of pond. The base perch may be an emergent or floating water plant or any object emergent from water. The females reached the breeding site later between 11.00 to 12.00 hrs.

Territoriality

The territorial area was determined by the range of distance beyond which the resident male didnot respond to intruder. The territorial area of male *A. pygmaea* was about 30 to 60 cm radius of a cirenlar area. Out of 213 marked males recoved in the field 166 individuals of males, *i.e.* 78% are territorial and 47 individuals of males *i.e.* 22% are non-territorial.

The non-territorial and were noticed cruised in the territory of males, instead of maintaining territories and at time they perch near territory. The non-territorial males preferred to establish territories on the vegetation near the bank and territorial males were on the vegetation mostly over water and thus competition sites were diminished. The non-territorial males often observed sharing their perches with 4 to 6 other conspecific and 1 to 2 heterospecific males like *A.lacteola* (Sel.) It thus appears that few males of *A. pygmaea* are non-territorial.

The territorial males exhibited five types of behaviour towards an intruder :

1. *Threat posture*

The resident male raising his entire abdomen without leaving its perch when a conspecific or hetero specific damselfly of the same size like *Enallagma parvum* (Sel.) approaches his perch. The male terminates this type of aggressive behaviour when the intruder retreats.

2. *Approach without aggressiveness*

These are like patrolling flights. If any moving object whether it is a conspecific or heterospecific male or female of any insect approaches towards the resident male, the resident male flew towards the moving object and returns to the base perch immediately.

3. *Approach-chase*

The chase display lasted for a few seconds. If any intruder enters into the boundary of the territorial male, the resident male leaves the perch and drives away the intruder to the boundary of its territory and comes back to its perch.

4. *Approach-threat-flighting (Including circle flights)*

When a conspecific male enters the territory and approaches his perch, the male performs alternate and repeated confrontation and vacillation display often moving in a cirrcular path and ceases with the withdrawal of intruder. During this display the resident male travels 1 to 2 m and duration of display is 3-6 min.

5. *Alarm reaction*

The resident male abandons his base perch if any large sized dragonflies like *Anax guttatus* or *Ictinogomphus rapax* approached. The resident male quits his perch and moves backwards upto a distance of 30-50 cm hovers for a few seconds there and returns to his perch. If the intruder persists in his territory for a long time the resident moves to a new perch and returns as soon as intruder vacates from the territory.

Thus, it is observed that territory was strongly defended against both conspecific and heterospecific males but not interferred with conspecific tandem pair and allowed it to oviposit within the territory. So, the aggressive displays of male is formal and not harmful physically.

Flight activities of territory : The territorial males performed the following flights during their stay in territories :

1. Patrol flights

The resident male performed short flights from the base perch to vegetation of the bank and returned back to the base perch. The distance travelled during this flight was 3 to 5 m and duration was about 15 to 20 Sec.

2. Shifting flights

The male shifts from one perch ot another nearby perch with in a radius of 1 to 2 m. The distance travelled in this flight was 1 to 2 m and duration of each flight was 6 to 12 Sec. These flights may be due to the disturbance caused to the base perch due to water current and wind.

3. Flight forwards

These flights were towards moving objects and if the intruder was conspecific male, the flight followed by an aggressive display. The male chase the conspecific male intruder upto the distance of 3 to 5 m and return back to the perch. The duration of this flight was 9 to 18 Sec.

4. Avoidance flights

The male performed this type of flights due to the disturbance caused by the large sized dragonflies, when any large sized dragonfly passing close to the territory of a resident, the male left the perch and moves by jerking backwards and returned back to its perch when the dragonfly flew away from his territory. The distance travelled during this flight was 50 to 70 cm and duration 12 to 30 Sec.

5. Feeding flights

These flights occurred whern a male separated from ovipositing female after oviposition at around 16.00 hrs. The male moved towards the vegetation of the bank and search for small insects for feeding. The male directly captured the insect with its mouth or with its fore-legs and then pass on to the mouth. During the morning hours 7.00 to 9.00 males were not observed feeding among the vegetation. In the territory also males were not observed on feeding.

Territorial attachment

The resident male stayed most of the daytime in the territory and it left the territory at around 15.00 hrs, followed by tandem link with female if it succeeds. Out of six males marked, on 3rd Sept.'97, three were found on the successive days of 4th, 5th and 6th Sept.'97 in their territory and disappeared. But one of them found on 10th and 12th Sept.'97. Out of 10 males marked on 10th Oct'97 only two occupied the territory on the same day, one appeared on 15th and 17th Oct'97 in their original territory. Out of 15 males marked on the 18th Oct'97 four returned to the original territory on the same day, two each returned, on 19th and 20th Oct'97, while one appeared on 24th Oct'97. Thus, males occupied their original territories for not more than 2 days. The males change their territories every day within a distance of 6 to 10 m of their previous territories.

Precopulatory Courtship

As soon as a female arrived in its territory, the resident male at once left the base perch and followed her. No specific courtship behaviour was displayed by the male except that he flew along the side of the female for some distance. The female when receptive, responded by flying along with the male 4-7 Sec ($\bar{X}$=4.2, N=9) and formed tandem link during the flight. Otherwise she escaped swiftly away. If any other conspecific male approached the female of the tandem, she curved her abdomen upwards as a negative signal. The tandem pair flew for sometime over a period ranging between 10-15 min ($\bar{X}$: 12, 25, N = 19). The territorial and non-territorial males recognise both isochromatic and heterochromatic forms of females in an equal consideration and forms tandem link.

Intra Male Sperm Translocation

The pair in tandem settled down on some vegetation, usually similar to the base perch growing in the water or on the bank of the pond. Only the male provide support on the plant whereas the female body hanged vertically down and the female did not provide any support to the male while the sperm translocation lasted. The intramale sperm translocation process lasted for 12 to 18 Sec ($\bar{X}$: 15.35, N : 8).

Copulatory Wheel

After intramale sperm translocation took place, the male relaxed his abdomen and in this process the female slightly pushed backwards. The abdomen of the female then curved ventrally forward so that its gonopore which is situated between the eighth and ninth sterinites was brought in juxtaposition with the male copulatory apparatus. A copulatory wheel was thus formed when the tandem always in perched condition and never it was in flight. More often than not, the female was successful in execting the copulatory wheel formation in its very first attempt. In *A. pygmaea* only two stages of wheel position observed. In the first stage rhythemic up and down movements of the male abdomen were noticed. In this stage the male alternately curved and straightened his abdomen. The duration of this first stage was 20-42 min ($\bar{X}$ = 28.05; N = 14). It is presumed that in this stage removal of sperm deposited previously by another male has taken place. The second stage followed by the first stage of the abdomen of male stopped for 2-6 min ($\bar{X}$: 4.75; N = 14). In this stage it is to be presumed that transfer of sperms to the female occurred. The wheel position was observed for 14 pairs, 13 reached wheel in first attempt, one after a short trail. In wheel position only male clapped his wings, when intruder approached his perch and also at time sin the absence of intruder.

Exploratory Flight

After the disruption of the copulatory wheel the tandem link maintained and the pair took rest for about 1-2 min ($\overline{X}$-1.52; N = 17) at the site of copulation. The tandem then performed post copulatory flight for 20-30 min ($\overline{X}$ = 22.40; N=18). One pair shifted about 32 times in 24 min and another pair 31 times in 28 min. The pair shifted for 2 to 4 m from the site of copulation to the effective oviposition site. During this period the female explored for suitable oviposition sites by touching the vegetation with her ovipositor.

Oviposition

The oviposition was always in tandem and was at the surface of the water. On finding a suitable site, the female alighted and perched on the horizontal stem of leaf of the water plant. The male remained verically up-right in the air still grasping the female thorax by his anal appendages. Occasionally male leaned forward to rest on the plant. The female then arched and extended her abdomen to insert her ovipositor into the plant tissue for laying eggs. The female alone flapped her wings when she changed sites. Apparently the female controlled the oviposition process while male only escorted her in tandem. Only on two occasions the wing beats were observed by male when a conspecific pair aproached them. The female oviposited at a site for 2-3 min and moved to another site. The other site could be some other spot of the same plant or an altogether a different plant. The total period of oviposition including interruptions has lasted for 21-80 min ($\overline{X}$ = 56.50; min N=22). Neither male nor female got submerged under water as the eggs were laid on the stem or leaf.

It was noticed that the tandem pair allowed 2 to 3 of other conspecific pairs to lay eggs in a circular area of 40-60 cm diameter. This species preferred horizontal aquatic plant material in the study area for oviposition. These species preferred *Hydralla verticillata* and *valisneria spiralis* plants for oviposition even though different aquatic plants growing in the same area.

The male often spearated from the ovipositing female and guarding her by hovering near the site or perching nearby until the termination of oviposition. On the occasion another male grabed the ovipositing female and formed tandem link.

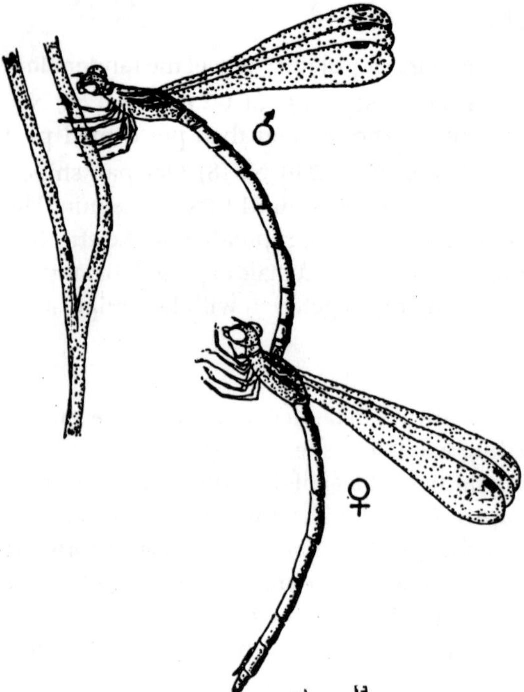

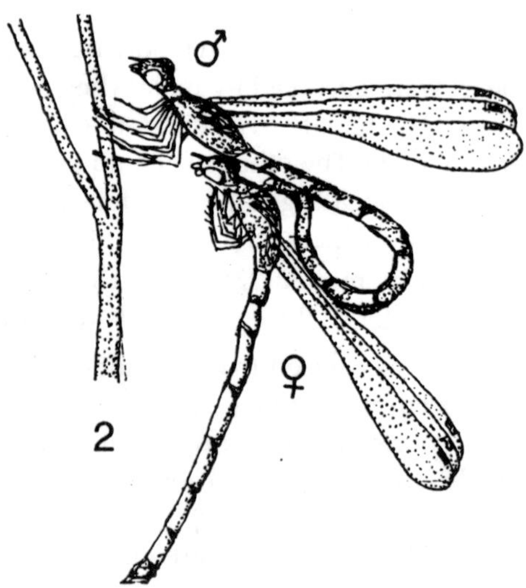

2

Figs. : Reproductive behaviour of *Agriocnemis pygmaea* :
(10.1) Male and female in tandem link;
(10.2) Intramale sperm transfer

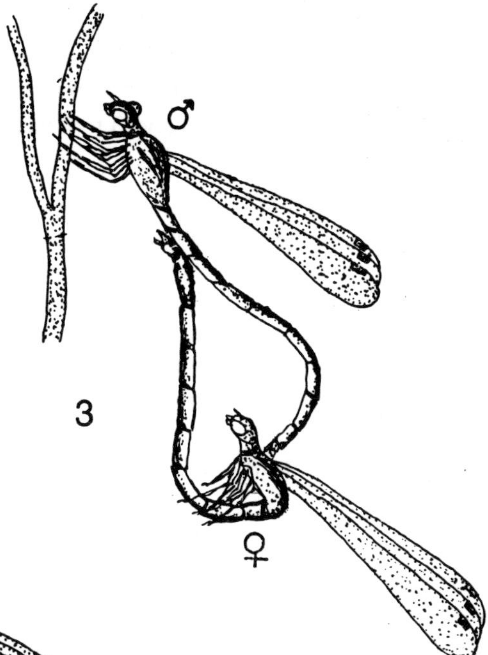

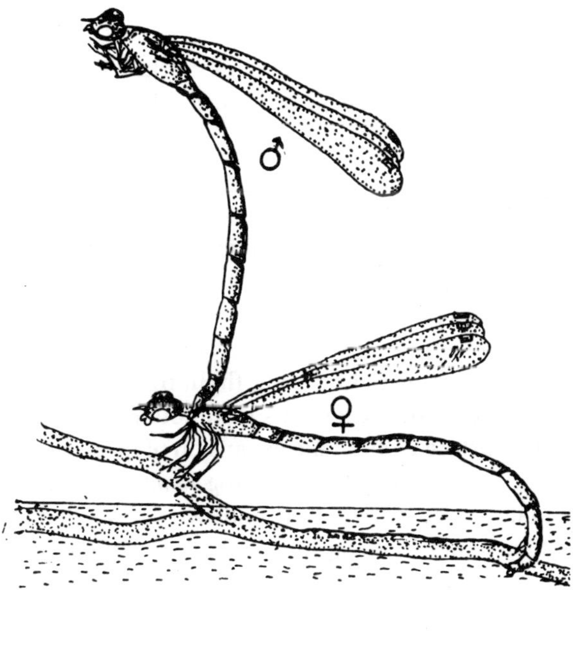

Figs. : (10.3) Tandem wheel position;
(10.4) Oviposition female with male in upright tandem.

Pre-reproductive period

The pre-reproductive period was estimated for 12 marked males and 13 females. The maturation period lasted for males was 4 to 6 days ($\bar{X}$:4.52; N = 12) and 5 to 7 days for females. ($\bar{X}$: 6.22; N = 13).

An information on the flight season was obtained by direct observation of emergence and presence of adult population during one year period. Flight season started from the first week of March and lasted upto the middle of November, 1997. The population was abundant during postmonsoon period, *i.e.* September and October and started decreasing on wards. No adults were seen at biotopes during the month of December, January and February.

Mating frequencies

It is clearly understood from the present study that males and females of *A. pygmaea* mate more than once in their lifespan. Out of 15 marked males M 18, M 37, M 52, M 63, M 81, M 92, M 118 mated twice, M21, M59, M102, M141 mated 4 times, M92, M121, M156, M167 mated 5 times. Out of 10 females marked F18, F47, F63, F123 mated twice, F09, F37, F83, mated thrice, F122 mated 4 times, F153 mated 5 times.

Mortality at breeding site

The mortality of male and female of *A. pygmaea* at breeding stie was observed due to several causes and predation was the main cause. Three tandem pairs were killed by a spider when they were traped in a web of spider in the bushes on the bank of the pond. One tandem pair was caught during flight by a Libellulid and another by an Aeschinid. Six tenerals were found dead and floating on the water surface without the expansion of their wings. Out of 9 pairs observed 7 were eaten by a frog *Rana* spp during oviposion when male settled horizontally and two pairs killed in upright tandem. One female was captured by a fish while it was ovipositing alone.

Discussion

The phenomenon of territoriality exhibited by males has been reported for many odonates (Corbet, 1962), but its absence has been recorded amongst members of the family Platycnemididae (Heymer, 1966), most Lestidae (Bick, 1972) and many Coenagrionidae (Fincke, 1982; Srivastava and Suri Babu, 1984) of the suborder Zygoptera. The behaviour of the mature male of the same species may differ due to its locality and other environmental conditions. The behaviour of males of *Ceriagrion coromadelianum* was reported as non-territorial by Srivastava and Suri Babu, 1984 at Sagar (M.P.) India at 23° 50′ N and 78° 35′ E. where as Prasad (1990) observed that the males of the same species as territorial at Paikpara, Calcutta , India at 88° 30′ E and 22° 55′ N. In the present study also the phenomenon of territoriality was exhibited by 78% males of *A. pygmaea* and 22% males were non-territorial. Waage (1979b) has also reported territorial and non-territorial males in the population of *Calopteryx maculata.*

Corbet (1980) stated that aggressive behaviour of the mature male at rendezvous are predominently against the conspecific males. In the present study *A. pygmaea* displays aggressive behaviour not only towards the conspecific males, but also towards the heterospecific intruders and in an equally strong terms. The aggressive behaviour is formal and not harmful physically, while *Pseudagrion decorum* (Ramb.) a coenagrid harms the intruder physically (Srivastava, *et. al.* (1994). The males of *A. pygmaea* keeps away from such combat in the territory.

As per Higashi and Ueda (1982) the aggressive behaviour of males of *Calopteryx cornelia* was classfied to the degree of response to intruders are four levels : (1) approach-without aggressiveness, (2) approach-chase, (3) approach-threath-chase and, (4) approach-threat-fighting (including circular flight). The territorial males of *A. pygmaea* exhibited above mentioned 1st, 2nd and 4th levels of response against intruders along with threat postures. Meskin (1989) reported that males of *Pseudagrion. i. inconspicuum* and *P. citricola* also exhibited the fourth level of aggressive behaviour.

Corbet (1962) stated that distinctive colours of the males may perhaps be of adoptive values in enabling them to recognise males of their own species and thereby reduce unnecessary interaction with those of others. The males of *A. pygmaea* also appear to be distinguished by their facial colouration of pale green. During aggressive confrontation flights males facing each other were not observed as they have no brilliant facial colours. The non-territorial males also have pale coloured faces which does not display aggressive behaviour. A similar observation has been made by Meskin (1989) that males of *Pseudagrion salisburyense* has no colourful facial areas and does not display to the conspecific males.

Meskin (1986) reported that males of *P. hageni tropicanum* occupy the same territories for upto 39 consecutive days and are most agressive and males of *P. kersteni* establish new territories each day. The males of *A. pygmaea* also forms new territories each day but occasionally stays on the same territory for a maximum of two days.

Meskin (1989) reported that male of *P. citricola* is more sedentary on its territory and feeds there, where as *P. i. inconspicuum* is more active and does not feed while in the possession of the territory. The territorial males of *A. pygmaea* does not feed while they are in territory and feeds only in the evening hours after the termination of guarding of ovipositing female.

Corbet (1980) believes that precopulatory courtship is displayed by the males of only those species who are defending their territories. So far three families Calopterygidae (Kumar and Prasad, 1977), Coenagriidae (Robet, 1958) and Protoneuridae (Srivastava and Suri Babu, 1985) were known in which precopulatory courtship is displayed by males. The male of *A. pygmaea* seems to be exception on both the counts. Being a member of Coenagriidae and strongly showing the territoriality, its males do not display prominent precopulatory courtship behaviour.

In some species the intramales sperm translocation is accomplished before the formation of tandem link (Kumar and Prasad, 1977) and in other species it occurs immediately after the assumption of tandem link (Bick, 1972). In this respect *A. pygmaea* comes under second category, in which the translocation occurs only in the perched condition after the tandem pair has settled down.

Cordero and Miller (1992) distinguished three stages in the copulation of *Ischnura graellsii.* They suggested that stage I concerned with sperm removal from female and that sperm was transferred during stage II. The stage III was thought to follow sperm transfer and represented a resting stage. In the present study of *A. pygmaea* only two clear phases of whell position could be distinguished. In the first was rhythmical abodominal movement may be typical of sperm removal and in the second stage was without movements may be indicative of spermtransfer. As reported by of Waage, 1979a and b for *Calopteryx maculata.*

Cordero (1989) reported that andromorph females of *Ischnura graellsii* (Ramb.) which are identical to the male in colouration are not recognized as females by males and these females mimic the behaviour of males. Contrary to this, males of *A. pygmaea* recognized both isochromatic and heterochromatic females as females and copulate with them in equal consideration.

The Zygopteran tandem which oviposit at small, shallow waters often exposed to a high pressure from predators hunting on the water surface (Rehfeldt, 1991). The author agreed with this statement as predation risk was greatly increased in tandem oviposition if male settle horizontally rather than with upright position of male which make a speedier escape as observed the tandem in the case of *A. pygmaea.*

Acknowledgements

The author expresses his gratitude to Prof. B.K. Srivastava, Sagar (M.P.), Dr. V.K. Srivastava, Allahabad, Dr. Arun Kumar, Dehra Dun and Dr. M. Prasad, Calcutta, India for providing literature and encouragement through the course of this study and to the Supdt. of Police, Dist. Bastar (M.P.) for providing facilities. The author indebted to the staff of Sciene of Crime Unit, Jagdalpur for their assistance in the field.

References

Bick, G.H. 1972. A review of territorial and reproductive behaviour in Zygoptera. *Contact. Br. ned. Libellen Onderz* 10 (Suppl.) : 1-14.

Corbet, P.S. 1962. A biology of dragonflies. Witherby. London.

Corbet, P.S. 1980. Biology of Odonata. *Ann. Rev. Ent.* 25 : 189-217.

Cardero, A. 1989. Reproductive behaviour of *Ischnura graellsii* (Rambur) (Zygoptera : Coenagrionidae), *Odonatologica* 18(3) 237-244.

Cardero, A. and Miller, P.L. 1992. Spermtransfer, displacement and precedence in *Ischnura graellsii* (Odonata : Coenagrionidae) *Behav. Ecol. Sociobiol.* 30 (3/4) : 261-267.

Fincke, O.M. 1982. Life time mating success in a natural population of the damselfly, *Enallagma hagnei* (Walsh) (Odonata : Coenagrionidae). *Behav. Ecol. Sociobiol.* 10(4) : 293-302.

Heymer, A. 1966. Etudes comparees du comportement inne de *Platycnemis acutipennis* selys 1841 et de *Platycnemis latipes* (Rambur 1842 (Oden. Zygoptera). *Annls Soc. ent. fr.* (N.S.) 2:39-73.

Higashi, K. and Ueda, T. 1982. Territoriality and movement pattern in a population of *Calopteryx cornelia* (Selys) (Zygoptera : Calopterygidae). *Odonatologica* 11(2) : 129-137.

Kumar, A. and Prasad, M. 1977. Reproductive behaviour in *Neurobasis chinensis chinensis* (Linn.) (Zygoptera : Calopterygidae). Odonatologica 6:163-171.

Meskin, I. 1986. Territorial behaviour in *Pseudagrion hageni tropicanum* Pinhey (Zygoptera : Coenagrionidae) *Odonatologica* 15:157-167.

Meskin, I. 1989. Aspects of territorial behaviour in three species of *Pseudagrion* Selys (Zygoptera : Coenagrionidae) *Odonatologica* 18(3) : 253-261.

Meskin, I. 1993. Territorial behaviour in *Pseudagrion kersteni* (Gerstaeker) (Zygoptera:Coenagrionidae) *Odonatological* 22 (1) : 63-70.

Prasad, M. 1990. Reproductive behaviour of *Ceriagrion coromandelianum* (Fabricius) and *Pseudagrion rubriceps* selys (Zygoptera : Coenagrionidae) *Ann. Entomol* 8(2) : 35-38.

Rehfeldt, G.E. 1991. The upright male position during oviposition as an anti-predator response in *Coenagrion puella* (L.) (Zygoptera : Coenagrionidae). *Odonatologica* 20(1) : 69-74.

Robert, P.A. 1958. Les libellules (Odonates). Delachaux and Niestle, Neuchatel.

Srivastava, B.K. and Suri Babu B. 1984. Some observations on oviposition of *Ischnura aurora* (Brauer) in Indian biotopes (Zygoptera : Coenagrionidae) *Fraseria* 6:24.

Srivastava. B.K. and Suri Babu B. 1985. A Reproductive behaviour of *Ceriagrion coromandelianum* (Zygoptera : Coenagriidae) *Proc. First Indian Symp Odonatol* : 209-216.

Srivastava B.K. and Suri Babu B. 1985 b. On some aspects of reproductive behaviour in *Chloroneura quadrimaculata* (Rambur) (Zygoptera : Protoneuridae). *Odonatologica* 14(3) : 219-226.

Srivastava V.K., Srivastava B.K. and Suri Babu B. 1994. The behaviour of Reproduction and oviposition in *Pseudagrion decorum* (Rambur) Zygoptera : Pseudagriinae) in central India. *Adv. Oriental. Odonatol.* 77-84.

Utzeri, C., Falchetti, E and Carchini. G. 1983. The reproductive behaviour in *Coenagrion lindeni* (Selys) in Central Italy (Zygoptera : Coenagrionidae) *Odonatologica.* 12(3) : 259-278.

Waage J.K. 1979a. Dual function of the damselfly penis : sperm removal and transfer. *Science.* 203:916-918.

Waage, J.K. 1979b. Adaptive significance of postcopulatory guarding of mates and non-mates by male *Caloperyx maculata* (Odonata) *Behav. Ecol. Sociobiol.* 6:147-154.

Glossary

Technical Terms

aggressive
conspecific
copulatorywheel
diamorphic
endophitic
heterochromatic
heterospecific
isochromatic
precopulatory courtship
rendezvous
reproduction
spermtranslocation
territoriality
oviposition

Genus & Species

Agriocnemis
Anax
Aurora
Brachythemis
Calopteryx
Cercion
Ceriagrion
Chinensis
Chloroneura
Citricola
Coenagrionidae
Cornelia
Contaminata
Coromandelianum
Crocothemis
Decorum
Enallagma
Graellsii
Guttatus
Hegeni
Ictinogomphus
Intermedia
Ischnuva

Inconspicuum
K. ersteni
Lactoela
Malayanum
Maculata
Neurobasis
Neurothemis
Pallidinervis
Parvum
Parthenope
Platycnemididae
Pseudagrion
pygmaea
Rapax
Salisburyense
Senegalensis
Servilia
Trithemis
Tropicanum
Quadrimaculata
Zygoptera

11

Comparative Studies on Eco-energetics of Nymphs of Anisoptera and Zygoptera (Odonate) in the River Mayurakshi in Santal Pargana, India

• *Arvind Kumar*

Abstract

Ecological energetics of certain dragonfly and damselfly nymphs of the river Mayurakshi in Santal Pargana (Jharkhand) have been estimated from January to December, 2000. The calorific values of anisopteran nymphs was recorded minimum in January (6.004±0.080 K cal/g dry wt.) and maximum in July (6.510±0.555 k cal/g drywt.). The zygopteran nymphs also showed variations and found maximum in June (5.955±0.080 k cal/g dry wt.) and minimum in January (5.515±0.105 k cal/g dry wt.). The possible factors responsible for these fluctuations in energetics have also been extensively discussed in the text.

Key words : Eco-energetics, Trophic level, Calorific value, Ash-fraction, River Mayurakshi.

Introduction

Eco-energetics concerns with the fixation of atmospheric carbon and its elaboration into organic compounds by primary producers, the subsequent flow and partitioning of fixed carbon between and within trophic levels of consumers and its eventual return to the atmosphere in the form of carbon dioxide. One of the main concerns of International Biological Programme (IBP) was to quantify parameters in ecological energetics in order to clarify such relationships within ecosystems, to enable objective comparisons to be made between ecosystems and hopefully to provide a greater understanding of natural complexity *via* an assessment of its functional interplay (Luxton, 1975). In freshwater ecosystems, the net primary production, *i.e.* the energy rich organic materials stored in aquatic autotrophs in excess of respiratory utilization by the phytoplankton, may be consumed by the herbivores living above in food chain. The relative contributions of the aquatic biota in allocating this energy differ considerably from place to place.

Ecological energetics being concerned with energy transformation by ecological units and is one of the important aspect of ecology, as it makes a significant contribution to our understanding of how nature works. The effective running of ecological systems depends upon the transfer of energy from organism to organism and energy transfer of this type constitute the food chains. The flow of energy along the food chain through the ecosystem and the manner and efficiency with which it is used is an ecological process of utmost importance. The basic unit studied in this type of work may vary - it may be species throughout its life cycle, the species population, the community, the ecosystem, a single trophic level or a single food chain.

Odonate larvae form an important component of zoobenthos of natural eutrophic wetlands. It acts as biological control agent for many known and unknown noxious insects and on the other hand devoures the fingerlings and fish-fries due to its predatory habits and it imposes a trenchent problem for aquacultural strategies. Though many reports are now available indicating the community structure (Kumar, 1973a and b; Kumar and Khanna, 1983), population dynamics (Kumar, 1995, Kumar and Kumari, 1999), secondary productivity (Roy, 1984; Kumar, 1995 and 1996) food and feeding habits (Kumar, 1994 and 1996), no attempt has been made to understand the ecological energetics in odonate larvae. Therefore, the present investigation was carried out to estimate the energy contents of certain odonate larvae in a wetland of Santal Pargana (Jharkhand).

Materials and Methods

A monthly collections of odonate larvae, from January to December of 2000 was made with an insect collecting net made of nylon cloth. The larvae of each hauling were sorted out in laboratory and were placed in an oven at 60°C till constant weight of material were reached. The higher temperature was applied at the begining to kill the mobile larvae. The loss in weight by drying was taken as the weight of water. The dried larvae homogenized in a mortar and the powder was pestled with distilled water to form pellets of approximately one gram under a pellet forming maching. The pellets were dried again in the oven for atleast before burning in the Oxygen Bomb Colorimeter. Three determinations were made for each larvae. In the drying process of the material, precaution was taken not to overheat the larvae since this method may cause under-estimation of the energy values of the material due to liberation of fatty acids from the decomposing lipids.

The maximum temperature difference was obtained in 6-7 min after the explosion. The ash content of the larvae was determined by weighing the residual in the cup after burning. Before pressurization, the bomb was flushed with oxygen (3 times) to remove nitrogen from the bomb. Taking into account, the ash content of the samples, the ash free energy values (organic fraction) of the materials was calculated. Finally, the energy contents of the larvae were determined according to Golley (1960).

Results

The larvae belonging to Anisoptera and Zygoptera were taken into account for estimation of energy contents. The common and most dominant species of larvae in a pond of each suborders are as follows :

Anisoptera

1. *Potamorcha obscura*
2. *Zyxomma petiolatum*
3. *Mesogomphus lineatus*
4. *Pentala* sp.

5. *Anax* sp.
6. *Cordulegaster* sp.

Zygoptera

1. *Rhodischehnura nursei*
2. *Ischnura senegalensis*
3. *Ischmura delicata*
4. *Lestis* sp.
5. *Agriocnemis* sp.

The seasonal changes in the energy contents of the above larval froms from January to December, 2000 have been shown in Table 11.1 and 11.2.

Discussion

The present study revealed that the energy content of the odonate larvae varies from season and also from month to month. The increase in the calorific values in the winter and summer may have been related to changes in the physiological conditions of these larvae. In winter season, most of these larvae remain in an inactive state or in diapause due to short photoperiod and low temperature. They usually store large amount of fat in their bodies during this phase. In summer, most of these larvae attain their sexual maturity and breed. Therefore, there was an increase in new individual in their population in this season. These young ones have comparatively more calorific values than old ones and thus high calorific values were recorded. The variation in the calorific contents of these larvae in present studies was also influenced by the calorific equivalent of daily ration available. This confirms the views of Rodgers and Quadri (1977).

It has also been found that the changes in the calorific contents were affected largely through variation in relative size of the ash-fraction. However, the calorific contents of the organic fraction remained relatively more or less constant. Changes in the calorific contents of the organic fraction may reflect changes in the amount of lipid contents (Gerking, 1955, Schindler *et. al.*, 1971). The odonate larvae having sclerotised exoskeleton have low calorific value and have a high ash-contents principally in the form of $CaCO_3$. In the present investigation it was found that the species of Anisoptera have relatively high ash-content and comparatively lower calorific value. This view was also supported by Atkinson and Wacasey (1976). Thus, the calorific values of an animal depends, to a considerable degree, on the fat content and to a lesser extent on the mineral content of the body.

Table 11.1 : Monthly variations in calorific contents of Anisopteran larvae from January to December, 2000

Months	Calorific value (Kcal/g. dry wt.)	Ash fraction (% dry wt.)	Organic fraction (Kcal/g dry wt.)
Jan	6.004 ± 0.080	0.075 ± 0.020	5.831 ± 0.085
Feb	6.310 ± 0.008	0.079 ± 0.000	5.525 ± 0.006
Mar	6.220 ± 0.155	0.065 ± 0.000	6.450 ± 0.155
Apr	6.260 ± 0.450	0.066 ± 0.000	6.220 ± 0.051
May	6.194 ± 0.375	0.080 ± 0.000	6.210 ± 0.315
Jun	6.430 ± 0.108	0.068 ± 0.000	6.525 ± 0.120
Jul	6.510 ± 0.555	0.700 ± 0.000	6.610 ± 0.555
Aug	6.325 ± 0.175	0.066 ± 0.001	6.215 ± 0.178
Sep	6.225 ± 0.015	0.069 ± 0.001	6.315 ± 0.002
Oct	6.470 ± 0.800	0.053 ± 0.008	6.223 ± 0.065
Nov	6.470 ± 0.185	0.078 ± 0.000	6.415 ± 0.180
Dec	6.005 ± 0.075	0.076 ± 0.001	6.425 ± 0.140

Table 11.2 : Monthly variations in calorific contents of Zygopteran larvae from January to December, 2000

Months	Calorific value (Kcal/g. dry wt.)	Ash fraction (% dry wt.)	Organic fraction (Kcal/g dry wt.)
Jan	5.515 ± 0.105	0.000 ± 0.000	5.515 ± 0.105
Feb	5.545 ± 0.035	0.000 ± 0.000	5.545 ± 0.035
Mar	5.520 ± 0.034	0.000 ± 0.000	5.520 ± 0.034
Apr	5.530 ± 0.115	0.005 ± 0.060	5.530 ± 0.050
May	5.530 ± 0.135	0.004 ± 0.001	5.526 ± 0.001
Jun	5.995 ± 0.080	0.000 ± 0.000	5.955 ± 0.080
Jul	5.670 ± 0.012	0.085 ± 0.000	5.550 ± 0.120
Aug	5.885 ± 0.125	0.056 ± 0.002	5.802 ± 0.002
Sep	5.335 ± 0.980	0.000 ± 0.000	5.335 ± 0.028
Oct	5.410 ± 0.150	0.025 ± 0.001	5.388 ± 0.001
Nov	5.340 ± 0.155	0.042 ± 0.001	5.225 ± 0.001
Dec	5.550 ± 0.015	0.000 ± 0.000	5.550 ± 0.015

The availability of abundant food materials in the habitat of these larvae is another factor regarding the seasonal variations in the calorific value. The calorific values also depend upon the relative size of the animals (Golley, 1959).

Therefore, literature available on the calorific values of larvae suggests that it exhibits considerable seasonal variations. These variations are to some extent changes in the fat content of the body and physiological state of the larvae and also change in the ecological conditions of the environment.

References

Atkinson, E.G. and Wacasey, J. W. 1956. Calorific values of zoobenthos and phytobenthos from the Canada, Fish and Mar. Ser. Technical Report No. 632. Environ. Canada.

Gerking, S.D. 1955. Influence of rate of feeding on body composition and protein metabolism of blue gill sunfish. Physiol. Zool. 28: 267-282.

Golley, F.B. 1959. Table of caloric equivalents. Mimeogr. University, Georgia, 74 p.

Golley, F.B. 1960. Energy dynamics of a food chain of an old field community. Ecol. Mongr. 30: 187-206.

Kumar, A. 1973a. Description of the last instar larvae of odonata from the Dehra Dun Valley (India) with notes on biology. I. (Suborder-Zygoptera). Oriental Insects. 7: 83-118.

Kumar, A. 1973b. Description of the last instar larvae of odonata from the Dehra Dun Valley (India) with notes on biology. II. Suborder-Anisoptera. Oriental insects. J. 2 : 291-331.

Kumar, A. 1994. Analysis of gut content and dimension of food items of certain dragonfly nymphs of the river Mayurakshi in Santal Pargana, Bihar, Proc. Nat. Aca. Sci. B64: 269-273.

Kumar, A. 1994. Population density, species diversity and evenness of dragonfly larvae in a freshwater pond of Santal Pargana, India. J. Freshwat. Biol. 6: 231-235.

Kumar, A. 1995. Species composition and population dynamics of odonata larvae in a fish farming pond of Santal Pargana, Bihar. Proc. Nat. Aca. Sci. B65: 401-411.

Kumar, A. 1996. Comparative studies on stomach content and forage ratio of zygopteran and anisopteran larvae in a fish pond of Santal Pargana, Bihar. Proc. Nat. Aca. Sci. B66 : 315-323.

Kumar, A. and Khanna, V. 1983. A review of the taxonomy and ecology of odonata larvae from India. Oriental Insects. 17 : 127-157.

Kumar, A. and Kumari, P. 1999. Seasonal population dynamics of zygopteran larvae in a fish pond. Fraseria. 6 : 3-8.

Luxton, M. 1975. Studies on the oribatid mites of a Danish beech wood soil II. Biomass, calorimetry and respirometry. Pedobiologia. 15: 161-200.

Rodgers, D. W. and Quadri, S.U. 1977. Seasonal variations of calorific values of some littoral benthic invertebrates of the Ottawa River. Ontario Canad. J. Zool. 55: 881-884.

Roy, S.P. 1984. Studies on gut content analysis of odonate nymphs in a freshwater fish pond at Bhagalpur (Bihar). Entomon. 9 : 25-29.

Schindler, D.W.; Clark, A.S. and Grey, J.R. 1971. Seasonal calorific values of fresh water zooplankton as determined in a phillipson bomb calorimeter for small samples. J. Fish. Res. Board. Canada. 28 : 559-564.

12

Synopsis of Progress in Taxonomical Studies on Odonata in India

• *A.R. Lahiri*

Publication of the Fauna of the British India, Odonata, volumes 1-3 by Late Dr. F. C. Fraser, (1933-1936) is one of the landmarks in Indian Odonatology. In these Fauna volumes for the first time references to all earlier works were consolidated and put together with descriptions of all taxa the then known from the country and keys for their identification under one cover.

However, much additional information has generated in all fields of Indian Odonatology since the publication of the Fauna volumes, as a result of contribution of several workers both from India and abroad. In this article an attempt to summarise the progress achieved in the field of taxonomical studies (adults) on Indian Odonata has been made.

Outline of Progress in Taxonomical Studies

Generally speaking, trends of progress in taxonomical studies on Indian Odonate fauna has involved addition of new taxa, some changes in generic nomenclature, synonymisation of some species and enhancement of existing knowledge of known taxa. Outline of more important of such studies for different family are discussed in the following :

Order **Odonata**
Suborder **Zygoptera**
Family **Coenagrionidae**

Sahni (1964b) described *Onychargia indica* and *Himalagrion pithoragarhicum* from Uttar Pradesh. Chowdhary and Das (1975) Described *Coenagrion kashmirus* from Jammu and Kashmir. Kumar and Prasad (1978) described *Agriocnemis corbeti* from Uttar Pradesh while Peters (1981) described *A. keralensis* from Kerala. Tsuda (1991) reported *Pyrrhosoma elisabethi* Schmidt 1948 and *Enallagma risi* Schmidt 1961 from India.

Asahina (1967, 1970) provided detailed account of *Ceriagrion spp.* including those occuring in India with many illustrations. Mitra (1975) illustrated the posterior lobe of prothorax of females of *Agriocnemis pygmaea* (Rambur), *A. femina* (Brauer), *A. lacteola* Selys, *A. dabrui* Frasar, *A. splendidissima* laidlaw, *A. pieris* Laidlaw, and *A. clauseni* Fraser highlighting interspecific differences, Mitra (1975) also synonymised *A. nainitalensis* Sahni 1965 under *Ischnura sp.* Prasad and Ghosh (1987) discussed with illustration of thoracic and abdominal markings interspecific differences of *Pseudagrion spp.* Mitra (1975) considered *Agriocnemis nainitalensis* Sahni misidentified as belonging to the genus *Agriocnemis* Selys.

Lahiri (1987) restored species status of *Argiocnemis obscura* Laidlaw suppressed under *A. rubescens* Selys by Fraser (1933). Asahina (1967) treated *Ceriagrion cerinomelas* Lieftinck and *C. pendleburyi* Laidlaw as subspecies of *C. fallax* Ris but Lahiri (*op.cit.*) considered them as synonymous as subspecific differentiation was based on overlapping and unstable characters. Lahiri (1987) also illustrated the posterior lobe of prothorax (either sex) of *Aciagrion pallidum* Selys, *A. tillyardi* Laidlaw, *Agriocnemis clauseni* Fraser, *A. lacteola* Selys, *A. pygmaea* (Rambur) and *Argiocnemis obscura* Laidlaw together with male anal appendages and penile organ of these and of *Ceriagrion coromandelianum* (Fabricius), *C. fallax* Ris, *C. olivaceum* Laidlaw, *Pseudagrion australasiae* Selys, *P. rubriceps* Selys and *P. spencei* Fraser to show their usefulness to differentiate various species.

Family Platycnemididae

Taxonomical studies on Indian Platycnemididae has involved a change of generic nomenclature from *Calicnemis* Selys to *Calicnemia* Strand 1926, description of a number of new species and addition to existing knowledge of interspecific differentiation.

Lahiri (1976) described *Calicnemia mukherjeei* from Meghalaya, while Lieftinck (1984) described *C. carminea* from Uttar Pradesh and *C. c. pyrrhosoma* from India. Lieftinck (1984) also provided synonymic notes on *C. maheshi* Sahni 1964 and *C. miniata doonensis* Sangal and Tyagi 1985.

Lieftinck (1977a) published comparative account of a number of *Calicnemia spp.* including *C. mortoni* (Laidlaw) and *C. miniata* (Selys). Lahiri (1987) illustrated male anal appendages and penile organ in respect of *C. eximia* (Selys) *C. imitans* Lieftinck and *C. mukherjeei* Lahiri.

Asahina (1970, 1984a, b, 1985a) made extensive contribution on the genus *Coeliccia* Kirby from India, Burma and Thailand with many illustrations. Asahina (*loc. cit.*) published accounts of *C. didyma* (Selys), *C.d. loringae* Laidlaw, *C. renifera* (Selys) and described *C. svihleri* from Assam (1970), *rotundata* and *C. schmidti* from Manipur (1984b) and *C. rossi* from Assam. Lahiri (1985) described *C. prakritii* (from Namdhapa sanctuary) and Lahiri (1987) described *C. sarbottama* from Meghalaya. Lahiri (1987) illustrated posterior lobe of prothorax, thoracic markings, male anal appendages and penile organ in respect of *C. didyma* (Selys), *C. fraseri* Laidlaw and *C. sarbottama* Lahiri.

Family Platystictidae

Kennedy (1958) described *P. rotosticta fraseri* from Assam and Kimmins (1958) described *P. rufostigma* from Tamil Nadu.

Family Protoneuridae

Cowley's recommendation resulted in the changes of generic nomenclature of *Indoneura* Laidlaw, *Caconeura* Kirby, *Chloroneura* Laidlaw and *Disparoneura* Selys to *Caconeura* Kirby, *Prodasineura* Cowley, *Disparoneura* Selys and *Elattoneura* Cowley respectively. Mitra (1995) suppressed *Disparoneura bhatnagari* Sahni 1965 under *Copera marginipes* (Rambur).

Lieftinck (1937) described *Elattoneua coomansi* from India, Lahiri (1987) illustrated male anal appendage, penile organ and posterior lobe of prothorax of *Elattoneura atkinsoni* (Selyse), *E, campioni* (Fraser) and *Prodasineura autumnalis* (Fraser) from Meghalaya.

Family Lestidae

Ander (1944) described *Sympecma annulata kashmirensis* from Jammu and Kashmir. Lieftinck (1960b) described *Burmargiolestes laidlawi* from Assam and provided comparative account of *Lestes umbrinus* Selys and *L. concinnus* Selys. Lahiri (1987) described *Lestes garoensis* and *Orolestes durga* from Meghalaya. Lahiri (1987) illustrated male anal appendages and penile organ of *Indolestes inidca* Fraser, *Orolestes durga* Lahiri, *Lestes concinnus* Selys, *L. dorothea* Fraser and *L. garoensis* Lahiri to show their interspecific differences.

Family Chlorolestidae

Lahiri (1979) described *Megalestes lieftincki* (allied to *M. irma* Fraser) from Arunachal Pradesh and Lahiri (1987) described *M. raychoudhuri* (allied to *M. major* Selys) from Meghalaya with illustrations of male penile organ and anal appendages. Tsuda (1991) reported *M. kurahashii* Asahina 1985 from India.

Family Amphipterygidae

Lahiri (1987) studied the only representative of the family, *Philognaga montana* (Selys) from Meghalaya and illustrated genitalia of either sex and penile organ from studied material.

Family Calopterygidae

Lahiri (1987) studied examples of *Caliphaea confusa* Selys, *Echo margarita tripartita* Selys, *Matrona basilaris nigripectus* Selys and *Vestalis g. gracilis* (Rambur) from Meghalaya and illustrated male penile organ and anal appendages for all of them from the studied material.

Family Chlorocyphidae

Fraser (1949) and Laidlaw (1950) proposed a number of new genera out of the genus *Rhinocypha* Rambur 1842 involving species occurring in India, *viz.*, *Heliocypha* Fraser, *Paracypha* Fraser, *Heterocypha* Laidlaw and *Aristocypha* Laidlaw. The chief diagnostic characters for the proposed new genera were differences in relative width of wing and the extent of mesothoracic triangle on synthorax. However, Lieftinck (1954) suppressed the genera *Heliocypha* Fraser and *Aristocypha* Laidlaw under *Rhinocypha* Rambur while Fraser (1957) also did not upheld *Aristocypha* Laidlaw and *Heterocypha* Laidlaw.

Davies and Tobin (1984) and Prasad and Vershney (1995) treated the proposed genera of Fraser and Laidlaw as subgenera of the genus *Rhinocypha* Rambur. However, Lahiri (1987), Lahiri and Sinha (1991) and Mitra (1999) have used the genus *Rhinocypha* Rambur in its original broad concept.

Cowley (1937) made a detailed study of penes of Chlorocyphidae. Chhotani, Lahiri and Mitra (1983) treated *Libellago andamanensis* Fraser as a subspecies of *Libellago lineata* (Burmeister). Singh and Prasad (1976) and Lahiri and Sinha (1985) suppressed *Rhinocypha bifasciata* Selys and *R. bifenestrata* Selys under *R. trifasciata* Selys and *R. cuneata* Selys respectively. Lahiri (1987) illustrated penile organ of *Libellago lineata* (Burmeister), *Rhinocypha beatifica* Fraser, *R. biforata delimbata* Selys, *R. ignipennis* Selys, *R. immaculata* Selys, *R. quadrimaculata* Selys, *R. spuria* Selys and *R. vitrinella* Fraser and highlighted the importance of the shape and extent of mesothoracic triangle in interspecific differentiation (either sex). Lahiri and Sinha (1991) made a review of the Indian Chlorocyphidae. A total of 24 species were listed, a new subspecies *Rhinocypha hilaryae miaoa* was described and critical study on *R. spuria* Selys and *R. quadrimaculata* Selys was made in this article.

Family Euphaeidae

Cowley (1936) described *B. kali* from Meghalaya, Asahina (1978b) described a new genus (allied to *Bayadera* Selys) *Schmidtiphaea* with *S. schmidti* spec. nov. from Manipur. Prasad and Ghosh (1984) described *Anisopleura lieftincki* from Meghalaya. Tsuda (1991) reported *Dysphae gloriosa* Fraser 1938 from India.

Lahiri (1987) studied *Anisopleura subplalystyla* Fraser, *Bayadera hyalina* Selys, *B. indica* (Selys) and *Allophaea ochracea brunnea* Selys from Meghalaya and illustrated male genitalia an anal appendages for each one of them. Mitra (1995) suppressed *Anisopleura kusumi* Sahni under *A. lestoides* Selys.

Suborder Anisoptera

Family Gomphidae

Cowley (1934) changed the generic names *Mesogomphus* Forster and *Ictinus* Rambur with *Paragomphus* Cowley and *Ictinogomphus* Cowley respectively. Fraser (1937) described *Onychogomphus schmidti* from Eastern India. Fraser (1940) established the genus *Dubitogomphus* with *Leptogomphus bidentatus* Fraser 1930 as Type-species from Meghalaya and published comparative account of penes for the family. Fraser (1953) described *Gomphidia platyceps* from India while Chhotani, Lahiri and Mitra (1983) described *G. ganeshi* from South Andaman island. Asahina (1985b) established the genus *Asiagomphus* involving some species known under the genus *Gomphus* Leach including *G. personatus* Selys from India. Lieftinck (1969) referred *Onychogomphus aureus* laidlaw to the genus *Phaenandrogomphus* Lieftinck. But Lahiri (1987) failing to observe any substantial reason for such change, restored original status of *aureus* under the genus *Onychogomphus* Selys. Lahiri (1987) reported the genus *Nihonogomphus* Oguma for the first time from India with *N. indicus* spec nov. from Meghalaya. Lahiri (*op. cit.*) also described *Onychogomphus meghalayanus* from Meghalaya and illustrated male genitalia, anal appendages, penile organ and/or female genitalia in respect of *Anisogomphu caudalis* Fraser, *Burmagomphus? vermicularis* (Martin), *Asiagomphus personatus* (Selys), *Merogomphus martini* (Fraser), *Niihonogomphus indica* Lahiri, *Nepogomphus modestus* (Selys), *Onychogomphus aureus* Laidlaw, *O.? maculivertex* (Srelys), *O. meghalayanus* Lahiri, *O. saundersi duaricus* Fraser, *Paragomphus lineatus* (Selys), *P. echinoccipitaslis* (Fraser), *Perissogomphus stevensi* Laidlaw and *Stylogomphus inglisi* Fraser. Tsduda (1991) reported *Onychogomphus thienmanni* Schmidt 1934 from India. Mitra (1995) suppressed *Acrogomphus mohni* Sahni under *Anisogomphus bivittatus* (Selys).

Family Aeshnidae

There was an addition of a number of new species from India to this family. These are : *Aeshna flavifrons* from India by Lichtenstein (1976); *Anaciaeschna kashmirense* from Kashmir by

Singh and Baijal (1954); *Anax indicus* from India by Lieftinck (1942); *Gynacantha rotundata* from India by Navas (1930), *G. dravida* from India by Lieftinck (1960b), *G. rammohani* from West Bengal by Mitra and Lahiri (1975); *Cephalaeschna klapperichi* from India by Schmidt (1961); *Oligoaeschna andamani* from South Andaman island by Chhotani, Lahiri and Mitra (1983) and *O. khasiana* from Meghalaya by Lieftinck (1968); *Periaechna assamensis* from Assam by Asahina (1978).

Lieftinck (1953, 1960b) provided comparative account of some *Gynacantha spp.* with illustrations of male anal appendages and second abdominal segment. Lahiri (1987) illustrated male anal appendages, penile organ and genitalia of *Austoaeschna intersedens* Martin, and female genitalia for *Periaeschna? biguttata* (Fraser), *Tetracanthagyna waterhousei* McLachlan, *Anax sp.* and *Gynacantha bayadera* Selys from Meghalaya. Lahiri and Das (1991) discussed with illustration interspecific variation of the epicranioorbital suture of some Aeshnini spp.

Tsuda (1991) reported *Aeshna juncea mongolica* Bartenef 1929 and *Gynacantha arundi* Asahina 1984 from India.

Family Cordulegasteriadae

Cowley (1934) proposed the new generic name *Neallogaster* for *Allogaster* Selys, Asahina (1982) described *N. ornata* from Himachal Pradesh and *N. schmidti* from India. Ischida (1964) described a new subspecies *Chlorogomphus brevistigma okinawensis* from India. Asahina (1986) described *C. precious ferandi* and *C. schmidti* from Assam with illustrations and comparative notes on *C. atkinsoni* (Selys) and *C. fraseri* St. Quentin. Lahiri (1987) illustrated male genitalia, penile organ and genitalia (either sex) of *C. atkinsoni* (Selys) and the female genitalia of *Anotogaster sp.indet.* from Meghalaya.

Family Corduliidae

Fraser (1939) described *Idionyx periyashola* from Western Ghats. Lieftinck. (1977b) reported the genus *Somatochlora* Selys for the first time from India by describing *S. daviesi* from Meghalaya. Tsuda (1991) recorded *Epopthalmia vittata bellicosa* Lieftinck 1948 from India.

Lahiri (1987) illustrated male genitalia and penile organ of *Macromia moorei* Selys, *Idionyx optata* Selys and *Hemicordulia asiatica* Selys as well as the fremale genitalia of *M. moorei* Selys from Meghalaya.

Family Libellulidae

Some new species were added to the family Libellulidae. These are : *Sympetrum himalayanum* from India by Navas (1934); *Nannophya katrainensis* from Himachal Pradesh by Singh (1955); *Bradinopyga saintjohanni* from Uttar Pradesh and *Crocothemis misrai* from Madhya Pradesh by Baijal and Agarwal (1955); *Orthetrum guptai* from Himachal Pradesh by Baijal (1955); *O. chandrabali* and *O. ganeshi* from Bihar by Mehrotra (1961); *O. martensi* from Jammu and Kashmir by Asahina (1978) and *Zygonyx iris intermedia* from Meghalaya by Lahiri (1987). Tsuda (1991) reported *Libellula quadrimaculata grigorievi* Schmidt 1961 from India.

Mitra (1993) suppressed *Sympetrum tandicola* Singh under *Pantala flavescens* (Fabricius) while Mitra (1995a) suppressed *Orthetrum gangi* Sahni under *O. glaucum* (Brauer).

Lahiri (1979) described isochrome female of *Neurothemis t. tullia* (Drury). Lahiri (1987) studied and illustrated penile organ, male genitalia and/or female genitalia of a number of Libelluloid dragonfly species from Meghalaya. These are : *Tetrathemis platyptera* Selys, *Lyriothemis bivittata* (Rambur), *Orthetrum galucum* (Brauer), *O. japonicum internum* McLachlan, *O. luzonicum* (Brauer), *O. pruinosum neglectum* (Rambur), *O. s. sabina* (Drury), *O. t. triangulare* (Selys), *Potamarcha congener* (Rambur), *Palpopleura s. sex, maculata* (Fabricius), *Brachydiplax sobrina* (Rambur), *Nannophya pygmaea* (Rambur), *Acisoma p. panorpoides* (Rambur), *Brchythemis contaminata* (Fabricius), *Crocothemis s. servilia* (Drury), *Diplacodes trivialis* (Rambur), *D. nebulosa* (Fabricius), *Neurothemis intermedia atlanta* Ris, *N. i. intermedia* (Rambur), *N. t. tullia* (Drury), *Sympetrum hypomelas* (Selys), *S. orientale* (Selys), *Trithemis aurora* (Burmeister), *T. festiva* (Rambur), *T. pallidinervis* (*Kirby*), *Zygonyx iris intermedia* Lahiri, *Onychothemis testacea* Laidlaw, *Rhyothemis v. variegata* (Linnaeus), *Tholymis tillarga* (Fabricius), *Hydrobasileus croceus* (Brauer), *Pantala flavescens* (Fabricius) and *Urothemis s. signata*

(Rambur). The study exemplified further the interspecific differences of allied species not only for males, but also for females (in respect of the nature of vulver aperture) specially in *Neurothemis spp* and *Orthetrum spp.*

Prasad (1994) discussed intraspecific variation in wing markings of *Neurothemis fulvia* (Drury). Mitra (1998) and Lahiri Walia (1998) reported the occurrence of preapical spot in hindwing in females of *Palpopleura s. sexmaculata* (Fabricius). Hamalainen *et.al.* (1999) working on Andaman and Nicobar fauna, recorded for the first time from India *Neurothemis r. ramburi* (Brauer), *Tramea transmarina euryale* Selys and *Zyxomma obtusum* Albarda.

Progress in Faunistic Survey

There has been substantial progress in existing knowledge of regional and state fauna as a result of extensive surveys carried out by scientists of the Zoological Survey of India in particular. These works not only extended the known range of distribution of many species but also provided useful information on habits and habitats of some of them. More important of such survey reports are :

Andaman and Nicobar islands : Chhotani *et. al.* (1983), Lahiri and Mitra (1993), Mitra (1995b), Lahiri (1998) and Hamalainen *et. al.* (1999).

Western Himalayas (Jammu and Kashmir, Himachal Pradesh and Uttar Pradesh) : Kumar and Prasad (1981).

Uttar Pradesh : Ram *et. al.* (1983)., Prasad and Singh (1994), Mitra (1998).

Rajasthan : Bose and Mitra (1976), Prasad and Thakur (1981).

Central India/Madhya Pradesh/Maharashtra : Mitra (1986), Prasad (1996a, b).

Kerala : Rao and Lahiri (1982), Madhaban and Miller (1989).

Orissa : Srivastava and Das (1987).

Estuarine fauna : Prasad and Ghosh (1998)

Bihar : Prasad and Vershney (1988).

West Bengal : Mitra (1983), Srivastava and Sinha (1993).

Sikkim : Prasad and Ghosh (1984b).

Arunachal Pradesh : Lahiri (1979), Lahiri (1985), Bhargava (1989).

Meghalaya : Lahiri (1983), Mitra (1999)

Mizoram : Lahiri (1979)

Manipur : Lahiri (1977b), Mitra (1996a).

Tripura : Mitra and Sen (1975), Lahiri (1977a), Mitra (1966b)

Discussion

Taxconomical studies outlined under different family gives a overall picture of advancement of knowledge on odonate fauna of India since the publication of the Fauna volumes on Odonata. A rise of nearly 15% of species composition through an addition of nearly 70 species is in itself quite remarkable and this evidently reflects the potential richness of the odonate fauna of India. However, rise in species composition was not uniform in all genera and was surprisingly higher in a few, *e.g.*, *Calicnemia* Strand, *Coelliccia* Kirby and *Megalestes* Selys. Most of the newly described species have come from hilly regions of the country. This perhaps corroborates the fact that the hilly areas with the background of changing altitude, rainfall and vegetation supports range of differential habitat suitable for variety of species with specific preference of habitat. There has also been advancement in our understanding of interspecific difference as a result of some useful comparative and revisional studies specially on the genera *Agriocnemis* Selys, *Ceriagrion* Selys, *Calicnemia* Strand, *Coeliccia* Kirby, *Rhinocypha* Rambur, *Chlorogomphus* Selys, *Gynacantha* Rambur etc. Since, in the Fauna volumes most of the keys are based on males, identification of a solitary female specimen is often rendered difficult. In this regard, the usefulness of posterior lobe of prothorax of females as diagnostic for *Agriocnemis* spp. as pointed out by Mitra (1975) is quite interesting. Lahiri (1987) has also shown similar usefulness of the posterior lobe of prothorax (for a number of species under the families Protoneuridae, Platycnemididae and Coenagrionidae), that of thre mesothoracic triangle (for many *Rhinocypha* spp.) and that of vulver scale and vulver aperture (for many species under the families Gomphidae,

Libeullulidae etc.). Taxonomical studies may continue in these lines to cover remaining taxa and even new vistas of study may possibly open up. However, with the wealth of knowledge already accumulated one can perhaps look into the future of a more extensive and purposeful taxonomical study on Indian odonate fauna in future.

Acknowledgement

The author is thankful to the Director, Zoological Survey of India, Calcutta, for providing an opportunity to carry out this work and to Dr. S.K. Mitra, Scientist SF and Dr. M. Dutta, Scientist E of the same department for various helpful suggestions.

References

Ander, K. 1944. Odonaten aus Kashmir und Ostrurkestan. *Forh. K. Fvsiogr Sallsk,* **14** : 168-178.

Asahina, S. 1967. A revision of Asiatic species of *Ceriagrion* (Odonata : Agrionidae). *Jap. J. Zool.,* **15** : 255-334.

Asahina, S. 1970. Burmese material collected by Dr. Arthur Svihla with supplementary notes on Asiatic *Ceriagrion* species. *Jap. J. Zool.* **16** : 99-126.

Asahina, S. 1978a. A new and some known species of Odonata from Kashmir *Senckenbergiana Biol.* **59** : 115-120.

Asahina, S. 1978b. A remarkable new damselfly allied to *Bayadera* (Odonata : Euphaeidae). *Proc. Japanese Soc. Syst. Zool.* **14** : 43-46.

Asahina, S. 1978c. Notes on the Chinese Odonata, VII. Further studies on the Graham collection preserved in the U.S. National Museum of Natural History, Suborder Anisoptera. *Contyu,* **46** : 234-252.

Asahina, S. 1982. A revision of the Himalayan dragonflies of the genus *Neallogaster* (Odonata : Cordulegasteridae). *Bull Natl. Sci. Mus., Tokyo* (A), **8** : 153-171.

Asahina, S. 1984a. A list of the Odonata recorded from Thailand. Part VI. Platycnemididae - genus *Coeliccia. Tombo,* **27** : 1-19.

Asahina, S. 1984b. Assamese and Burmese *Coeliccia* species in the collection of Dr. Erich Schmidt (Odonata : Platycnemididae). *Trans. Shikoku Ent. Soc.,* **16** : 1-9.

Asahina, S. 1985a. Further contribution to the taxonomy of Southasiatic *Coeliccia* species (Odonata : Platycnemididae). *Cho-Cho,* **8**: 1-13.

Asahina, S. 1985b. Contribution to the taxonomic knowledge of the *Megalestes* species of continental South Asia (Odonata : Synlestidae). *Cho-Cho,* **8** : 2-18.

Asahina, S. 1985c. A revisional study of Japanese and East Asiatic '*Gomphus*' species with description of *Asiagomphus* gen nov. *Gekkon Mushi.,* **169** : 6-17.

Asahina, S. 1986. Revisional notes on Nepalese and Assamese dragonfly species of the genus *Chlorogomphus* (Odonata : Corgdulegasteridae). *Cho-Cho,* **9** : 11-26.

Baijal, H.N. 1955. Entomological survey of the Himalayas. Part XIV. Notes on some insects collected by the second Entomological Expedition to the North West Himalayas. (1955) with description of three new species of Odonata. *Agra Univ. Res (Sci.) Suppl.* **4** : 741-766.

Baijal, H.N. and Agarwal, J.P. 1955. Opuscula Libellulogica. I. *Agra Univ. J. Res.* (Sci.) **4** : 453-470.

Bhargava, R.N. 1989. On a collection of Odonata from West Kameng district, Arunachal Pradesh. *Indian Odonatol,* **2** : 45-46.

Bose, G. and Mitra, T.R. 1976. The odonate fauna of Rajasthan. *Rec. zool. Surv. India* **71**, 1-11.

Chowdhury, S.K. and Das, S.M. 1975. Notes on the life history of *Coenagrion kashmirus* n. sp. (Odonata : Zygoptera : Coenagriidae). *3rd All India Congr. Zool.* : 60-61. (Abstr.)

Cowley, J. 1934. Changes in the generic names of Odonata. *Entomologist,* **67** : 200-205.

Cowley, J. 1936. A new species of *Bayadera* (Odonata). *Ann. Mag. Nat. Hist.* (10), **18** : 477-482.

Cowley, J. 1937. The penes of Chlorocyphidae (Odonata) as a group character. *Trans R. ent. Soc. Lond.,* **86** : 1-18.

Davies, D.A. L. and Tobin, P. 1984. The dragonflies of the world : A systematic list of the extant species of Odonata. *Soc. Internat. Odonatol. Rapid. Comm. (Suppl.) No 3* : 1-127.

Fraser, F.C. 1933. 1934 and 1936. Odonata, Vol 1-3, *Fauna Brit. India,* Taylor & Francis Ltd.

Fraser, F.C. 1937. A note on the confusion of *Onychogomphus bistrigatus* (Hagen) with a new species of *Onychogomphus* (Odonata). *Proc. R. ent. Soc. Lond.* (B), **6** : 161-164.

Fraser, F.C. 1940. A comparative study of the penes of the family Gomphidae (Order Odonata). *Trans. R. ent. Soc. London,* **90** : 541-550.

Fraser, F.C. 1949. A revision of the Chlorocyphidae with notes on the differentiation of the Selysian species *rubida, glauca, cyanifrons* and *curla. Bull. Inst. R. Sc. Nat. Belg.*, **25** : 1-50.

Fraser, F.C. 1953. Notes on the family Gomphidae with descriptions of a new species and the female of another (Order Odonata). *Proc. R. ent. Soc. Lond.* (B). **22** : 189-194.

Hamalainen, M., Parsanath, M., Mohanraj, M. and Veenakumari, K. 1999. Further additions to the odonate fauna of the Andaman and Nicobar islands, Indian Ocean. *Notul Odonatol*, **5 (3)** : 27-29.

Kennedy, C. H. 1936. *Protosticta fraseri*, a new species of dragonfly (Platystictidae : Odonata) from Assam, India. *Proc. R. ent. Soc. Lond.* (B), **5** : 67-71.

Kumar, A. and Prasad, M. 1978. On a new species of *Agriocnemis* Selys, 1839 with description of its larva from Dehra Dun Valley, India. *J. Bombay nat. Hist. Soc.* **75** : 174-179.

Kumar, A. and Prasad, M. 1981. Field ecology, zoogeography and taxonomy of the Odonata Western Himalaya, India. *Rec. zool. Surv. India, Occ. Pap.* No. **20** : 1-118.

Laidlaw, F.F. 1950. A survey of Chlorocphidae (Odonata : Zygoptera) with diagnosis of a proposed new genera and description of a new geographical subspecies. *Trans. R. ent. Soc. Lond,* **101** : 253-280.

Lahiri, A.R. 1975. A note on behaivour of *Bayadera hyalina* Selys in the rain (Zygoptera : Euphaeidae). *Odonatologica,* **4** : 255-257.

Lahiri, A. R. 1976. *Calicnemia mukherjeei* sp. nov. from Khasi hills, India. *Odonatologica,* **5** : 273-276.

Lahiri, A.R. 1977a. New records of Odonata from Tripura and Arunachal Pradesh. *Proc. Zool. Soc. Calcutta,* **30** : 45-49.

Lahiri, A.R. 1977b. On a collection of Odonata from Manipur with new records. *Rec. zool. Surv. India* **72** : 409-418.

Lahiri, A.R. 1979. Odonata (Insecta) from Different States of North Eastern India. *Orient. Ins.* **12** : 119-132.

Lahiri, A.R. 1985. Odonata : Insecta. *Rec. zool. Surv. India,* **82** : 61-67.

Lahiri, A.R. 1987. Studies on Odonate fauna of Meghalaya. *Rec. zool. Surv. India, Occ. Pap.* No. **99** : 1-402.

Lahiri, A.R. 1998. New record of Odonata (Insecta) from Little Andaman islands. *Fraseria* (N.S.), **5** : 57-59.

Lahiri, A.R. and Das, S. 1991. Observation on the epicranio-orbital complex of some Indian Aeshninae (Odonata : Aeshnidae). *Rec. zool. Surv. India,* **89** : 155-166.

Lahiri, A. R. and Mitra, B.M. 1993. New records of dragonflies (Insecta : Odonata) from Bay islands. *J. Andaman Sci. assoc.,* **9** : 96-99.

Lahiri, A. R. and Sinha, C. 1985. A new synonym in India *Rhinocypha* Rambur with revision of species groups *Fenestrella* and *Bifasciata* (Odonata : Chlorocyphidae). *Rec. zool. Surv. India* **7** : 33-36.

Lahiri, A. R. and Sinha, C. 1991. A revision of Indian Chlorocyphidae (Insecta : Odonata) with additional notes on taxonomy and distribution of some species and description of a new subspecies. *Rec. zool. Surv. India,* **89** : 257-268.

Lahiri, A. R. and Walia G. 1998. On the status of female *Palpopleura sexmaculata* (Fabricius) (Anisoptera : Libellulidae) marked by preapical spot *Fraseria (N.S.),* **5** : 61-62.

Lieftinck, M. A. 1937. Descriptions and records of South East Asiatic Odonata. *Treubia,* **16** : 55-119.

Lieftinck, M. A. 1942. The dragonflies (Odonata) of New Guinea and neighbouring islands, Part VI. The results of the 3rd Archbold Expedition 1938-39 and of the Roux Expedition 1939 to new Guinea. (I Anisoptera). *Treubia,* **18** : 441-608.

Lieftinck, M.A. 1953. Additions to the Odonate fauna of the Indo Australian Archipelago. *Treubia,* **22** : 233-269.

Lieftinck, M.A. 1954. Handlist of Malaysian Odonata. *Treubia 22 (Suppl.)* : xiii = 1-202.

Lieftinck, M. A. 1960a. Considerations of the genus *Lestes* Leach with notes on the classification and description of new Indo-Australian species and larval forms (Odonata : Lestidae). *Nova Guinea,* **8** : 127-171.

Lieftinck, M. A. 1960b. On the Identity of some little known South-East Asiatic Odonata in the European Museums described by E de Selys Longchamps with descriptions of new species. *Mem. Soc. ent. Ital,* **38** : 229-256.

Lieftinck, M.A. 1968. A review of the genus *Oligoaeschna* Selys in south-east Asia. *Tijdschr.Ent.* **111** : 137-186.

Lieftinck, M.A. 1969. Two new species of Odonata from south-east Asia with comments on previously described species. *Dt. Ent. Z.* **16** : 205-215.

Lieftinck, M. A. 1977a. Ergebnisse der Bhutan expedition 1978 des Naturhistorischen Museum in Basel Odonata. *Entomologica Basiliensia,* **2** : 11-37.

Lieftinck, M.A. 1977b. New and little known Corduliidae (Odonata : Anisoptera) from the Indo-Pacific regions. *Oirent. Ins.* **11** : 157-179.

Lieftinck, M.A. 1984. Further notes on specific characters of *Calicnemia* Strand, with a key to the males and remarks on some larval forms (Zygoptera : Platycnemididae). *Odonatologica,* **13** : 351-375.

Madhaban, S. and Miller, P.L. 1989. A collection of dragonflies (Odonata) made in the Periyar National Park, Kerala, South India in January, 1988. *Soc. Internat. Odonatol. Rapoid Comm.* (Suppl)., No. **10** : 1-10.

Mitra, A. 1998. A note on the heterochromous female of *Palpopleura* s. *sexmaculata* (Fabricius) (Anisopter : Libellulidae). *Notal. Odonatal.*, **4** : 151-159.

Mitra, T.R. 1973. *Sympetrum landicolo.* Singh, 1955a synonym of *Pantala flave scam* (Fabricius). (Odonata : Libellulidae). *Ent. Rec.*, **85** : 30-31.

Mitra, T.R. 1975. A review of Indian species of *Agriocnemis* Selys (Insecta : Odonata : Zygoptera : Coenagrionidae), with a note on *Agriocnemis nainitalensis* Sahni. *B.S. Chauhan Comm. Vol., Zool. Soc., India,* Bhubaneswar : 403-409.

Mitra, T.R. 1983. A list of Odonata of Calcutta, India. *Ent. Mon. Mag.*, **119** : 29-31.

Mitra, T.R. 1986. Note on the odonate fauna of Central India. *Rec. zool. Surv. India,* **83** : 69-81.

Mitra, T.R. 1988. On a collection of Odonata (Insecta) from Mirzapur district of Uttar Pradesh, India. *J. Bengal nat. Hist Soc., N.S.,* **7** : 64-68.

Mitra, T.R. 1995a. Taxonomic status of four Odonate species described by D.N. Sahni from Western Himalaya. *J. Bengal nat. Hist. Soc. (N.S.)* **14** : 71-72.

Mitra, T. R. 1995b. Additions to the odonate fauna of the Great Nicobar island, Indian ocean. *Opusc. Zool. Flumin.,* **129** : 1-6.

Mitra, T.R. 1996a. Additions to odonate fauna of Manipur, eastern India (Odonata). *Opusc. Zool. Flumin.,* **141** : 1-6.

Mitra, T.R. 1996b. Additions to odonate fauna of Tripura, eastern India (Odonata). *Opusc. Zool. Flumin.,* **141** : 7-9.

Mitra, T.R. 1999. Geographical distribution and Zoogeography of Odonata (Insecta) of Meghalaya, India. *Rec. zool. Surv. India, Occ. Pap.,* No. 170, pp. 1-63.

Mitra, T.R. and Lahiri, A. R. 1975. A new species of *Gynacantha* Rambur, 1842 (Odonata : Aeshnidae) from India. *Ent. Rec.,* **87** : 148-149.

Mitra, T.R. and Sen. G.C. 1975. First records of deagonflies (Odonata : Anisoptera) from Tripura (India). *Ent. Rec. J. Var.* **87** : 119-120.

Navas, R. P.L. 1930-1934. Insectos de la India. *Rev. Acad Cienc. Zaragoza,* 13 : 29-31 (1930); 15 : 11-41 (1932); **17** : 29-33 (1933-34).

Prasad, M. 1994. Variation in wing markings of *Neurothemis fulvia* (Drury) collected from various regions of India. *Fraseria,* **1** : 5-7.

Prasad, M. 1996a. Studies on the odonate fauna of baster, Madhya Pradesh. *Rec. zool. Surv. India* **95** : 165-213

Prasad, M. 1996b. An account of the Odonate fauna of Maharashtra state, India. *Rec. zool. Surv. India,* **95** : 305-327.

Prasad, M. and Ghosh, S.K. 1984a. A note on the Odonata (Insecta) from Khasi Hills, Meghalaya with description of a new species. *Bull. Zool. Surv. India,* **6** : 37-44.

Prasad, M. and Ghosh, S.K. 1984b. A note on the Odonata from Sikkim, India. *J. Bombay nat. Hist. Soc.,* **80** : 435-438.

Prasad, M. and Ghosh, S.K. 1987. Notes on the Indian species of the genus *Pseudagrion* Selys (Coenagrionidae : Odonata). *Bull. Zool. Surv. India,* **8** : 313-321.

Prasad, M. and Ghosh, S.K. 1988. A contribution on the estuarine Odonata of eastern India. *Rec. zool. Surv. India,* **85** : 197-216.

Prasad, M. and Singh, A. 1994. Fauna of Conservation Area 5 : Rajaji National Park. Odonata : 195-215. *Zoological Survey of India,* Calcutta.

Prasad, M. and Thakur, R.K. 1981. Further additions to the Odonata (Insecta) from Rajasthan, India. *Jantu,* **1** : 26-28.

Prasad, M. and Vershney, R.K. 1988. The Odonata of Bihar, India, *Rec. zool. Surv. India, Occ. Pap,* No **110**. 45 p.

Prasad, M. and Vershney, R.K. 1995. A Checklist of the Odonata of India including data on larval studies. *Orient. Ins.,* **29** : 385-428.

Rao, K.R. and Lahiri, A.R. 1982. First records of Odonata (Arthropoda : Insecta) from Silent valley and new Amarambalam Reserve forests. *J. Bombay nul. Hist Soc.,* **79** : 557-562.

Sahni, D.N. 1964. Some new species of Odonata from Kumaon Hills (India). *Agra Univ. J. Res. (sci.),* **13** : 89-98.

Schmidt, E. 1961. Ergebuisse der Deutschen Afganistan Expedition 1956 der Landessamlungen Fur Naturkunde Karlsruche Sowie der Expedition J. Klapperich Bonn 1952-53 und Dr. K. Lindberg, Lund (Schweden) 1957-1960. Libellen (Odonata). *Beitr. Naturk Forsch Sudw. Dtl,* **19** : 399-435.

Singh, S. 1955. Entomological survey of the Himalayas, Part V. On two new species of Odonata. *Agra Univ. J. Res. (Sci.)* **4** : 171-174.

Singh, S. and Baijal, H.N. 1954. Entomological Survey of Himalayas, II. On a collection of Odonata. *Agra Univ. J. Res. (Sci.)* **3** : 385-400.

Singh. A. and Prasad, M. 1976. Odonata of Doon valley, I. Anisoptera. *Rec. zool. Surv. India,* **70** : 21-38.

Srivastava, V.D. and Das, S. 1987. Odonata in Fauna of Orissa. **1** : 135-159. State Fauna series. *Zoological Survey of India, Calcutta.*

Srivastava, V.D. and Sinha, C. 1993. Odonata in Fauna of West Bengal. 3 : 51-168. State Fauna series. *Zoological Survey of India, Calcutta.*

Strand, E. 1926. Miscellaneous nomenclatoria zoologica at palaeontologica. *Arch. naturg, 92A,* 8 : 30-75.

Tsuda, S. 1991. A distributional list of World Odonata, Osaka. 362 p.

13

Odonate Diversity in Jharkand State with Special Reference to Niche Specialization in Their Larval Forms

• *Arvind Kumar*

Abstract

Altogether 109 species belonging to 14 families of odonata have been recorded from the different regions of the Jharkhand state. They exhibited remarkable diversity in their distribution and habitat preference. Few species were found confined to a specific niche and some were seasonal while few were also migratory. Niche specialisation among larval forms of certain dominant dragonfly and damselfly has also been extensively discussed in the text.

Introduction

Odonates are one of the scintillating white wonder, most conspicuous, beautifully coloured, and fascinating insects found flying swiftly around waterbodies. They are characterised by predaceous habit, palaeopteran wings, large bulging prominent eyes, rectal and caudal gills, secondary copulatory apparatus and the complex reproductive and feeding behaviours. Odonates were flourished during upper carboniferous and pernian periods, about 255 million years ago and are still surviving as descendants of one of the most ancient insect stock – "the protodonata". They lead a dual life, *i.e.* their larvae are purely aquatic whereas the adults lead an aerial mode of life. Now-a-days they are extensively used in controlling causative agents of malaria and filaria on the global basis.

About 5000 species containing 500 genera have so far been recorded (Fraser, 1933, 1934, 1936 and 1957; Kumar, 1973a and b; Srivastava, 1998). First of all, Johannson (1768) recorded *Rhyothemis variegata* Linn. from Eastern India. That was the first taxon of odonata recorded from India. During the last century, 494 species and sub-species have been recorded from the country. But ill today, no record of odonate fauna was recorded from Jharkhand state except Kumar (1982) and Kumari (1990). Therefore, the present article makes a modest attempt to update the existing fauna of Jharkhand state.

Study Area

South Bihar has been separated from the Bihar on 15 November, 2000 as separate state named as Jharkhand. Jharkhand means land of plants and trees. This state has been blessed with rich minerals and a beautiful topography (Fig. 13.1). Topography of this state shows that it may be divided into three distinct parts: (1) the hilly portion, which covers about three-eight of the entire area; (2) the rolling country covering half of it; and (3) the flat country on the north-east of the state. The hilly part of this state stretches continuously for about 250 miles from the Ganga at Sahibganj to the southern boundary. The hills are covered with luxuriant forests and savannah of considerable sizes. The rolling country includes whole of the west and south-west of the division. It contains long ridges of rocky intervening depressions with scrub jungle. The third division consists of a fringe of lowland between the Ganga and the hills, which is largely cultivated for winter and summer crops. But this part of state is subjected to annual inundation of flood of river Ganga during monsoon months. Therefore, there are large number of perennial water bodies for fish culture in this area. The displaced fisherman of East Pakistan, now Bangladesh rehabilated in this area and as these people are expert personnel in the field of fishing, as a result large quantum of fishes landed and transported to large cities specially Calcutta daily. Thus, this area of Jharkhand provides a vast resource for aquaculture for the tribal people.

Fig. 13.1 : Jharkhand State of India

The flora of alluvial tract of south-east resembles that of the Indo-Gangetic region. The upland countries have not uniform hilly systems but at places are found low areas for paddy cultivation. In the southern hills on way to Dumka, there are woodland forests which are protected by Forest Department of the State Government.

Geology

The Jharkhand state is chiefly an upland of ancient crystalling rocks which are covered with thick flows of volcanic lava in the east. The later forms the Rajmahal hills. In between these two main geological formations, is a narrow strip of lower Gondwana rocks which fringe the lava formation along its western margin. The ancient crystalline rocks cover the greater part of the division in the western and south-western parts.

Although the Jharkhand state is not very rich in minerals, however, a number of economic minerals occur in the divison. The Santhal Parganas is also rich in natural resources and huge belt of coal deposits has been recorded along Rajmahal hill ranges. There are two coal fields; Eastern Coalfield Limited, Lalmatia and Chitra. Due to uncountrolled mining operation, the flora and fauna of the region have been adversely affected (Datta Munshi, 1989). The Rajmahal hills have a great significance in plant and animal distribution as well as palaeobotany because various fossils of plants have been recorded in the rock strata.

Meteorological Status

The meterological parameters have been obtained from the Meterological Observatory, Dumka and Ranchi, presented in Figs. 13.2 and 13.3. The climate of the area is characterised with hot dry summer, a good rainy season and a cool pleasant winter season. The year may be divided into three main seasons from November to February – winter season; from March to May – summer season and from June to September – monsoon season. October is a transitory month from monsoon to winter.

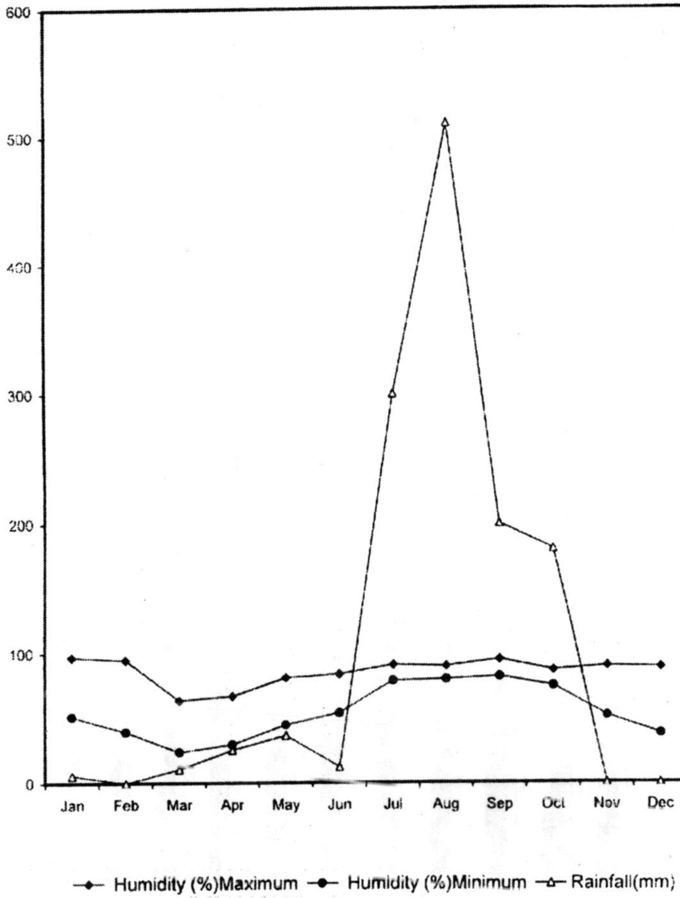

Fig. 13.2 : Meteorological data of Jharkhand State

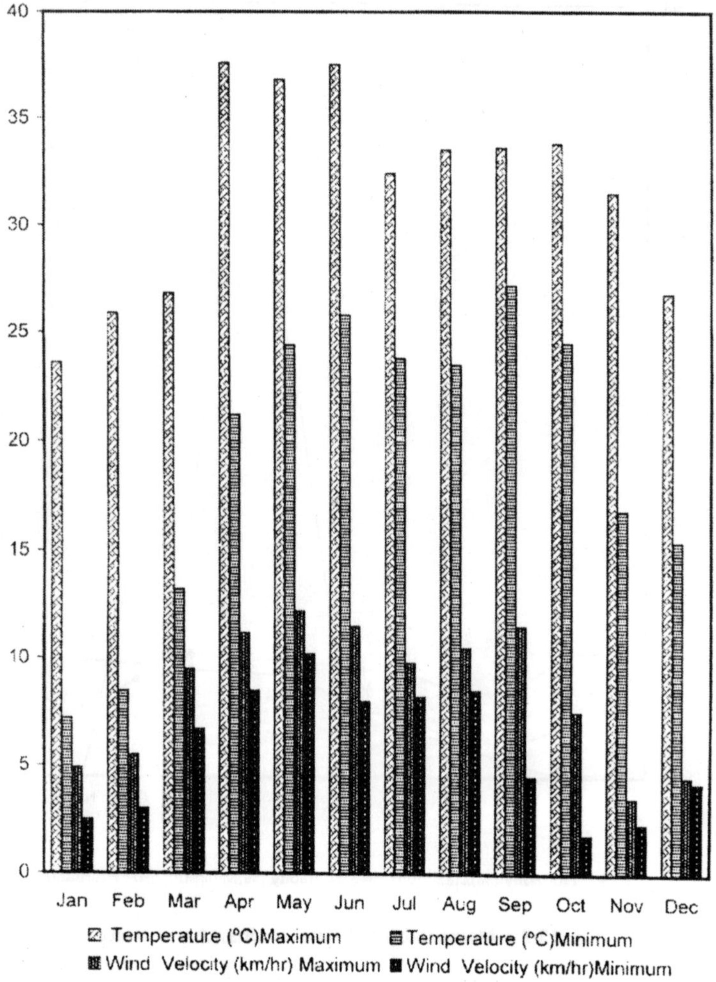

Fig. 13.3 : Meteorological data of Jharkhand State

Humidity

Humidity is an important environmental factor because it is essential for controlling the loss of water from the body of animals on the land. If the humidity is too low, the animals die from dessication at high temperature (Ananthakrishnan and Vishvanathan, 1976) and if the humidity is too high, it may harm them in a number of ways including the development of fungal and bacterial epizootics. Thus, high atmospheric humidity brings about a high incidence of insect diseases and is an important factor of natural control. Humidity and temperature are closely related and a combination of them gives an optimum conditions for propagation of these insects populations in a given habitat. Since odonates are amphibiotic in nature, thus, atmospheric humidity plays an important role for only adult stages which spent aerial mode of life.

The minimum humidity varied from 24.0-82.0% whereas the maximum humidity varied from 64.0-95.0% March, April and May are the driest months of the whole year. Humidity increases from the June and in monsoon months, *i.e.* from September to March. The maximum humidity remains between 64.0-97.0% usually. The periods of higher temperature hav e appreciably meagre percentage of humidity and *vice-versa*. However, in the present study humidity was found comparatively in lower values in February, March and April during the study period.

Rainfall

The rainfall varied from nil - 510.5 mm with the maximum record in August. The annual rainfall was recorded 1268.9 mm showing slightly higher precipitation with respect to the Bihar state.

The rainfall is monsoonic, cyclonic as well as torrential. It was found that the high rainfall and high water discharge displace and destroy the habitat of these benthic larvae causing a reduction in number of these insects in monsoon months. Rainfall is pre-requisite for these larvae as they need water throughout the year for their growth and propagation. The rain water get accumulated in water bodies which are in the form of perennial rivulets, derelict and managed ponds in the area that serve suitable habitat from these larvae.

Temperature

There is gradation of temperature on land, in water and even in the soil. Since the surface of lands being directly exposed to the changes in the atmospheric temperature thus, there is great variation in the diurnal and seasonal temperature. But the variation is comparatively small in water and soil.

During the period of study, the average monthly maximum temperature varied from 23.8-37.5°C and minimum temperature varied from 7.2-25.8°C. January was the coldest month. From the end of February, temperature begins to rise. Day temperature are almost equally high in April and May. During the later period of the summer until the onset of monsoon by the middle of June, days are extremely disagreable on account of the hot dusty winds. April to June were recorded as the hotter months.

Atmospheric temperature is inevitable components of the environment, penetrates into every region of the biosphere, profoundly influencing all forms of life and exerts its action by increasing or descreasing some of the vital activities of insects life such as reproduction, development, growth and death. The temperature affects the life in combination with various physical factors. Usually normal life activities go on smoothly at a specific temperature or at a specific range of temperatures. This is named as the optimum temperature or the optimum range of temperature. The insects react to any rise and fall of the optimum temperature range and behaviour of insect is often conditioned by it. The insects usually possess a narrow range of tolerance of temperature, since there is no great difference between maximum and minimum temperature in the aquatic medium (Ananthakrishana and Vishwanathan, 1976).

Wind Velocity

The minimum wind velocity varied form 1.8-10.2 km/h. The maximum wind velocity varied from 3.5-12.2 km/h.

Winds are generally highest with slight increase enforced in the late summer and monsoon months. In May the winds are mainly from direction between north-east to south-east. In monsoon, the winds are easterly or south-easterly. Winds are usually variable in October. In the remaining part of the year, the winds are south-westerly to westerly in the morning and westerly to north-westerly in the afternoon. Wind velocity plays an important role in the disporsal of these insects and thus helpful in preventing them to colonize in over-crowded habitat.

Intensity of Light

The response of the organisms to environmental rhythms of light and darkness is called photoperiodism. The length of the day is called as photoperiod. Regularly occurring daily cycles of light and darkness have been known to exert a synergetic effect on the biology of insects. Therefore, seasonal rhythms of insects are sometime due to change in climatic regime of the environment while the seasonal periodicity may be influenced by the light factor. The intensity of light affect the morphological, physiological, ecological and behavioural adaptations in insects.

Cloudiness

The cloudness increases from June during the north-east and south-east monsoon seasons. The skies are mostly heavily clouded or overcast.

Special Weather Phenomena

Thunder, storms and depressions particularly in the later monsoon and post-monsoon months which originate in the Bay of Bengal and rush in north-western to northern direction, affect Jharkhand and cause heavy rain with strong winds. Thunder and hailstroms are observed during the period Febraury-October, but their frequencies were observed highest in the monsoon months. Occasionally fogs also occur in the winter.

Materials and Methods

The adult odonates were collected from the different parts of teh Jharkhand state during day-time and mid-night with help of hand nets. The live specimens were brought to the laboratory, killed in cyanide bottle, dried at 36-40°C in incubator and mounted on insect trays. The dried specimens along with alcohol-formaline-glacial acetic acid fixed specimens were sent to Zoological Survey of India for proper identification.

Larval forms were collected from the littoral, limnetic and benthic zones of lentic and lotic freshwater ecosystems by an aquatic insect collecting net made up of nylon cloth (mesh size : 40 to 80/cm^2). The benthic and littoral larvae were collected by separate technique with specific sampling method. The benthic forms were collected by modified surber sampler while littoral fauna were collected by D-shaped aquatic net (Meritt and Cummins, 1978). The larvae were brought to the laboratory and preserved in 70% alcohol. The habitat preference and niche specialisation of these larvae have also been taken into account.

Result and Discussion

During the present investigation following odonate species have been recorded from different collecting sites in Jharkhand state :

Sub order : Anisoptera

I. Family : Libellulidae

(1) *Potamarcha obscura*

(2) *Potamarcha congener*

(3) *Potamarcha* sp.

(4) *Pantala flavescens*

(5) *Orthetrum chrysis*

(6) *O. luzonicum*

(7) *O. glaucum*

(8) *O. sabina sabina*

(9) *O. pruinosum neglectum*

(10) *O. brunneum brunneum*

(11) *O. taeniolatum*

(12) *Orthetrum* sp.

(13) *Tramea basilaris*

(14) *T. virginia*

(15) *Trithemis festiva*

(16) *T. aurora*

(17) *T. pallidinervis*

(18) *Acisoma panorpoides panorpoides*

(19) *Neurothemis tullia tullia*

(20) *Tholymis tillarga*

(21) *Zyxomma petiolatum*

(22) *Brachydiplax contaminata*

(23) *B. sorbina*

(24) *Diplocodes trivialis*

(25) *D. nebulosa*

(26) *Bradinopyga germinata*

(27) *Crocothemis servilla servilla*

(28) *Brachydiplax* sp.

(29) *Epithemis* sp.

(30) *Latherecista* sp.

(31) *Nannophya* sp.

(32) *Psendotramea* sp.

(33) *Rhodothemis* sp.

(34) *Libellula* sp.

(35) *Sympetrum* sp.

(36) *Indothemis* sp.

(37) *Urothemis* sp.

(38) *Zygonyx* sp.

(39) *Nesoxenia* sp.

(40) *Tholymis* sp.

II Family : Gomphidae

(41) *Paragomphus lineatus*

(42) *Paragomphus* sp.

(43) *Ictinogomphus rapax*

(44) *Gomphidia* sp.

(45) *Anisogomphus* sp.

(46) *Cyclogomphus* sp.

(47) *Macrogomphus* sp.

(48) *Aerogomphus* sp.

(49) *Megalogomphus* sp.

(50) *Ophiogomphus* sp.

(51) *Nepogomphus* sp.

(52) *Perissogomphus* sp.

(53) *Burmogomphus* sp.

III Family : Corduliidae

(54) *Macromidia* sp.

(55) *Macromia moorei*

(56) *Macromei* sp.

(57) *Somatochlora* sp.

(58) *Epophthalmia* sp.

IV Family : Aeshnidae

(59) *Anax* sp.

(60) *Anax guttatus*

(61) *A. immaculifrons*

(62) *Polycanthagyna* sp.

(63) *Hemianax* sp.

(64) *Tetracanthagyna* sp.

(65) *Planaeschna*

(66) *Oligoaeschna*

(67) *Periaeschna*

V. Family : Cordulegasteridae

(68) *Cordulegaster* sp.

(69) *Anotogaster* sp.

(70) *Chlorogomphus* sp.

Sub order - Zygoptera

I. Family : Platycnemididae

(71) *Copera marginipes*

(72) *Calicnemia* sp.

(73) *Copera* sp.

(74) *Indocnemis* sp.

(75) *Platycnemis* sp.

II. Family : Lestidae

(76) *Lestes* sp.

(77) *Indolestes* sp.

(78) *Orolestes* sp.

III. Family : Protoneuridae

(79) *Disparoneura quadrimaculata*

(80) *Melaneura* sp.

(81) *Phylloneura* sp.

(82) *Caconeura* sp.

(83) *Prodasineura* sp.

IV. Family : Coenagriidae

(84) *Pseudagrion rubriceps*

(85) *P. laidlawi*

(86) *P. decorum*

(87) *P. hypermelas*

(88) *Ceriagrion coromandelianum*

(89) *Ceriagrion dyeri*

(90) *Agreocnemis lacteola*

(91) *A. pygmaea*

(92) *Ischnura delicata*

(93) *I. senegalensis*

(94) *I. aurora aurora*

(95) *Rhodischnura nursei*

(96) *Enallagma parvum*

(97) *Mortonagrion varalli*

(98) *Archibasis* sp.

(99) *Ceriagion* sp.

(100) *Onychargia* sp.

V. Family : Chlorocyphidae

(101) *Rhinocypha unimaculata*

(102) *Indocypha* sp.

(103) *Calocypha* sp.

VI. Family : Epallagidae

(104) *Bayedera indica*

VII. Family : Agriidae

(105) *Neurobasis chinensis ehinensis*

VIII. Family : Calopterygiadae

(106) *Caliphaea* sp.

(109) *Vestalis* sp.

IX. Family : Euphaeidae

(107) *Anisopleura* sp.

(108) *Bayadera* sp.

(109) *Dysphaea* sp.

The above list indicates that odonates are very diversified in this zone of India. Altogether 109 species belonging to 14 families were collected. The anisopteran species were predominantly collected at night by switching on light lamps. The forests, hills, streams, reservoirs and other geological factors have divided the zoogeography of odonates into small niche with their own habitat ecology.

Niche Specialisation in Larval Odonates

The larval forms of odonata are purely aquatic and habitat ecology is very remarkable. Table 13.1 and 13.2 summarise the habitat preferences of anisopteran and zygopteran larvae.

Table 13.1 : Niche specialisation in Anisopteran larvae in natural waterbodies of Jharkhand

Species	Lentic waters				Lotic waters		
	A	B	C	D	E	F	G
1. *Anax guttatus*	+	-	-	+	+	+	-
2. *I. Immaculifrons*	-	-	+	+	-	+	-
3. *Brachythemis* sp.	-	-	-	-	+	-	-
4. *B. contaminata*	-	-	-	-	+	+	-
5. *Crocothemis servilia servilia*	+	+	+	-	+	+	+
6. *Macromia* sp.	-	-	-	-	-	+	-
7. *M. moorei*	-	-	-	-	-	+	+
8. *Mesogomphus lineatus*	-	-	-	-	-	+	+
9. *Orthetrum* sp.	-	-	-	-	+	-	-
10. *O brunneum brunneum*	-	-	-	-	+	+	+
11. *O. pruinosum neglectum*	+	-	+	-	+	+	+
12. *O. sabina sabina*	+	+	-	+	+	+	+
13. *Potamarcha obscura*	+	+	-	-	+	-	-
14. *Potamarcha* sp.	-	-	-	-	+	-	-
15. *Trithemis aurora*	-	-	-	-	+	+	-
16. *T. festiva*	-	-	+	+	-	+	-
17. *Tholymis tillarga*	-	-	-	-	-	+	+
18. *Zyxomma petiotum*	+	-	-	-	+	-	-

A = Deposite of plant, B = Bottom mud, C = Bottom waters, D = Among weed, E = Deposite of plant debris, F = Clinging to weeds and roots, G = Burried in sandy, + = Present, - = Absent.

In all dominant larvae of 19 species belonging to anisoptera and 17 species of zygoptera were observed. Most of anisopteran larvae were found in shallow lotic water having a semi-sandy bottom and abundance of macrophytes. *Potamarcha obscura, Potamorcha* sp. *Crocothemis servilia servilia, Zyxomma petiolatum, Orthetrum pruinosum, O. brunneum brunneum, O. servilia servilia* were observed at lower as well as higher altitudes. *Anisogomphus occipitalis, Mesogomphus lineatus, Macromia moorei, Trithemis aurora, Trithemis festiva, Orthtrum taeniolatum,* and *Tholymis tillarga* were found in waters at higher altitude, whereas *Brachythemis contaminata* and *Macromia* sp. were recorded in waterbodies of lower altitude. Besides these, *Macromia moorei* preferred cool waters with a sandy bottom. *Anax* sp. and *Potamarcha* sp. were also found in organically polluted waterbodies. *Zyxomma petiolatum* through present at higher altitude and in plains also preferred decaying plant debris whereas *Tholymis tillarga* were specially observed in highly alkaline water.

Table 13.2 : Niche specialisation in zygopteran larvae in natural waterbodies of Jharkhand

Species	Lentic waters				Lotic waters		
	A	B	C	D	E	F	G
1. *Agriocnemis pygmaea*	-	-	-	+	-	-	-
2. *Bayedera indica*	-	-	-	-	-	-	+
3. *Ceriagrion coromandelianum*	-	-	-	+	+	-	-
4. *Copera marginipes*	-	-	-	+	-	+	-
5. *Ischnura delicata*	-	-	+	+	-	+	-
6. *I. senegalensis*	-	-	-	+	-	-	-
7. *I. aurora aurora*	-	-	+	-	-	+	-
8. *Lestes* sp.	-	+	+	-	-	-	-
9. *L. praemrosa praemrosa*	-	+	-	+	-	-	-
10. *Neurobasis chinensis chinensis*	-	-	-	-	+	+	-
11. *Pseudagrion laidlawi*	-	-	-	-	-	+	-
12. *P. rubiceps*	-	+	-	+	+	-	-
13. *P. decorum*	-	+	-	-	-	-	-
14. *P. hypermelas*	-	-	-	+	-	+	-
15. *Rhinocypha animaculata*	-	-	-	-	-	+	-
16. *Rhodischnura nursei*	-	-	-	+	-	-	-
17. *Enallagma parvum*	-	-	-	+	-	+	-

A = Deposite of plant, B = Bottom mud, C = Bottom waters, D = Among weed, E = Deposite of plant debris, F = Clinging to weeds and roots, G = Burried in sandy, + = Present, - = Absent.

On the other hand zygopteran larvae were recorded in a great variety of habitats and in association with an abundance of algae and macrophytes. The larvae of *Ischnura delicata* and *I. senegalensis* were found in different localities in the valley. They were recorded from slow running streams weedy banks of large rivers and in temporary village ponds. Mature larvae are found in abundance in these ponds towards the decline of monsoon while in perennial streams these larvae are found attached to the submerged aquatic vegetation during spring and summer. *Neurobasis chinensis chinensis* larva occurs in shallow water of river Mayurakshi attached to submerged vegetation. They also inhabit slow-running marshy streams and channels such as Narganj. *Rhodischneura nursei* larve were recorded in the temporary monsoon ponds attached to the submerged vegetation. The larva of *Bayedera indica* and *Rhinocypha unimaculata* were found in shallow streams having currentful and clear water. The streams bottom was stony and devoid of vegetation, so larvae was found attached to the lower surface of the pebbles and small stones. Larvae of *Agriocnensis pygmaea* were found in temporary pools and ponds and generally remain attached to the submerged aquatic vegetations. *Desparoneura compioni* larva occurred in rivers, fast-flowing streams with stony substratum and with little vegetation. The larval froms of *Pseudagrion rubiceps* were recorded in turbid and muddy ponds and streams. The larvae of *Pseudagrion laidlauri* and *Ceriagrion coromandelianum* were commonly inhabited in slow running streams near the banks of river.

Taking into account the above observation, it may be concluded that odonate nymphs are adapted in different ecological niche. Tonapi (1980) also remarked that the odonatan nymphs are found in freshwater habitats when there is an abundance of oxygen and no pollution. However, the present study indicates that these insects were adapted in polluted, semi-polluted and unpolluted waters having luxurient growth of macrophytes.

References

Ananthakrishnan, T.N. and Visvanathan, S.A. 1976. General Animal Ecology. The MacMillan Co. of India Ltd., Madras., p. 324.

Datta Munshi, J.S. 1989. Ethnozoological survey and ecology of tribal areas of Santal Pargana (Bihar). Final Technical Report, S. No. 14/20/86; MAB/RE.

Fraser, F.C. 1933. Fauna of British India, Odonata 1. London. Taylor and Francies Ltd.

Fraser, F.C. 1934. Fauna of British India, Odonata 2. London. Taylor and Francies Ltd.

Fraser, F.C. 1936. Fauna of British India, Odonata 3. London. Taylor and Francies Ltd.

Fraser, F.C. 1957. A reclassification of the order odonata. R. Zool. Soc. N.S.W. Hand Book No.

Kumar, A. 1973a. Description of the last instar larvae of odonata from the Dehra Dun Valley (India) with notes on biology. I. (Suborder-Zygoptera). Oriental Insects. 7 : 83-118.

Kumar, A. 1973b. Descriptions of the last instar larvae of odonata from the Dehra Dun Valley (India) with notes on biology. II. Suborder-Anisoptera. *Oriental Insects. J.* 2 : 291-331.

Kumar, A. 1982. Ecophysiology of odonate larvae in Santal Pargana. Ph. D. Thesis, Bhagalpur University, Bhagalpur. p. 320.

Kumari, P. 2000. Checklist of Anisopteran larvae of Santal Pargana. *Indian J. Environ. and Ecoplan.* 3 : 20-27.

Meritt, R. W. and Cummins, K.W. 1978. *An introduction to the aquatic insects of North America. Hunt.* Publishing Co. p. 427.

Shrivastava, B.K. 1998. The marvellous odonata. *Fraseria.* 5 : 1-8.

Tonapi, G.T. 1980. *Freshwater animals of India.* I.B.H. Publishing Co., New Delhi. p. 258.

14

TROPHIC BIOLOGY AND ENERGY CONTENTS OF LARVAL ODONATES WITH SPECIAL REFERENCE TO THEIR ROLE IN THE MANAGEMENT OF AQUATIC ECOSYSTEM

• *Surendra Prasad Roy*

Abstract

The present article deals with the seasonal variations in the energy contents, productivity in terms of g/m^2/month/year, food and feeding biology and foraging ratio (FR) of *Mesogomphus lineatus* Selys, *Cordulegaster* sp. (Anisoptera : Odonata) and *Ischnura* sp. (Zygoptera : Odonata) of a fish pond of Bhagalpur (Bihar, India). The maximum calorific values of *M. lineatus* Selys and *Cordulegaster* sp. (5.487±0.003 Kcal/g dry weight and 5.430±0.00 Kcal/g ash-free dry weight) was recorded in September. However, the minimum calorific value (4.133+0.09 Kcal/g dry weight) was recorded in January. The calorific values of *Ischnura* sp. (3.706±0.032 Kcal/g dry weight and 3.470±0.034 Kcal/g ash-free weight) was measured. It was investigated that the calorific values of these larvae varied from instar to instar, month to month and upon the physiological state of the animals. The annual productivity was measured as 2.414g dry wt/cub. met/year with monthly productivity being 0.210 g dry wt/cub. met/month.

The gut content analysis of the larval odonates revealed that the Rotifera. Cladocera, Rhizopoda and aquatic insects form the maximum percentage of food items in *Ichnura* sp. but in *M. lineatus* Selys and *Cordulegaster* sp. Rhizopoda were recorded in very small quantity and other food items such as Rotifera, Cladocers, Copepoda, aquatic insects and other animal tissues were found in maximum percentage in the foregut.

Due to the utilization of food present at various trophic levels of the food chains, they have regulatory impact in the management of the aquatic ecosystems as well as their mere presence indicates healthy and non-contaminated environments.

Key words : *Energy contents, Productivity, Food and feeding biology, Forage ratio, Management, Odonate larvae.*

Introduction

In animal taxa, odonate larvae are at the top of the aquatic food chain[1]. Literature survey reveals that considerable work has been done on the eco-energetics, productivity and food and feeding biology of ononate larvae[2-6] but work pertaining to the flow of energy and nutrients in the eco-system through them, prey destroying efficiencies, gut content analysis and food preferences potentialities under field conditions is rather scantly in India. Therefore, in the present study the energetics, standing corp/m^2/ day/month/year, prey destroying potentialities; food preferential capacities, prey turn-over rates and their role in the management of the aquatic ecosystems were undertaken with a view to assess their role in exploiting the fish food organisms.

Materials and Methods

The collection of odonate larvae was made with the help of aquatic insect collection net and samples through the method of Merritt and Cummins[7] after slight midifications. Monthly random sampling of larvae associated with the roots of free-floating macrophytes in the littoral and benthic zones of the selected water bodies was done by specially designed 1.6 m^2 lift net made of nylon cloth (mesh size- 40 to 80 m^2). Ekman-dredge has been used for the collection of odonate larvae found on the benthic zones of ponds. The collected materials with vegetational parts were brought to the laboratory in a plastic bucket and thoroughly washed. Larvae not freed by washing were removed manually. Larvae of each hauling were sorted out species-wise and were preserved in 70% alcohol. Finally, the larave of each sampling unit were counted and identified. The energetics, productvitiy, prey destroying efficiencies, food and feeding habits and forage ratio of odonate larvae were investigated in the laboratory with the help of methods noted below :

(i) Determination of Energetics

The energetics of odonate larvae was determinated by the oxygen bomb calorimetry method. Five-replicated determinations were made for each sample. The maximum temperature difference is obtained in 6-7 mins after the explosion.

The energy content of the samples was determined according to the formula derived from the heat balance equation[8].

(ii) Determinations of the Standing Crop and Productivity

Biomass method was adopted to evaluate the standing crop of the larvae and productivity was determined by Winberg method[9]. The ash free dry weight of the larvae was used as a measure for its biomass[10]. The data was expressed g/m^2 for standing crop g/m^2/month for net monthly productivity of the larval population.

(iii) Prey Destroying Capacity and Prey Turnover Rates of the Odonate Larvae

The three instars (2nd, 3rd and 4th instar) of dragonfly *Cordulegaster* sp. of different length and body weight groups were separately placed in various glass aquaria (20 x 15 cm) which were filled with fish pond water. The larvae were kept starved for 24 hours and then exposed to their prey. They were fed with spawn and fry of both *Catla* and *Lebeo* respectively at five different prey densities (5, 10, 20, 40 and 80). The prey-predator relationship was studied following "Ellis Borden" method[11].

(iv) Food and Feeding Biology of Odonate Larvae

The intact fore gut was removed and dissected further on a glass slide. Various techniques[12-13] were used with some modifications to characterize the feeding habits. The food items were counted using the method of Brown[14]. Approximate per cent by volume of food items (Zooplankton-Rotifera. Cladocera, Copepoda aquatic insects, algae and other matters) were recorded for each season. The plankton of foraging medium were collected by a plankton net mady of bolting silk no. 25 (mesh size 0.3 - 0.4mm) and counted by drop micro-transect counting by Lackey method modified by Edmondson[15].

(v) Determination of Forage Ratio (FR) of Odonate Larvae

The foraging ratio (FR) of odonate larvae was determined as per Usinger method[16].

Results

The seasonal change in the energy contents and productivity of *Mesogomphus lineatus* Selys and *Ischnura* sp. are shown in Table 14.1 and Fig. 14.1 respectively.

Table 14.1 : Seasonal variation in calorific values of dragonfly larvae (*Mesogomphus lineatus* (Selys) and *Cordulegaster* sp.)

Months	Calorific value	Ash fraction	Org. fraction
Jan	4.133 ± 0.009	0.00 ± 0.00	4.133 ± 0.009
Feb	4.647 ± 0.042	0.00 ± 0.00	4.647 ± 0.042
Mar	4.656 ± 0.010	0.00 ± 0.00	4.656 ± 0.010
Apr	5.103 ± 0.003	0.006 ± 0.06	5.098 ± 0.001
May	5.163 ± 0.001	0.003 ± 0.001	5.157 ± 0.013
Jun	5.204 ± 0.003	0.00 ± 0.00	5.204 ± 0.003
Jul	5.388 ± 0.00	0.06 ± 0.002	5.306 ± 0.001
Aug	5.18 ± 0.01	0.82 ± 0.00	5.039 ± 0.01
Sep	5.487 ± 003	0.057 ± 0.005	5.43 ± 0.00
Oct	5.261 ± 0.321	0.00± 0.00	5.261 ± 0.321
Nov	6.416 ± 0.01	0.024 ± 0.002	5.389 ± 0.004
Dec	5.413 ± 0.00	0.053 ± 0.001	5.359 ± 0.001

(i) Energy contents

The maximum calorific value of *Mesogomphus lineatus* Selys (5.487 ± 0.003 Kcal/g dry weight) was recorded in September and minimum calorific values (4. 133 ± 0.009 Kcal/g ash-free dry weight) was recorded in January. The ash fraction varied from 0.082±0.00 to 0.00±0.00 with maximum and minimum recorded in August and January, March, June and October respectively (Table 14.1). The calorific values of *Ischnura* sp. (3.706±0.032 Kcal/g dry weight and 3.470±Kcal/g ash free dry weight) was recorded.

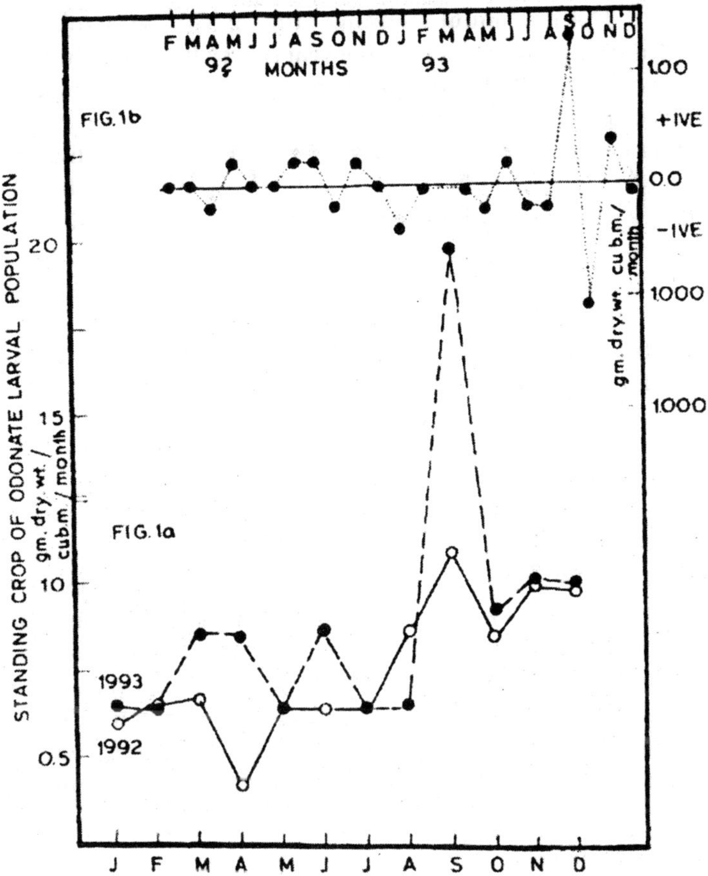

Fig. 14.1 : (a) Monthly standing crop of Odonate larval population; (b) Net monthly productivity of Odonate larvae.

The percentage ash fraction was 0.072±0.001. It was observed that *Mesogomphus lineatus* Selys had higher values of energy contents than the *Ischnura* sp.

(ii) Standing crop and productivity of odonate larvae

The standing crop, net daily productivity and net montly productivity of the larval population have been presented in

Fig. 14.1. The maximum standing crop (g/dry wt/haul) was recorded in September for both the years of study (0.55 g/haul for 1992 and 1.0 g/haul for 1993), whereas minimum standing crop was recorded in April of 1992 (0.21 g/haul). The standing crop of odonate larvae dry wt/cub met/day was obtained after dividing monthly net productivity of the population (g/dry wt/cub met/ month) by the days of intervals between two successive dates of sampling. The maximum net monthly productivity of the larval population was found in summer and monsoon months when usually positive increase in the standing crop of the larval population was recorded in 1992. In this year, however, negative productivity in the larval population was found in April, July, October, and December. In 1993 the negative produtivity of population was found in January-February, April-May, July, October and December. It was found that comparatively lesser increase in the standing crop was obtained in 1993 than 1992. In 1993, the maximum net montly productivity was found in September (1.36 g/dry wt/cub met/month) when positive increase in the standing crop was recorded. Usually negative productivity of the population was found in winter seasons of both the years of study. The net daily productivity of the population varied from 0.008 g. dry wt./cub. met/day to 0.004 g dry wt/cub met/day in 1992 and from 0.04 g dry wt/cub met/day to 0.008 g dry wt/cub met/day in 1993, having positive increase in the larval population (Fig. 14.1b).

The annual net productivity comes to 1.18 g dry wt/cub met/ year with montly productivity being 0.09 g dry wt/cub met/month of 1992. In 1993 the annual productivity was 2.4 g dry wt/cub met/year with monthly productivity being 0.2 g dry wt/cub met/ month.

The Product moment correlation co-efficient for the standing crop of dry wt/haul with water temperature, pH, DO_2, FCO_2, CO_3, HCO_3, and Cl^- had been worked out (Table 14.2) Among all the correlations existing therein, only DO_2 content of water had shown significant positive correlation with the larval odonate population (e = 0.898; $p > 0.05$; t = 0.282).

Table 14.2 : Correlation co-efficient between standing crop g dry wt/ haul with various physico chemical

Parameters	Correl coeff.	Regress	Intercept	T_R For
1992				
Temperature	0.01638	2.575-03	0.465	0.525
pH	0.206	4.652-03	1.360-03	0.666
DO_2	8.398-02	4.303	0.362	0.282
FCO_2	00.151	3.402-03	0.375	0.483
Corbonate	0.078	2.363-03	0.376	0.248
Bicarbonate	0.238	-1.348-04	0.419	0.776
Chloride	0.152	0.121	0.460	0.488
1993				
Temperature	2.425	8.886-04	0.445	7.664-02
pH	0.336	0.265	1.571	1.131
DO_2	0.392	3.105-02	0.244	1.350
FCO_2	0.366	5.637-02	0.431	1.325
Carbonate	-0.383	-1.823.02	0.589	1.314
Bicarbonate	-0.197	-8.114-05	0.487	6.244-02
Chloride	-0.259	-0.225	0.6121	0.350

(iii) Prey destroying efficiencies and prey turn over rates

The predatory efficiencies of *Cordulegaster* sp. on different prey densities (5, 10, 20, 40 and 80) of both spawn and fry of *Catla catla* Ham. and *Labeo rohita* Ham. in the laboratory were recorded. The mean values of maximum spawn and fry consumption (N/hr.) for larvae on their respective prey densily of spawn and fry was recorded for one hour of predation trial and presented in Fig. 14.2. thc maximum utilization (N/hr.) of spawn and fry of these major carps by the three instar stages (II, III, IV) of *Cordulegaster* sp. was found on maximum spawn and fry density (80).

The mean values of maximum consumption (N/hr.) by three larval stages (2nd, 3rd, 4th) of different length and body weight groups of *Cordulegaster* sp. on higher prey density (80) of *Catla* and *Labeo* spawn were 5.10 ± 0.88; 8.00 ± 1.27; 8.80 ± 1.08 and 5.60 ± 0.87; 6.10 ± 0.99; 10.20 ± 1.09 respectively. Analysis of variance shows that the consumption capacity of three larval stages of different length and body weight groups and different prey densities of spawn on their predation were significantly different at 1% level of significance. Interaction between predator and prey density was also found significant.

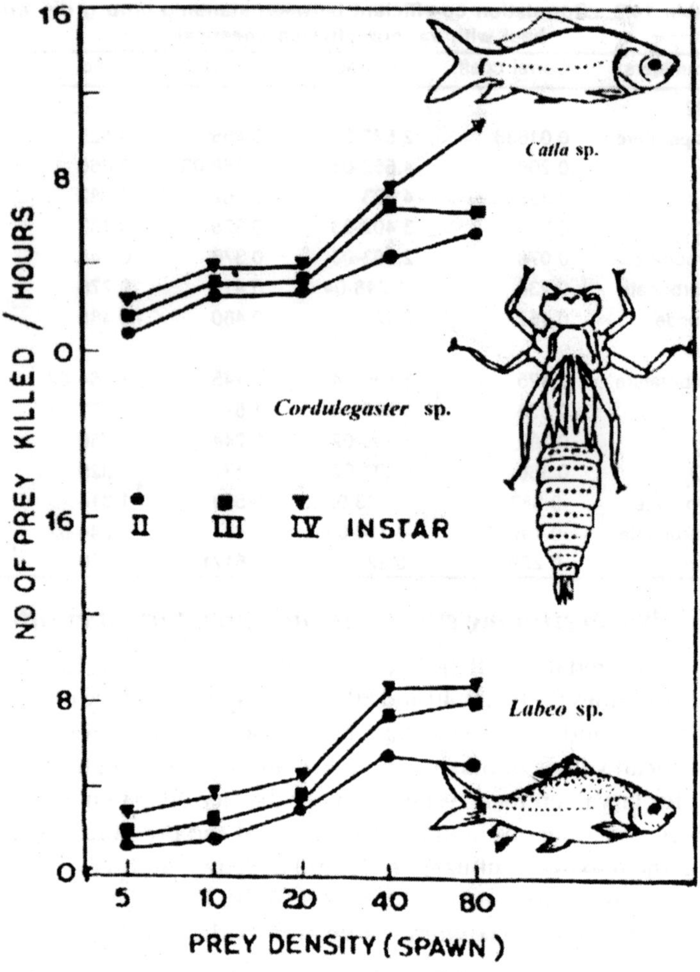

Fig. 14.2 : Prey destroying efficencies and prey turn over rates

(iv) Food and feeding biology

The seasonal variation in the percentage composition of the major dietary components of *Mesogomphus lineatus* (Selys) and *Cordulegaster* sp. and *Ischnura* sp. (Zygoptera) have been summarized in the Table 14.3 and 14.4 respectively. The gut content analysis of these larvae revealed that they are primarily

Table 14.3 : Percentage composition of food items in the gut of *Mesogomphus lineatus* and in the foraging medium, during the period 1992-1993

Season	Rotifera %		Cladocera %		Copepoda		Other %	Algae aquatic %		Muscular %		Other Tissue
	Gut.	Med.	Gut.	Med.	Gut.	Med.	insects	Gut	Med.	Gut.	Med.	Unidentified
Summer	20.79	52.44	18.81	21.77	14.85	25.77	09.90	15.38	13.85	16.23	08.31	12.87
Monsoon	14.85	44.31	21.78	25.14	11.88	30.53	12.87	25.24	07.92	12.46	16.83	13.86
Winter	19.80	26.00	28.70	54.57	13.86	19.42	14.85	23.43	03.96	06.58	06.93	11.88
Spring	23.76	36.47	32.67	31.96	15.84	31.55	12.87	17.77	02.97	09.42	04.95	06.93

Table 14.4 : Percentage composition of food items in the gut of *Ischnura* sp. (Zygoptera) and in the foraging medium, during the period 1992-1993

Season	Rotifera %		Cladocera %		Copepoda		Other %	Algae aquatic %		Muscular %		Other Tissue
	Gut.	Med.	Gut.	Med.	Gut.	Med.	insects	Gut	Med.	Gut.	Med.	Unidentified
Summer	15.84	53.07	22.71	21.49	16.83	25.43	12.87	15.46	08.91	14.25	11.08	10.00
Monsoon	19.41	40.70	26.21	29.64	13.59	29.64	15.53	19.29	10.67	20.46	05.82	08.73
Winter	19.80	27.35	29.70	43.58	10.89	28.05	20.79	22.48	04.95	08.82	05.95	07.92
Spring	29.24	37.50	32.03	36.11	19.59	26.38	10.67	13.95	04.84	07.81	05.82	07.76

carnivorous in habit, feeding on microcrustacea (Cladocera and Copepoda), Rotifera, aquatic insects and other animal life present in the water bodies. Only a small percentage of plant manerials could detected in the gut. They were mainly algae belonging to Chlorophyceae, Bacillariophyceae and Myxophyceae. The seasonal variation in the feeding habits showed that both dragon fand damsel fly larvae prefer to take Rotifera, Cladocera, Copepoda, and aquatic insects Mayfly larvae, *Chironomus* larvae, mosquito larvae). The forage ratio (FR) and food preferences have been presented in Table 14.5 for dragonfly and damselfly larvae.

(v) Foraging ratio (FR) and food preference of odonate larvae

The forage ratio (FR) of dragonfly larvae varied from 0.38 to 0.78% with the average of 0.58 for Rotifera; 0.54 to 1.06% with the average of 0.87% for Cladocera, 0.44 to 0.73% with the average of 0.57% for Copepoda; 0.49% to 0.78% with avaerage of 0.60% for aquatic insects and 0.38 to 0.89 with the average of 62% for algae. The forage ratio (FR) of damselfly larvae for Rotifera varied from 0.30% to 0.76% with the average of 0.55% for Cladocera 0.71 to 1.09% with the average of 0.94%; for Copepoda 0.39% to 0.68% with the average of 0.53%; for other aquatic insects 0.76% to 1.29% with the average of 0.96%; and for algae 0.28% to 0.82% with the average of 0.52%.

Table 14.5 : Foraging ratio (FR) of Mesogomphus lineatus and Ischnura sp. (Zygoptera)

Season	Rotifera		Cladocera		Copepoda		Aquatic insects		Algae	
Summer	0.47	0.30	0.90	1.09	0.60	0.68	0.63	0.86	0.72	0.68
Monsoon	0.38	0.51	0.98	0.95	0.44	0.49	0.48	0.94	0.51	0.82
Winter	0.78	0.76	0.54	0.71	0.73	0.39	0.78	1.29	0.38	0.31
Spring	0.69	0.66	1.06	1.02	0.52	0.56	0.51	0.76	0.89	0.28
Average	0.58	0.55	0.87	0.94	0.57	0.53	0.60	0.96	0.62	0.52

In the case of dragonfly larvae Cladocera and Rotifera form the maximum percentage of the food items followed by the other aquatic insects and algae. In the damselfly larvae Cladocera, and other aquatic insects were recorded maximum while copepods and algae were found in very small quantity.

Discussion

The variation in the energy contents of odonate larvae from season to season seems to be related with the calorific equivalent of daily ration available to them in their proper habitat. This view was supported by Rodgers[17]. Changes in the energy content of the organic fraction may reflect changes in amounts of lipid contents[18]. It may also change with age and with different development stages fo animals[19]. The gross energy contents for animal with heavy skeleton may be highly diverse. But when expressed terms of the ash free dry weight, becomes very similar to each other[20].

The availability of abundant food in the habitat of odonate larvae is the another factor regarding the seasonal variation in the energy contents of these insects. The energy contents also depend upon the relative size of the larvae. Winberg found that it varied from 2.0 to 8.0 $Kcal/g^{-1}$.

From Table 14.2 and Fig. 14.1(a) and 14.1(b) it was found that during October-December in both the years very little production took place in the odonate larval population in their natural habitat.

During monsoon season, *i.e.* in July-August, comparatively low standing crop was recorded in both the years. Monsoon rains and flooding found unfavourable to these sedentary larvae and benethic dragonfly larvae. As such in the winter season, specially in November-December, consistently large standing crop of these larvae was recorded. This view supports the work of Benke who studied dragonfly production and prey turn-over rates in Atlanta, Georgia. He strongly suggested that their annual production is directly related to the availability of food prey. It was found by Benke that highest production occurred from June to September in the dragonfly larvae thus suplementing the present work. Lawton[21] found the damselfly, *Pyrrhosoma nymphula* in an English pond to have an annual production of 0.7 g/met sq. The abundance of vegetation provided hiding places for their prey. The abundance of prey is one of the major factors is standing crop oscillation which in turn depends upon the seasonal appearance and disappearance of macrovegetations of the habitat. Thus, the production of odonate larvae is directly controlled by the density and abundance of the food (prey) of the habitat.

Of all the physico-chemical factors studied, DO_2 show marked correlation with the larval population ($r = 0.898$; $p>0.05$). These larvae live in such environment in which DO_2 content is the most important limiting factor. These larvae take DO_2 content of water in their respiration. The environment in which DO_2 content is meagre then the larval population becomes less represented. Thus, in the present study a significant correlation has been established with the standing crop g/cub met of water of the larval population with the DO_2 content of water.

In most predation trials, the experimental dragonfly larvae (*Cordulegaster* sp.) of different length and body weight groups and different consumption capacity of spawn of *Catla* and *Labeo*, the rate of predation is gradually increasing as the densities of prey increased. *Cordulegaster* sp. do not inflict any injuries to fry and fingerling stages of major carps. This might be due to the large size of prey that directly affects the predation. The increasing body size of prey is an effective mechanism for reducing predation[22]. Alikunh *et.al.*,[23] reported that a species of dragonfly (15 mm) had annihilated 17 carp spawn within 3 hours in the laboratory. Similar trends of results were observed during the present study of predation trials of *Cordulegaster* sp. on carp spawn.

The prey destroying efficiencies of this larvae thus, increases with the increase in the densities of prey (spawn). This is an accordance with the report that the predator prefers certain size proportion, as they are relatively abundant in nature[24-25].

The interaction of predatory dragonfly larvae with the spawn of major carps have been investigated on the basis of two way classification of analysis of variance (ANOVA). The present computation of data on predatory efficiencies and prey annihilation indicates that the prey density is the prime factor increasing the predation rate. The odonate larval population have capacity to annihilate and consume prey species approximately seven times than its body weight.

It was observed that the *Cordulegaster* larvae with lower length and body weight groups cause uniform destruction of spawn of carps than the larvae with higher length and body weight.

In the present investigation. Rotifera, Cladocera, Copepoda, other aquatic insects were recorded in the gut of odonate larvae.

A small percentage of algal community belonging to the *Chlorophyceae, Bacillariophyceae, Myxophyceae* were also recorded in different seasons of the year. The animal diet comprises the maximum percentage of dietary components whereas non-animal diet forms insignificant percentage. This view supported by Benke.

It was found that the percentage composition of Rotifers was low in the gut of dragonfly larvae than the environment water in all seasons. The dragonfly larvae have some preference for Cladocera as their forage ratio (FR) was found more than 1.0. In *Pyrrhosoma nymphula microcrustacea* always comprised off over 50% of the prey consumed by this species[21].

In case of damselfly larvae, percentage composition of Cladocera in the gut was found lower than the environmental water except in the summer.

The forage ratio (FR) for Cladocera was found more than 1.0 except in winter (0.71%) with average of 0.94%. Thus, the damselfly larvae have some preference for Cladocera and aquatic insects. It is interesting to note from the above data that the switchover to algal diet by these odonate larvae suggesting transition from an algal to a carnivorous diet vice versa similar to the larvae of Trichoptera and Plecoptera[26].

Thus, due to their mobility and utilization of food present at various trophic levels of the food chain, the odoante larvae can be considered promising organisms for the monitoring of the aquatic environment. As top carnivores, they can provide information about the organisms occupying lower trophic levels of the food chains and serve as biological indicators of the aquatic productivity.

References

[1] Alikunhi, K. H.; Choudhary, H. and Ramchandran, *V. Ind. J. Fish Soc.* 2 : 257.

[2] Benke, A. C. and Benke, S. S. 1975. *Ecology* 56 : 302.

[3] Benke, A. C. 1976. *Ecology*. 57 : 915.

[4] Roy, S. P. 1986. *Biol. Bull. India.* 8 : 169.

[5] Roy, S. P. 1990. *J. Freshwater Biol.* 23 : 257.

[6] Kumar, V. 1987. Eco-energetic studies of aquatic insects of certain freshwater ponds, Ph. D. Thesis, University of Bhagalpur.

[7] Merritt, R. W. and Commins, K. W. W. 1979. Kendall/Hunt Publishing Co. 2460, Kerper Boulevard, Dubugue, Lowa, 52001, U.S.A, Third Printing.

[8] Gorecki, A.; Grodzinski, R. Z. and Klekowski, A Duncan. 1975. Blackwel Scientific Publications. Oxford, London. pp. 275-292.

[9] Winberg, G.G. (ED) 1971. Academic Press, New York. 175 p.

[10] Beukema, J.J. 1976. *Netherland J. of Sea. Research.* 10 : 236.

[11] Ellis, R. A. and Borden, J. H. 1970. *Ann. ent. Soc. Ann.* 63 : 963.

[12] Mecon J. O. and Cummins, K. W. 1964. *Trans Amer. Microscop. Soc.* 83 : 233.

[13] Shapas, J. J. and Hilsenhoff, W.L. 1976. *Hilsenhoff Entomologist.* 94 : 175.

[14] Brown, D.D. 1961. *J. Anim. Ecol.* 38 : 55.

[15] Endmondson, W.T. 1974. (Ed. R.A. Vollenwider) 14.

[16] Usinger, R. L. (Ed.) (1956) : Univ. of California Press, Berkeley.

[17] Rodgers, D.W. 1976. Food habits, growth, energy transformation and pollutant accumulation of yellow perch (Perca flavescerus) in the Ottawa River, M. Sc. Thesis, Univ. of Ottawa, Ottawa.

[18] Schindler, D.W.; Chark, A.S. and Grey, *J. R. 1971. J. Fish Res. Board Can.* 98 : 559.

[19] Gokecki, *A. Acta Theriol.* 10 : 333.

[20] Slobodkin, L.B. and Richman, S. 1961. *Nature.* 191 : 299.

[21] Lawton, J. H. 1970. *J. Annim. Ecol.* 39 : 669.

[22] Semlitsch, R.D. 1990. *Can. J. Zool* 65 : 1027.

[23] Munshi, J. S. D.; Singh, O. N. and Singh, D.K. 1990. *Tropical Ecol.* 31 : 138.

[24] Peckarskey, B. T. 1982. *Bioscience.* 32 : 261.

[25] Fox, L.R. and Murnoch. 1978. *J. Animal Ecol.* 47 : 945.

[26] Winterboum, M. J. 1974. *Fresh Water Biol.* 4 : 507.

15

Species Composition and Seasonal Fluctuations in Biomass of Zygopteran Nymphs in a Wetland of Santal Pargana, India

• *Poonam Kumari*

Abstract

The present investigation has been made on the systematics, faunistic composition, and community structure and biomass estimation of zygopteran larvae. During the present observation, altogether 22 species belonging to 9 genera under 4 families were recorded. The maximum production of the larvae was found in summer and monsoon months when positive increase in biomass was recorded. The maximum production rate was recorded in October (0.060 g/cub m/day) and minimum (0.004 g/cub m/day) in January. It was also found that the production of these larvae is governed by complex extrinsic and intrinsic physical, chemical and biological factors discussed in the text.

Introduction

Insects are so fantastic that they often seem like creatures from another world. Insects of almost every imaginable shape, beautiful, bizarre, grotesque, ugly, or cosmopolitan distribution can be found. Insects apparently came into being more than 350 million years ago, during the Paleozoic era.

The surrounding environment determines the insect population of a region. In more recent time taxonomists recognized the significance of the related sciences like Anatomy, Cytology,

Histology, Embryology, and Ecology in defining a particular morphological character. Certain taxonomists felt that ecological niche and broad ecological factors play significant role in their taxonomical variations, and their life cycles.

Elton (1927) emphasized the importance of taxonomy regarding ecology as follows – "The extent to which progress in ecology depends upon accurate identification and upon the existence of a sound systematic ground work for all groups of animals". Thus, proper identification and taxonomical position of species living in an ecosystem becomes the essential basis for all ecological studies.

Aquatic insects do not comprise a systematic group but are scattered through a large number of insect orders. These insects are specific that are more or less closely associated with water at least for any one phase of their life cycle. They fall into two major groups, – (i) insects having aquatic larval and pupal stages, and (ii) those spending their entire life cycle in or near the free water area. The former group includes the Ephemeroptera, Odonata and Diptera. The second group includes aquatic Hemiptera and Coleoptera, whose entire cycle is passed in water.

Zygopteran larvae constitute an integral part of the fresh water communities and are fascinating in structure and biology but the taxo-eco-biology of these larvae has been studied scantly in the past except works of Sangal and Kumar (1970a, b) and Kumar and Khanna (1984). Since the publication of Fauna of British India odonata (Vol. 1-3) by Fraser (1933 and 1936), and Rhynchota (Vol. 3-5) by Distant (1906 and 1910), a number of important treatise on odonata and Rhynchota-Heteroptera have appeared. Though approximately 500 species of odonata have been recorded from different parts of our countries, no serious attempts has been made so far to study the larval stages of the dragonfly. Smith and Pritchard (1956) and Gloyd and Wright (1959) have studied extensively different aspects of these insects. However, these studies are concerned with the identification of adult zygoptera. The identification of larva still remains a puzzling problem for limnologists and aquaculturists. However, Kumar (1971a, b and c 1973a and b) and Kumar and Prasad (1977) have extensively studied the taxonomy and ecology of odonate larva of Doon Valley

(U.P.). Roy (1980) has also made an attempt to classify certain zygopteran larvae in a fishpond of Bhagalpur. Kumar (1985) has made a successful attempt in making a checklist of odonate larvae of the wetlands of Santal Pargana. Roy *et.al.* (1988) have studied the faunistic composition of aquatic insects of eastern Bihar with special reference to odonata with notes on some aspects of their ecology. Kaushik *et.al.* (1990) also studied the habitat ecology of odonatan nymphs in certain water bodies of Madhya Pradesh. Andrew and Tembhare (1997) collected and classified the odonates from Nagpur city (Maharashtra). Srivastava and Suribabu (1997) also made an exhaustive study on annotations on the damselfly of Central India. Kumar and Kumari (1999) have extensively studied the taxo-eco-biology of zygopteran larva of different lentic freshwater ecosystems of South Bihar.

However, no information is available regarding the systematic and community structure of dragonflies, larvae and other species inhabiting the freshwater resources of Santal Pargana (Dumka). Therefore, to fill this gap in the tribal zone of Bihar, the present investigation has been made on the systematic, faunistic composition, community structure and biomass estimation of zygopteran larvae of a wetland of Dumka (Santal Pargana).

Materials and Methods

Odonate larvae were collected from the littoral, limnetic and benthic zones of this pond. These larvae were collected by an insect collecting net made up of nylon cloth (mesh size : 40 to 80/cm^2). The benthic and littoral fauna were collected by separate techniques with specific sampling method. The benthic forms were collected by modified surface samplers while littoral fauna were collected by D-shaped aquatic net. The D-frame aquatic net is made according to Merritt and Cummins (1978) and the samples were taken by the method used by Allan (1975). The samples were brought to the laboratory and preserved in 70% alcohol for sorting. Identification of larvae was made in the laboratory. After the identification, the number of species in each hauling was counted and their relative abundance was determined. The habitat preferences and niche specialization of these larvae have also been taken into account.

For the estimation of productivity, Biomass method as suggested by Winberg (1971) was employed. The samples of each hauling were sorted in the laboratory and they were placed in the oven at 105°C for 24 hours and its weight was recorded until a constant weight of the materials was reached. The weight loss at this temperature is considered to represent the ash-free dry weight of the animal and is used as a measure for their biomass (Beukema, 1976). Biomass (g/cub m^{-1} day^{-1}) and net production (kcal/g/day) were calculated by estimating the calorific values of these insects in Oxygen Bomb Calorimeter. The product moment correlation co-efficient for biomass (g/cub m of water) with water temperature, humidity, rainfall and atmospheric temperature was also computed.

Results

During the present investigation, altogether 22 species belonging to 9 genera under 4 families (Figs. 15.1-15.13) were collected whose taxonomical characters are described below :

Order : Odonata

(1) The insects of this order are predatory in both larval and adult stages.

(2) Larvae are purely aquatic and predatory capturing prey from microcrustacea to various aquatic insects and other invertebrates to small fishes and tadpoles.

(3) They are characterized by having two sub-equal pairs of membranous wings, each with a complex reticulation of small cross veins and usually a conspicuous stigma, 3-segmented tarsi, short settiform antennae, strong chewing mouth parts and large compound eyes.

(4) The damselflies (Zygoptera) held the wings together above the body while the dragonflies (Anisoptera) held the wings horizontally in repose.

(5) The males have forcep-like appendages at the end of their abdomen used for grasping the head (Dragonflies) of the female during mating.

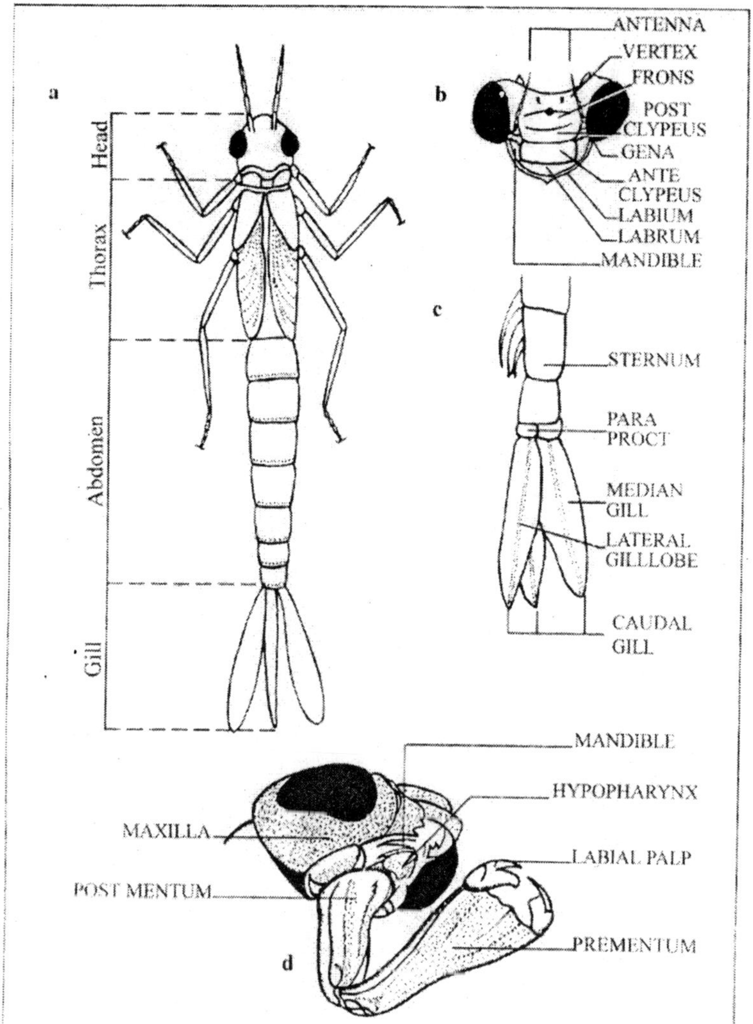

Fig. 15.1 : **(a) Detail structures of a larva of *Ischnura* sp. (Zygoptera); (b) Head portion; (c) Caudal parts; (d) Latero-ventral view of head and mouth parts.**

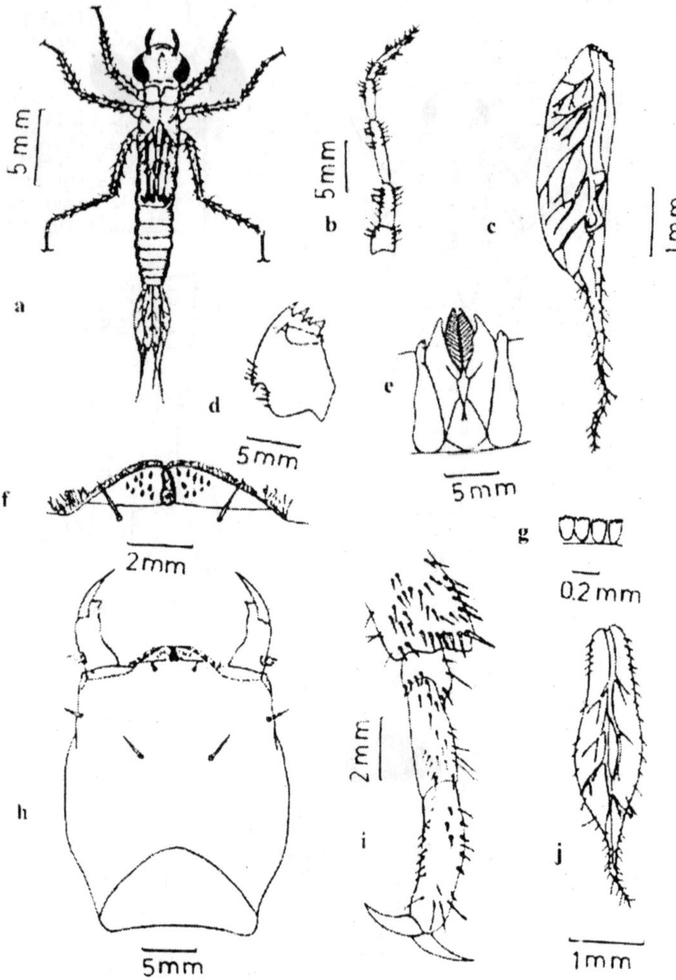

Fig. 15.2 : ***Desparoneura compioni*** **(Fraser) : (a) larva; (b) antenna; (c) paraproct; (d) mandible; (e) female gonapophyses; (f) enlarged view of distal margin of prementum; (g) palisade setae; (h) labium; (i) tibial comb and tarsi; (j) epiproct.**

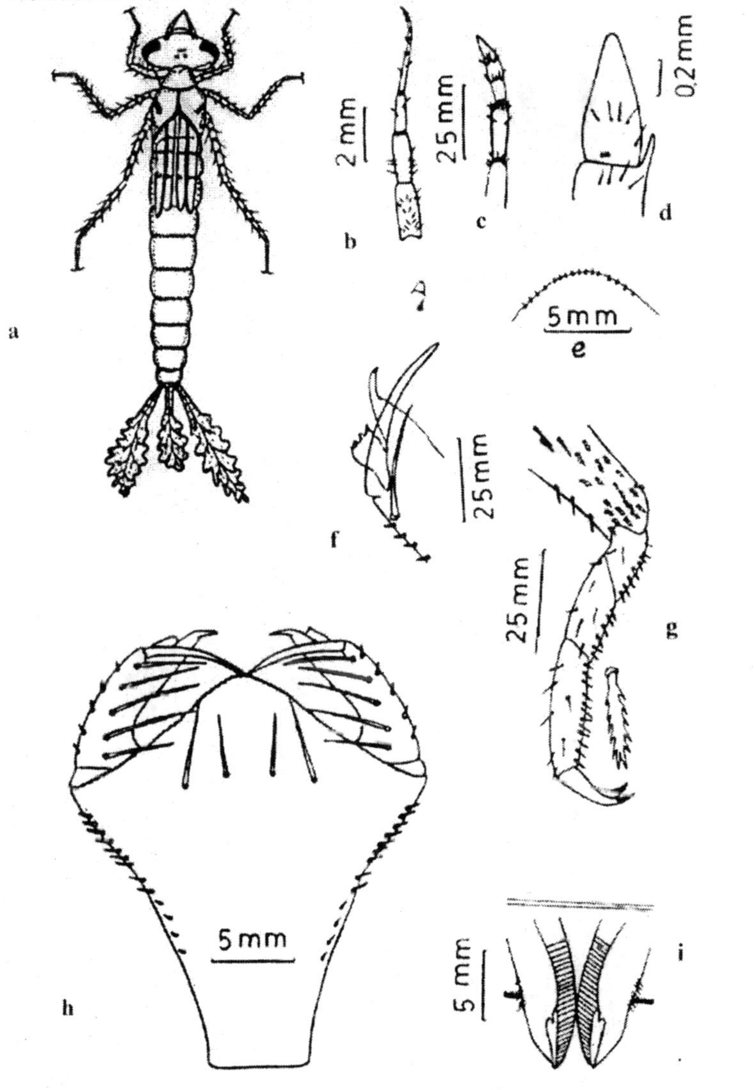

Fig. 15.3 : ***Copera marginipes*** **(Rambur) : (a) larva; (b) antenna; (c) enlarged view last 4 segments antenna; (d) apical spine on sixth segment of antenna; (e) enlarged view distal margin prementum; (f) enlarged view distal margin of palpus; (g) tibial comb and tarsi and a pectinate seta; (h) labium; (i) female gonapophyses.**

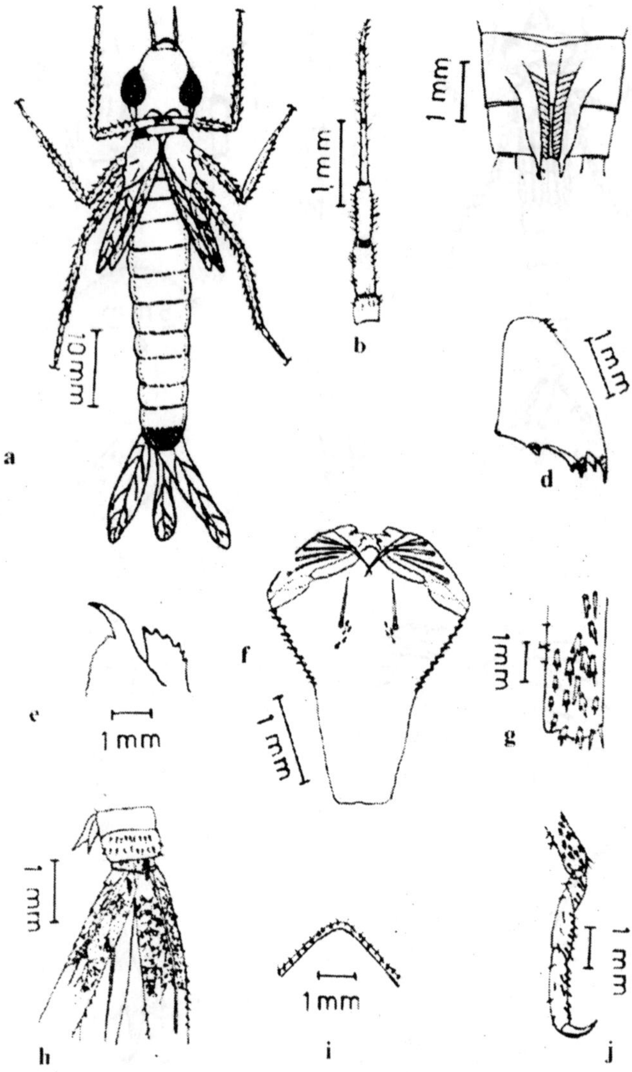

Fig. 15.4 : ***Pseudagrion rubriceps*** **(Selys) : (a) larva; (b) antenna; (c) female gonapophyses; (d) mandible; (e) enlarged view distal margin palpus; (f) labium; (g) enlarged view of tibial comb; (h) caudal lamellae; (i) enlarged view distal margin prementum; (j) tibial comb and tarsi.**

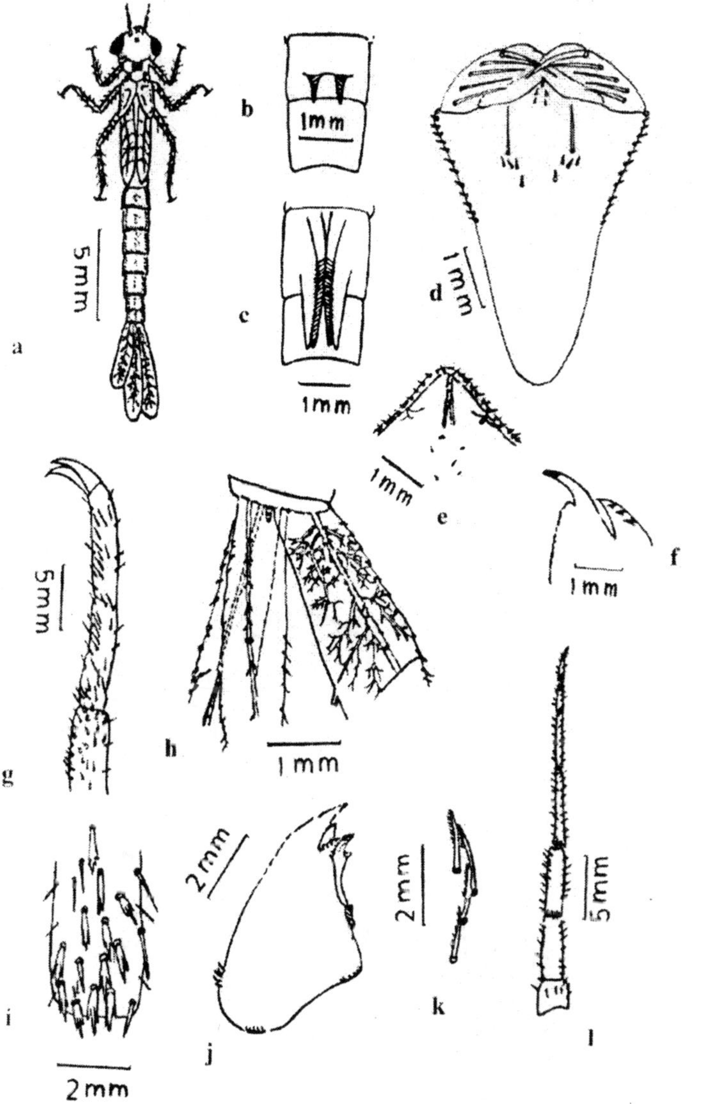

Fig. 15.5 : *Pseudagion laidlawi* (Fraser) : (a) larva; (b) male gonapophyses; (c) female gonapophyses; (d) labium; (e) enlarged view distal margin prementum; (f) enlarged view distal margin palpus; (g) tibial comb and tarsi; (h) caudal lamellae; (i) enlarged view tibial comb; (j) mandible; (k) pectinate sotae tarsi; (l) entenna.

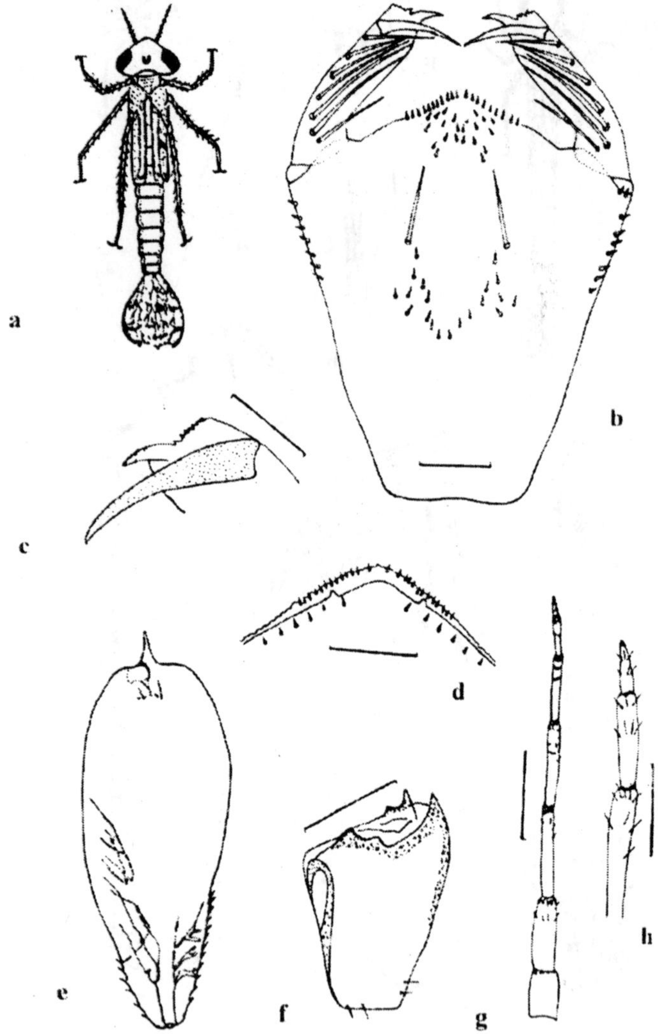

Fig. 15.6 : ***Ceriagrion coromandelianum*** **(Fabricius) : (a) larva; (b) labium; (c) enlarged view distal margin palpus; (d) enlarged view distal margin prementum; (e) epiproct; (f) mandible; (g) antenna; (h) enlarged view distal segments antenna.**

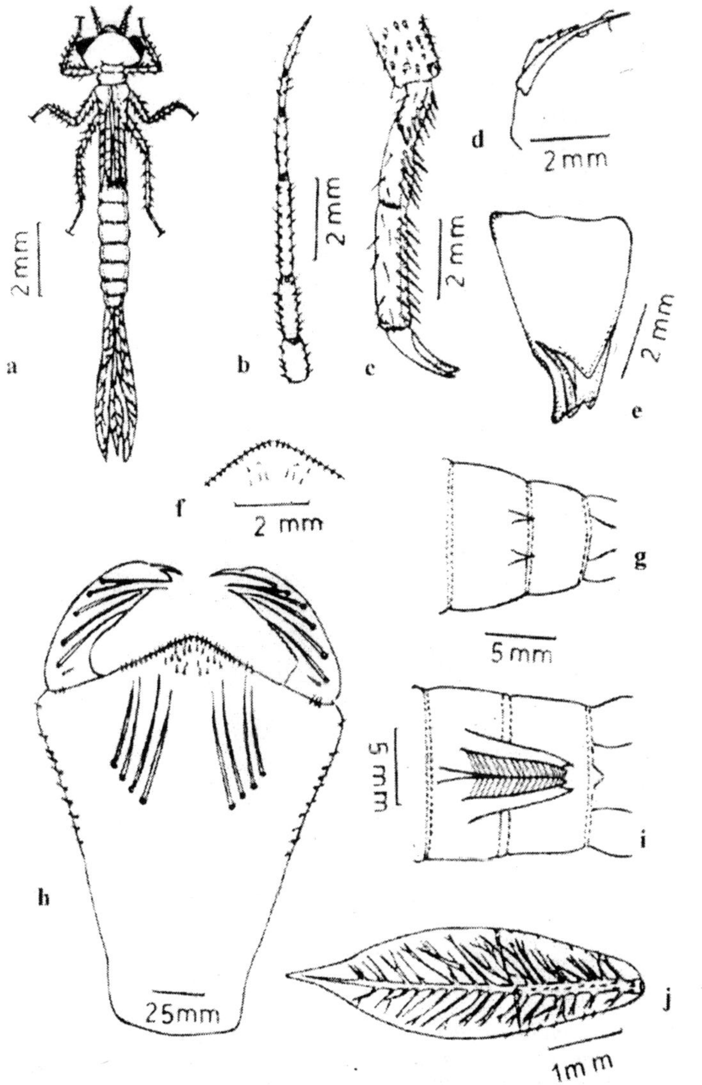

Fig. 15.7 : ***Ischnura delicata*** **(Hagon) : (a) larva; (b) antenna; (c) tibial comb and tarsi; (d) enlarged view distal margin palpus; (e) mandible; (f) enlarged view distal margin prementum; (g) male gonaophyses; (h) labium; (i) female gonapophyses; (j) epiproct.**

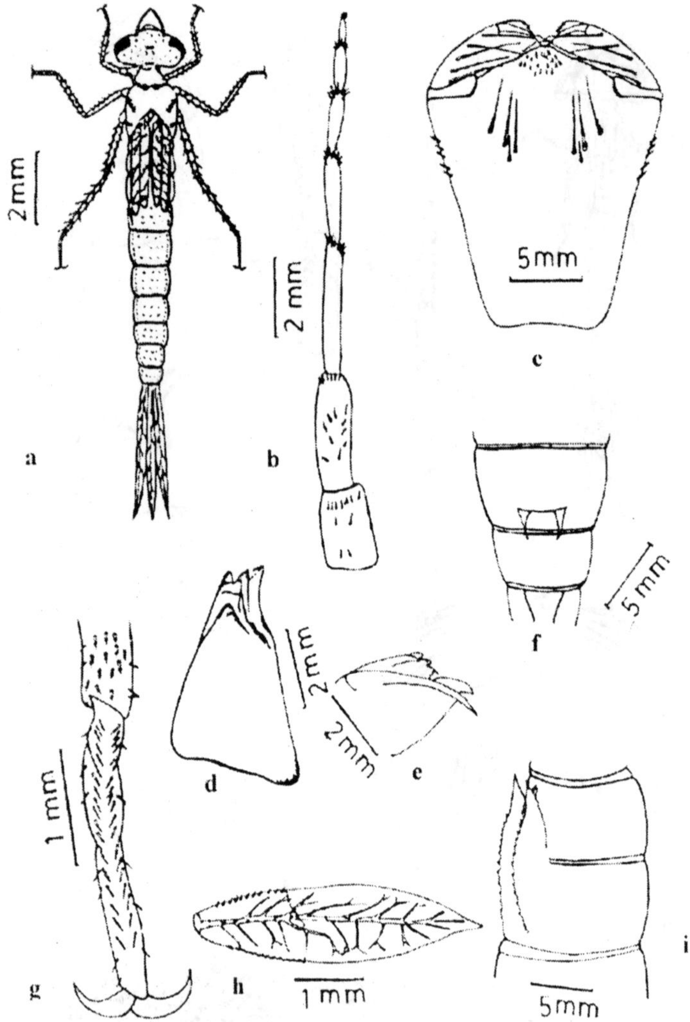

Fig. 15.8 : ***Ischnura senegalensis*** **(Rambur) : (a) larva; (b) antenna; (c) labium; (d) mandible; (e) enlarged view distal margin palpus; (f) male gonapophyses; (g) tibial comb and tarsi; (h) epiproct; (i) female ganopophyses.**

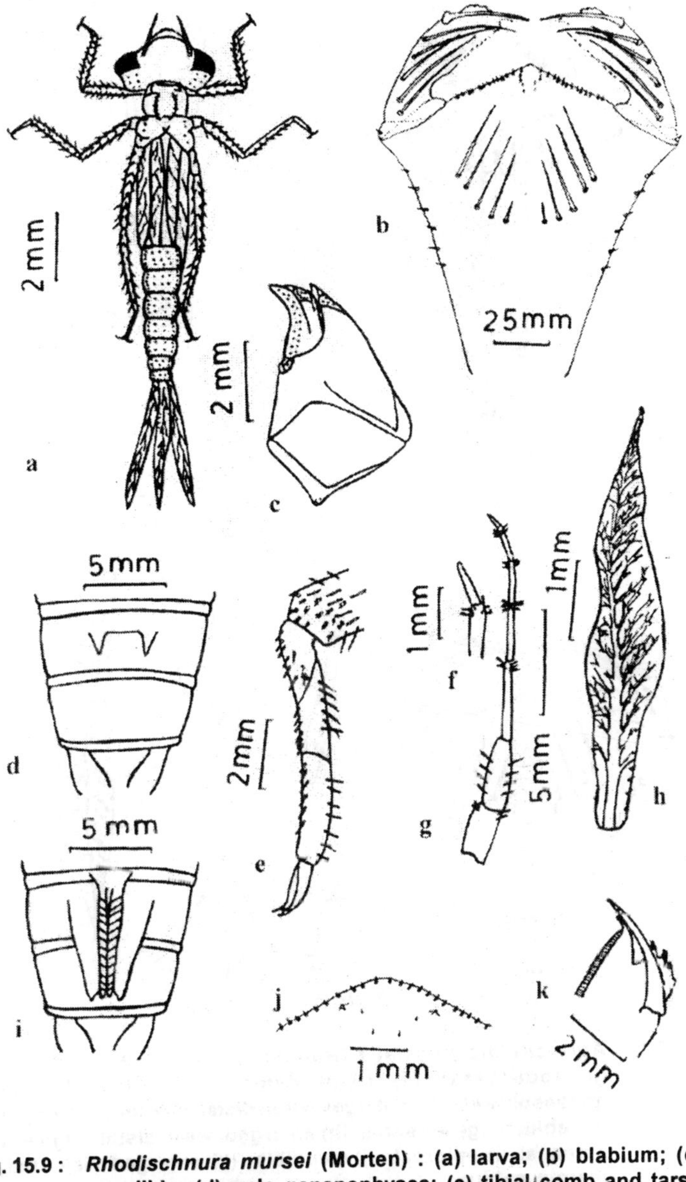

Fig. 15.9 : ***Rhodischnura mursei*** **(Morten) : (a) larva; (b) blabium; (c) mandible; (d) male gonapophyses; (e) tibial comb and tarsi; (f) apical spine on 6th segment antenna; (g) antenna; (h) opiproct; (i) female gonapophyses; (j) enlarged view distal margin premontum; (k) enlarged view distal margin palpus.**

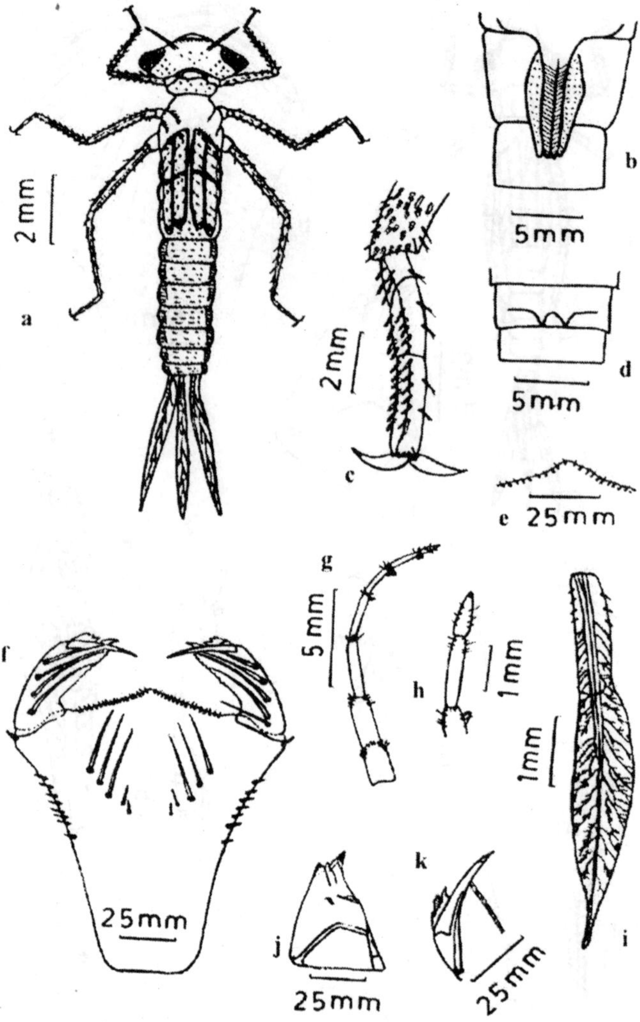

Fig. 15.10 : ***Agriocnemis pygmaea*** **(Rambur) : (a) larva; (b) female gonapophyses; (c) tibial comb and tarsi; (d) male gonapophyses; (e) enlarged view distal margin prementum; (f) labium; (g) antenna; (h) enlarged view distal segments antenna; (i) epiproct; (j) mandible; (k) enlarged view distal margin palpus.**

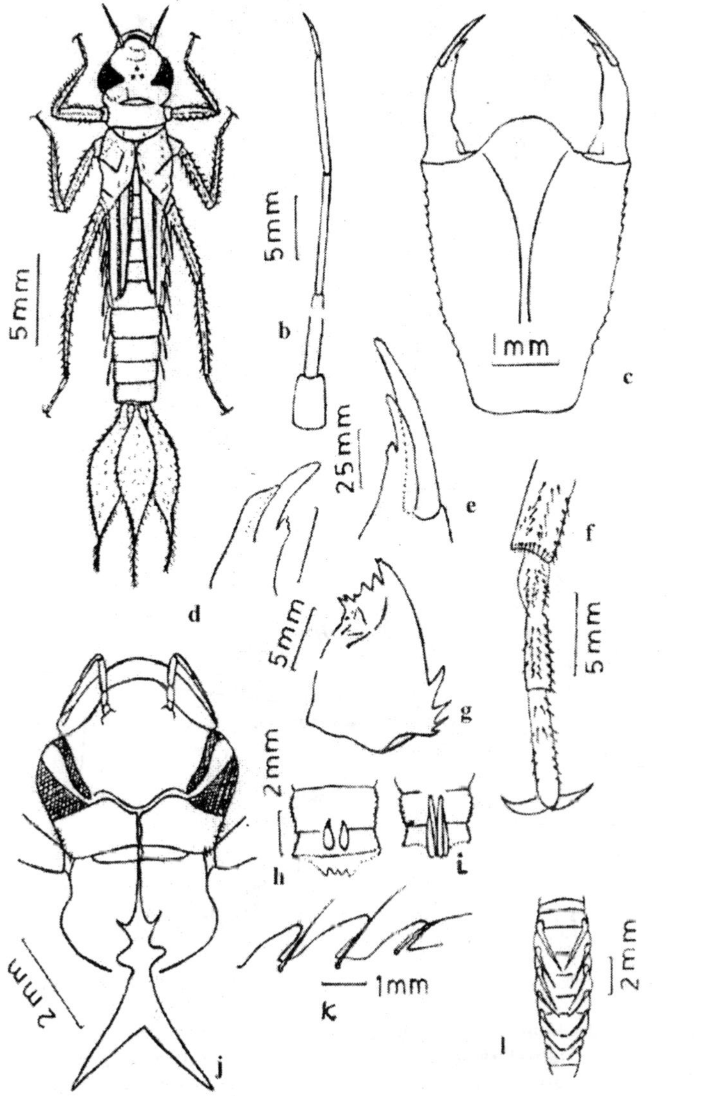

Fig. 15.11 : ***Bayedera indica*** **(Selys) : (a) larva; (b) antenna; (c) labium; (d) enlarged view distal margin palpus; (e) movable hook; (f) tibial comb and tarsi; (g) mandible; (h) male gonapophyses; (i) female gonapophyses; (j) enlarged view head (exuvia); (k) enlarged view lateral prementum; (l) paired ventral abdominal appendages.**

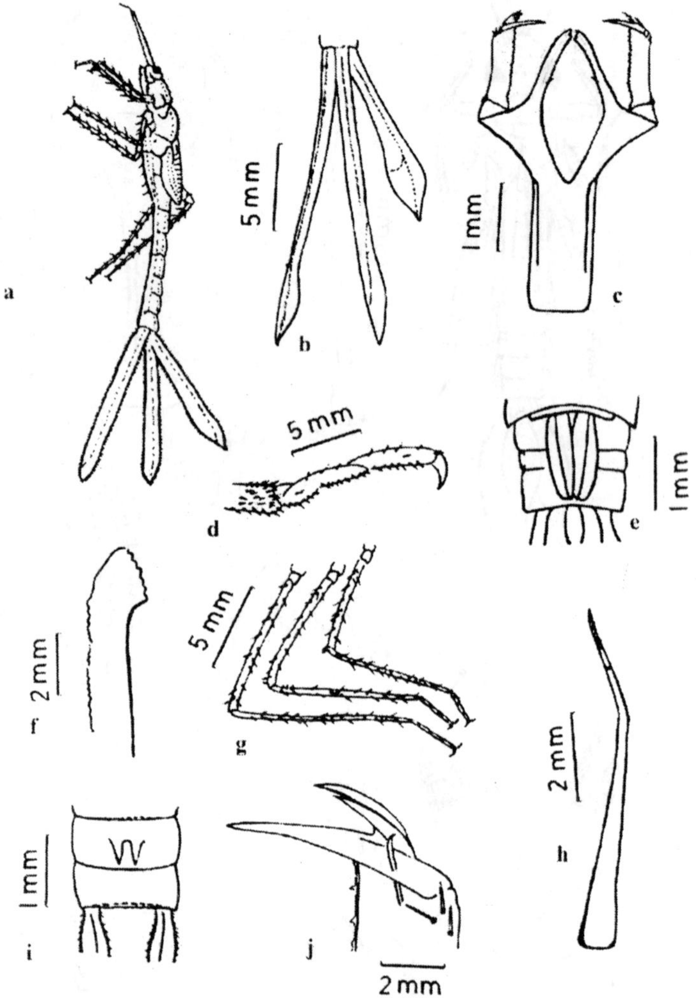

Fig. 15.12 : ***Neurobasis chinensis chinensis*** **(Linn) : (a) Larva; (b) caudal lamellae; (c) labium; (d) tibial comb and tarsi; (e) female gonapophyses; (f) apical lobe prementum; (g) fore, middle and hind legs; (h) antenna; (i) male gonapophyses; (j) enlarged view distal margin palpus.**

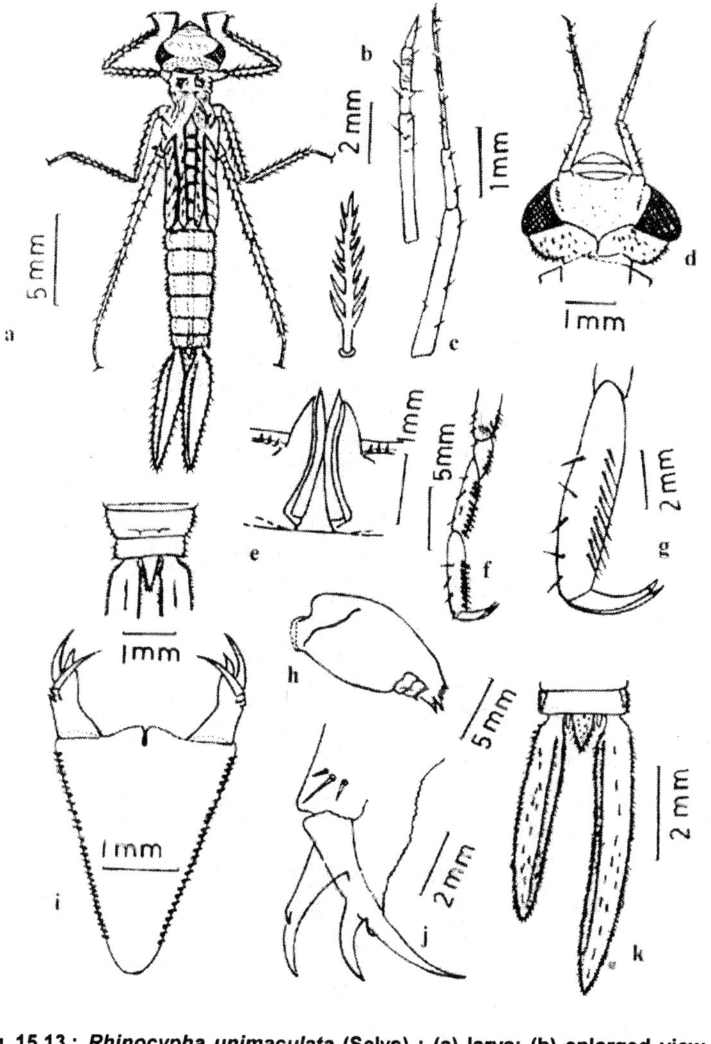

Fig. 15.13 : ***Rhinocypha unimaculata*** **(Selys) : (a) larva; (b) enlarged view distal segments antenna and a pectinate seta tarsi; (c) antenna; (d) enlarged view head; (e) male gonapophyses and female gonapophyses; (f) tibial comb and tarsi; (g) enlarged view last tarsal segment; (h) mandible; (i) labium; (j) enlarged view distal margin palpus; (k) caudal lamellae.**

(6) The mouthparts are remarkable because labium ending in a pair of spiked jaws, which can be extended or retraced by means of an elbow-like articulation. Once prey is captured, the victim is drawn to mandibles where it is chewed up and devoured.

(7) Dragonfly larvae respire by means of a rectal chamber which pumps water in and out and have jet-propulsive locomotive devices.

Sub-order : Zygoptera

1. Larvae long and slender.
2. Abdominal pleurites infolded
3. Abdomen terminating in three caudal lamellae.

Super family : Coenagrioidea

1. Labium almost triangular in out line.
2. Median lobe of prementum without left in the middle.
3. No major setae present on the movable hook.
4. Caudal lamellae with secondary branches oblique to the median trachea.
5. Tibial comb mostly consists of tridentate setae.

Family : Coenagriidae

1. Antenna with first segment of flagellum longer than the pedicel.
2. Caudal lamellae denodate or nodate type.
3. Marginal setae thick and differentiated into stout antenodal and fine postnodal setae. Larvae found in both running and still waters.

Sub-family : Pseudagriinae

1. Premental setae do not exceed 1 + 1.
2. Caudal lamellae lamellate type.
3. Rounded at the apices.

Genus : Pseudagrion (Selys)

1. Caudal lamellae long, typically lamellar type, rounded at the apices an divided distinctly into antenodal and post-nodal areas.

Species :

1. *P. rubriceps,* (Selys).
2. *P. hypermelas* (Selys).
3. *P. spencei* (Fraser).
4. *P. decorum* (Rambur).
5. *P. microcephalum* (Rambur).

(B) Genus : *Ceriagrion* Selys

1. Premental setae 1 + 1.
2. Caudal lamellae broad, obtusely pointed.
3. Found usually in temporary ponds (bivoltine species).

Species :

1. *C. coromandelianum.* (Fabr)
2. *C. olivaceum.*
3. *C. cerinonebellum.*

Sub-family : *Ischurinae*

1. Caudal lamellae broader in the middle.
2. Apex sharply pointed.
3. Spiniform setae present on the median trachea and laterally in the antenodal region.
4. Larvae occur in slow running streams, perennial and temporary ponds.

Genus : *Ischnura* (Char.)

1. Antennae without an apical spine at the fourth flagellar segment.
2. Caudal lamellae broader in the middle with apices sharply pointed.

Species :

1. *Ischnura delicata* (Hagen).
2. *I. senegalensis* (Rambur).
3. *I. nefostigma* (Selys).
4. *I. aurora aurora* (Brauer).

Genus : *Rhodischnura* (Laidlaw)

1. Premental setae 4 + 4.
2. Palpal setae 5 and 5.
3. Tibial comb consists of a number of scattered tridentate setae.

Species :

1. *R. nursei* (Morton).

Genus : *Enallagma*

1. Apex sharply pointed.
2. Caudal lamellae broader in the middle.
3. Larvae occur in perennial and temporary ponds.

Species :

1. *Enallagma parvum.*

Sub-family : *Agriocneminae*

1. Caudal lamellae elongated leaf – like, with apices ending into narrow processes.
2. Lateral bands on lamellae present.
3. Spiniform setae present on the lateral sides in antenodal region.

Genus : *Agriocnemis* (Selys)

1. Premental setae 3 +3.
2. Palpal setae 4 and 4.
3. Tibial comb consists of a number of scattered tridentate setae.
4. Tarsi with a double row of pectinate setae.

Species :

1. *A. pygmaea* (Rambur).
2. *A. dabreui* (Fraser).
3. *A. splendissima* (Laidlaw).

Super family : Lestinioidea

1. Labium greatly contracted basally.
2. Median lobe of pre-mentum with a cleft in the middle.
3. Movable hook armed with major setae.
4. Caudal lamellae with secondary branches at right angles to the median trachea.
5. Tibial comb consists of a number of bidentate setae.

Family : Lestidae

1. Labium spoon-shaped.
2. Distal margin of palp divided into two lobes by a deep cleft.
3. Movable hook armed with major setae.
4. Caudal lamellae lamellate.
5. Larvae plentiful in ponds during monsoon.

Sub family : Lestinae

Movable hook armed with major setae with lamellate caudal lamellae.

Genus : *Lestes* (Selys)

1. Premental setae 6 + 6.
2. Palpal setae 3 and 3 (2 setae on each movable hook).
3. Tibial comb and tarsi consists of bidentate setae.
4. Caudal lamellae simplex, lamellate type.
5. Epiproct with a median curvature.

Species :

1. *Lestes umbrima* (Selys).
2. *L. viridula* (Rambur).
3. *L. praemrosa praemorsa* (Selys).

Super family : Agrioidea

1. Antennae with scape longer than the total length of the remaining segments.
2. Labium without major premental or palpal setae.
3. Caudal lamellae saccoid or triqueteral type.

Genus : *Rhinocypha* (Rambur)

1. Light yellowish in colour with brown markings.
2. Characterized by forcep like anal appendages.
3. Remain concealed beneath the small pebbles.

Species : *Rhinocypha bisignata* Selys

Family : Agriidae

1. Large spider like larvae.
2. All three caudal lamellae large.
3. Antennae 6 - segmented.

Genus : *Neurobasis* Selys

1. Remain attached to the submerge vegetation.
2. Very large in size.
3. It is easily recognized because of their long thin look and spider like legs.

Species : *N. chinensis chinensis* (L.)

Biomass

The biomass and net production of zogopteran nymph during different months of the year have been shown in Table 15.1. The maximum biomass (4.058 g/cub m of water) was recorded in September and minimum in January (1.123 g/cub m). The maximum secondary production of the population was found in summer and monsoon seasons when positive increase in biomass was recorded. The maximum production the g/cub m/day was recorded in October (0.060 g/cub m/day) and minimum g/cub m/day (0.004 g/cub m/day) in January. The maximum net production was recorded in September (0.595 kcal/cub m/day) and minimum in January (0.219 kcal/cub m/day). The product moment correlation co-efficient for biomass (g/cub m) with water temperature, pH, humidity, rainfall and atmospheric temperature was established but interestingly the biomass was found to be significantly correlated with rainfall (r = 0.68; P>0.02). Other factors did not show any marked correlation, however, pH was negatively correlated (r = - 0.161).

Discussion

Insects are well suited for their use in environmental impact assessment because of their qualitative and quantitative abundance over other animal communities and occurrence in almost every imaginable habitat. In the post few years, great interest in aquatic insects has been shown regarding their habit and habitats in relation to the environment (Rosenberg, 1986).

The identification of the larvae is based on the combination of a number of characteristics, namely, the general body shape, distribution of permanental and palpal setae on the labium, and the disposition and grouping of other smaller setae on it. The shape of the labium and that of the distal margin of the prementum and palpals are reliable characters. The disposition and type of the setae on the tibiae and tarsi also help in distinguishing the larvae. Caudal lamellae are one of the important characters for identification. The shape form, tracheation and coloration of the lamellae very in different species but are constant for one species.

Although *Ischnura delicata, I. senegalensis, Rhodischnura nursei, Enallagma parvum, Agriocnemis pygmaea, and Lestes umbrina* were also previously listed by Kumar (1984) and Kumari (1999) from this locality. However, 16 species *viz. P. rubriceps, P. hypermelas, P. spencei, P. decorum, P. microcephalum, C. coromandelianum, C. olivaceum, C. cerinonebellum, I.nefostigma, I. aurora aurora, A. dabreui, A. splendissima laidlaw, L. viridula, L. praemorsa praemrosa, R. bisignata* and *N. chinensis chinensis* were recorded for the first time only from Jharkhand.

It is evident that South Bihar is quite rich in the zygopteran diversity. Considering the enormous number of species of the family coenagrionidae prevalent in the area of Santal Pargana, it may be conclusively state that the climatic conditions of Santal Pargana are very apt and suitable for fostering the members on this family.

Further, two species of the family coenagrionidae deserve additional comments. In *Ischnura senegalensis* and *Agriocnemis pygmaea*, the phenomenon of colour dimorphism has been observed. A part from the usual isochromatic (homochromatic) forms, heterochromatic females are also common.

In the present investigation, the maximum net growth (g/cub m) of the zygopteran nymph population was found in summer and monsoon seasons when positive increase in biomass was investigated (Table 15.1). the net production during specific time interval to be the sum of the new biomass productivity produced by individuals surviving from the beginning to the end of the interval. This total new production is, in fact, equal to the weight loss calculated plus the increase in standing stock biomass. Therefore, in the present investigation during the periods of weight loss, *i.e.*, from October to January (Table 15.1) very little production is going on in the aquatic system.

The weight loss from October to January may also be due to heavy mortality of zygopteran nymphs population, unavailability of their suitable food substances and adverse abiotic factors. Out findings support the views of Benke (1972 and 1976), who studied dragonfly production and prey turnover rate in Georgia. The periods of late October to January designated as winter season, which is thought to be unfavourable for the growth of these nymphs. Thus, these nymphs enter the non-growing phase in the winter and as such there was no consistent increase in their biomass during this season. In the spring (February onwards), when development resumed, biomass once again gets increased. Biomass of these nymphs dropped off rapidly as some of them emerged or attempted to emerge.

In the present study both positive and negative growth of the total nymph population was recorded. According to Boughey (1968) the rate of increase may have a negative as well as positive value. At any instant of time a population may be decining by the

Table 15.1 : Biomass and net production of zygopteran nymphs population from May, 1999 to May, 2000

Date	Biomass (g/haul* day)	Biomass (g/cub m)	Net growth of the population (g/cub m)	Period (days)	Biomass (g/cub m/ day)	Net Production (Kcal/ cub m/day)
1999						
18 May	0.92	1.872	-	-	-	-
19 Jun	1.22	2.439	0.577	32	0.027	0.397
17 Jul	1.55	3.046	0.617	27	0.032	0.575
16 Aug	1.86	3.653	0.617	31	0.029	0.589
19 Sep	2.05	4.058	0.415	35	0.022	0.595
16 Oct	1.28	2.641	1.427	28	0.060	0.529
20 Nov	0.78	1.629	1.022	35	0.038	0.276
18 Dec	0.58	1.224	0.415	28	0.024	0.275
2000						
15 Jan	0.56	1.123	0.111	28	0.004	0.219
15 Feb	0.74	1.447	0.334	31	0.020	0.237
20 Mar	0.86	1.669	0.232	33	0.007	0.277
19 Apr	0.98	1.933	0.274	29	0.009	0.365
17 May	1.39	2.742	0.819	29	0.029	0.490

(* 1 haul =494.0 litres of water)

loss of individual rather than growing by their addition. The change could be due to emigration or immigration, to the birth of the new individuals or the death of other (*i.e.*, natality and mortality).

Product moment correlation co-efficients for biomass (g/cub m) with pH, water temperature, humidity atmospheric temperature and rainfall were estabished in the present study. pH is negatively correlated (r = 0.16) to the biomass zygopteran nymph population, however, each species has been described to have its own tolerance of acidity and lakalinity. Water temperture, atmospheric temperature and humidity showed no marked correlation with the biomass. Off all the abiotic factors studied, rainfall is highly correlated (r = 0.68; P>0.02).

Acknowledgement

The author is grateful to Professor (Dr.) Sheodani Prasad Singh, Department of Chemistry, R. D. College, Sheikpura for his continuous encouragement.

References

Allan, J.D. 1975. The distributional ecology and diversity of benthic insects in cement creek, Colorado, *Ecology*. 56 : 1040-1053.

Andrew, R. J. and Tembhare, D. B. 1997. Collection of odonata from Nagpur City, Maharastra State, India. *Fraseria*. 4 : 1-4.

Benke, A. C. 1972. An experimental field study on the ecology of coexisting larval odonates. Ph. D. dissertations. Athens, Georgia, Univ. of Georgia. 122p.

Benke, A.C. 1976. Dragonfly production and prey turnover. *Ecology*. 57 : 915-927.

Benke, A. C. and Benke, S.S. 1975. Comparative dynamics and life histories of coexisting dragonfly population. *Ecology*. 56 : 302-317.

Beukema, J.J. 1976. Biomass and species richness of the macrobenthic animals living on the tidal flats of the Dutch Wadden Sea. *Netherland J. of Sea Research*. 10 : 236-261.

Distant, W. L. 1906. *The Fauna of British India*, Vol. III. Heteroptera, London, Taylor and Francis Ltd.

Distant, W. L. 1910. *The Fauna of British India*. Vol. V. Heteroptera, London, Taylor and Francis Ltd.

Elton, C. S. 1927. *Animal Ecology*. *Sidgwick* and Jackson Ltd., London.

Fraser, F. C. 1933. *Fauna of British India, Odonata* I. London, Taylor and Francis Ltd.

Fraser, F. C. 1936. *Fauna of British India, Odonata* 3. London, Tylor and Francis Ltd.

Gloyd, L. K. and Wright, M. 1959. "Odonata" In : *Freshwater Biology* 2nd. John Wiley and Sons. INC., New York. pp. 917-940.

Kumar, A. 1971a. Taxonomic studies of the last instar larvae of odonata from the Dehra Dun Valley (India), Ph. D. thesis, Univ. of Meerut.

Kumar, A. 1971b. The larval stages of *Orthetrum brunneum* with a description of the last instar larvae of *Orthetrum tacniolatum* (Schreidor) (Odonata : libellulidae). *J. Nat. Hist*. 5 : 121-132.

Kumar, A. 1971c. Studies on the phenology of dragonflies of the Dehra Dun Valley (India). *Ins. Europ. Symp. Odonatology Gent*. 29-30.

Kumar, A. 1973a. Description of the last instar larvae of odonata from the Dehra Dun Valley (India) with notes on biology. I. (suborder-zygoptera). *Oriental Insects.* 7 : 83-118.

Kumar, A. 1973b. Description of the last instar larvae of odonata from the Dehra Dun Valley (India) with notes on biology. II. Sub-order-Anisaptera. *Oriental insects. J. pp.* 291-331.

Kumar, A. and Prasad, M. 1977. On the larvae of *Tamea viggínia* (Rambur, 1842) from India, with notes on the larvae of India representatives of the genus *Tramea* Hagen, 1861 (Libelluloidea : odonata). *J. Bombay Nat. Hist. Soc. 74* : 199-202.

Kumar, A. and Khanna, V. 1983. A review of the taxonomy and ecology of odonata larvae from India. *Oriental Insects. 17* : 127-157.

Kumar, A. 1982. Some aspects of eco-physiology of larval odonates of wetlands of Santal Pargana, Ph. D. Thesis, Bhagalpur Univ. Bhagalpur.

Kumar, A. 1999. Impact of sewage pollution on the population dynamics of odonate larvae. Fraseria (in press).

Kumar, A. and Kumari, P. 1999. Impact of human interferences on certain abiotic and biotic factors of lentic freshwater ecosystem. *J. Env. and Poll.* (in press).

Kumar, A. and Kumari, P. 1999. Taxo-ecobiology of zygopteran larvae of different freshwater ecosystems of South Bihar, India. *J. Env. and Ecoplan.* (in press).

Kaushik, S.; Sharma, S.; Saxena, M.N. and Saksena, D.N. 1990. Habitat ecology of Anisoptera nymphs in certain water bodies of Madhya Pradesh. *Acta. Hydrochin. Hydrobiol. 18* : 443-447.

Laudin, J. 1976. Seasonal patterns in abundance of water beetles belonging to the Hydrophiloidea (Coleoptera) *Freshwater Ecol.* 6 : 89-108.

Merrit, R. W. and Cummins, K. W. 1978. *An introduction to the aquatic insects of North America.* Kendall/Hunt Publishing Company 2460 Keeper Boulevard Dubuque IUWA-52001, U.S.A. III Printing (1979) p. 427.

Roy, S. P. 1980. Studies on some aspects of ecology of the aquatic insects of Bhagalpur. Ph.D. Thesis, Bhagalpur Univ., Bhagalpur (Bihar), India.

Rosenberg, M. D. 1986. Importance of Insects in environmental impact assessment; *Environmental Management.* 10 : 773-783.

Roy, S. P.; Pathak, H. S. and Kumar, V. 1988. Faunistic composition of aquatic insects of eastern Bihar, with notes on their some aspects of ecology. *Rec. Zool. Sruv. India. 85* : 49-57.

Smith, R. F. and Pritchard, A. E. 1956. Chapter 4, odonata In O. R. L. Usinger (ed). Aquatic Insects of California with keys to North American Genera and California species. Univ. of California Press, Berkeley and los Angeles. pp. 106-153.

Sangal, S. K. and Kumar, A. 1970a. Studies on the taxonomy of larvae of Doon Valley. Odonata I. *J. Nat. Hist.* 4 : 33-38.

Sangal, S. K. and Kumar, A. 1970b. Studies on the taxonomy of larvae of Doon Velley. Odonata II. *J. Nat. Hist.* 4 : 305-313.

Srivastava, V. K. and Surbelu, B. 1997. Annotations on the damselfly collection from Sugar, Central India. *Fraseria.* 4 : 13-15.

Winberg, G. G. 1971. *Methods for the estimation of production of aquatic animals*. Academic Press, New York. 175p.

16

On the Structure and Life History of Three New Species of Septate Gregarines (Apicomplexa : Conoidasida) from Odonates of West Bengal

• *D.P. Haldar and S. Biswas*

Abstract

Descriptions of structures and life history of three new species of septate gregarines (Apicomplexa : Conoidasida) from different odonate hosts of West Bengal have been given. The gregarines are : *Odonaticola platyepimerita* sp. n. from *Ortehtrum sabina* (Drury), *O. bengalensis* from *Trithemis aurora* (Burmeister) and *O. contaminata* from *Brachythemis contaminata* (Fabricius), all belonging to the family Libellyulidae under the order Odonata.

Information regarding the deposition of type materails have also been furnished.

Key Words : *Septate gregarines; Apicomplexa; Conoidasida; Three new species of Odonaticola.*

Introduction

The septate gregarines (Apicomplexa : Conoidasida) are endoparasitic protozoans and are commonly found inside the digestive tract of various arthropod hosts. Interest in the study of the group has gained greater importance in the recent years because of their possible role as biological control agents of harmful and noxious insects (Lipa, 1967; Levine, 1988; Haldar *et.al.*, 1988).

In course of our investigation, we have obtained several septate gregarines in the midgut of odonates commonly found in the state of West Bengal, India, with particular reference to the district of Nadia adjoining the neighbouring country of Bangladesh. Detailed information on the structure and life history of three of these parasitic species, all belonging to the genus *Odonaticola* Sarkar and Haldar, 1981 have been incorporated in this communication. All have been designated as new species.

Odonates are one of the most elegant and beautiful creatures of nature and, according to Srivastava and Agarwal (1992), these have no parallel in the whole of the insect world with regard to their prevalence and abundance in nature. These are beneficial insects and are wonderful predators for many known and unknown insect pests. Despite their much economic values, they have touched the life and activities of man only very lightly. It is likely that if they disappear from the earth, most of us would feel their loss in the slightest. Ecologically, odonates are most powerful determining factors in preserving the balance of life in the ponds, rivers, lakes and their surroundings (Tillyard, 1917).

The slides containing the new species have been deposited at the Department of Zoology, University of Kalyani.

Material and Methods

The host insects were collected in and around the localities of a small town of Karimpur situated between 23° - 25°N latitude and 88°-89°E longitude in the district of Nadia adjoining Bangladesh. Since the Tropic of Cancer passes through the district, the average summer temperature rises up to 42°C while it comes down to around 4°C during winter. During the monsoon months, the average rainfall in the locality is recorded to be around 6.7 mm a day and the relative humidity ranges from 87.7% to 92.2%. The condition is thus very favourable for the gregarines to flourish during the monsoon.

Odonates do not survive under captivity and therefore these were examined on the very day of collection in order to reduce the chances of defaunation due to starvation or some other reasons. The insects were dissected under dissecting binocular and the entire gut was carefully taken out and placed on a glass slide with a little drop of 0.5% saline solution. The presence of parasites within the gut could be detected by microscopic examination. The gut was gently teased with needles for the parasites to come out of the gut lumen. Thin smears were prepared on glass slides and were allowed to be semidried. The slide containing gregarines were fixed in Schaudinn's fluid for 20 m, stained in Heidenhain's haematoxylin, differentiated, dehydrated, cleared in xylene and finally mounted in DPX. Cytes were collected form the midgut as well as hindgut of the infected hosts and cultured according to a technique of Sprague (1941) with slight modification. Oocysts were stained with Lugol's iodine for the detection of the number and arrangement of sporozoites. Illustration presented in this paper have been drawn with the aid of a mirror type camera lucida. The measurements were done by an ocular micrometer calibrated with a stage micrometer.

Observation

Odonaticola platyepimerita sp. n.

Trophozoite : Only three trophozoites have so far been encountered in smear preparations. These are solitary and elongated whose epimerite consists of a slender neck bearing a small flattened hat-shaped knob at its anterior extremity. The epimeritre is drawn into eight small, slender hook-like projections at its periphery. The projections are downwardly recurved and are very sharp at their apices. Teh protomerite is ovoidal to dome-shaped and its width is generally greater than length. The deutomerite is cylindrical to cylindro-conical and the single, vesicular nucleus occupies its anterior portion. There is a distinct nuclear membrane.

The cytoplasm of the trophozoite is usually lightly granulated; however, any epicyteal striations are not observed.

Sporadin : These are found in various sizes and shapes. The smallest sporadin found in smears measures 146.5 μm in length and 71.0 μm in width. Its protomerite is very samll and conical to dome-shaped. In young sporadin, the deutomerite is conical in shape being wider at its anterior end and tapers to a sharp posterior

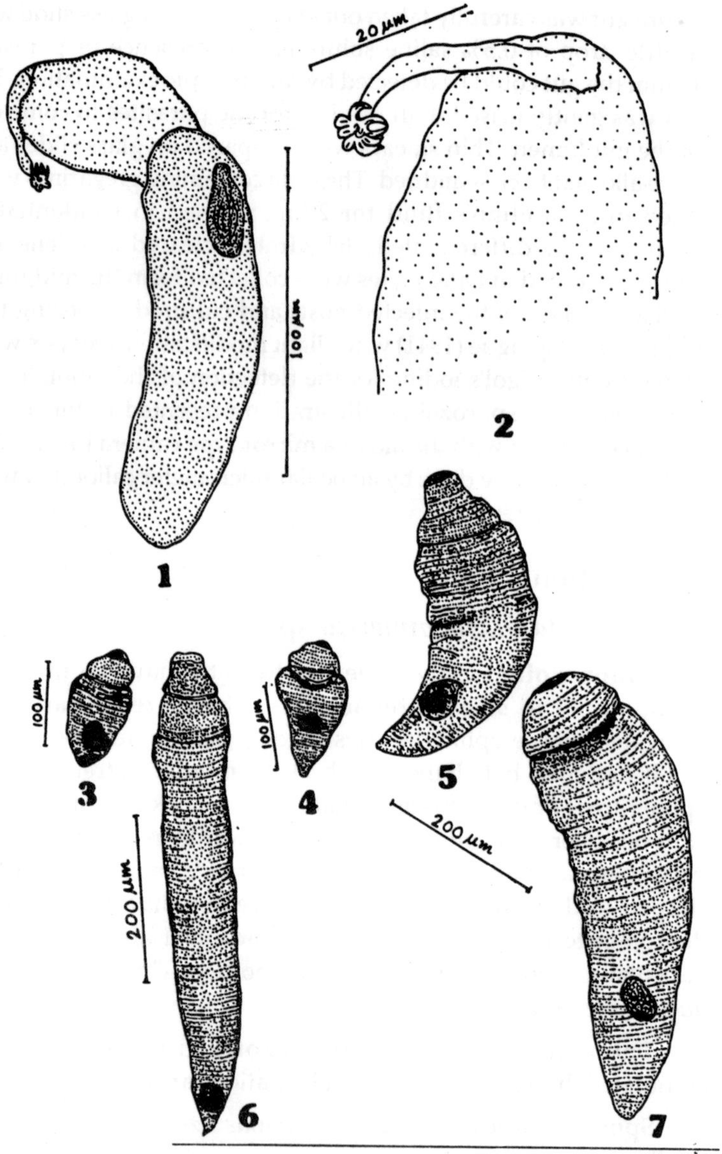

Plate I

Figs. 16.1-16.7 : Camera lucida drawings of *Odonaticola* sp. I n.sp. (16.1) A fully grown trophozoite; (16.2) An epimerite enlarged; (16.3-16.6) Young sporadins of various shapes and sizes; (16.7) A mature sporadin.

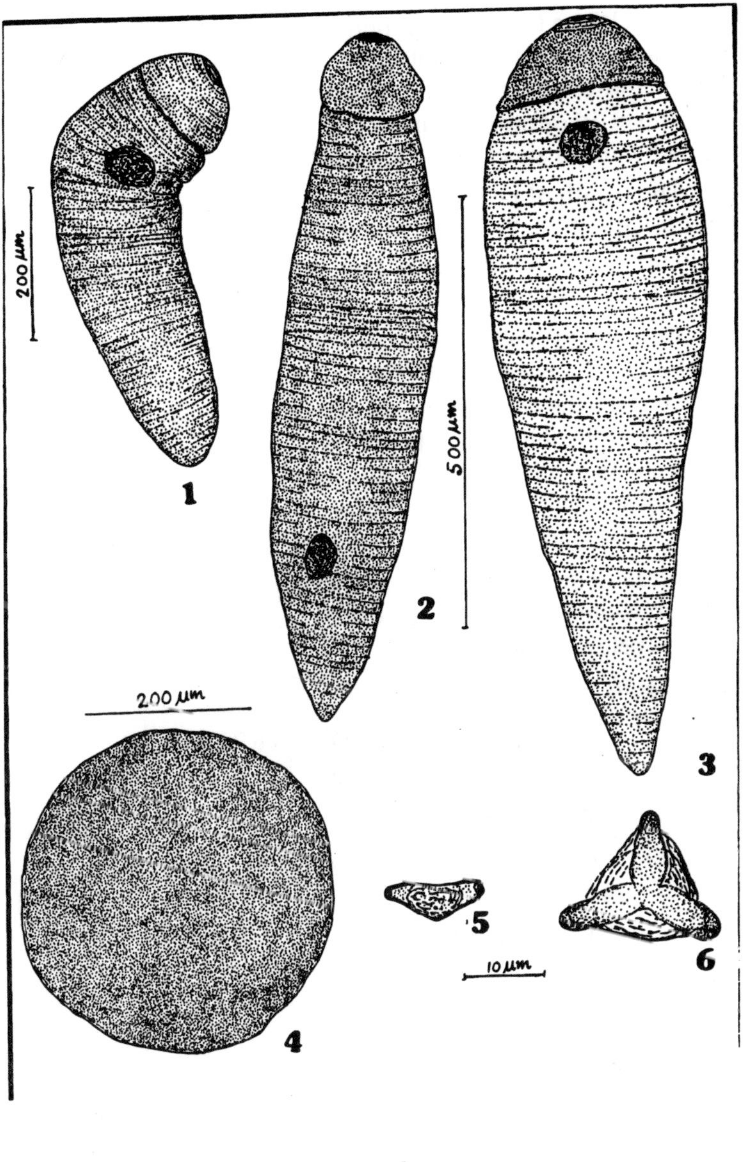

Plate II

Figs. 16.1-16.6 : Camera lucida drawings of *Odonaticola* sp. I n. sp. (16.1-16.3) Mature sporadins of various shapes and sizes; (16.4) A mature gametocyst; (16.5) A oocyst; (16.6) Three oocysts in cluster.

extremity. When fully grown, sporadins assume a napiform shape which are widest at their anterior end. The nuclei are of variable shapes and may assume any position in the deutomerite. The cytoplasm is densely granulated and distinct striations of myoneme fibres are clearly discernable.

The length of protomerite is five to 18 times small than the total length (LP : TL = 1 : 4.8 - 18.4), while the width of the protomerite is four times lesser than that of the deutomerite (WP : WD = 1 : 0.6 - 3.9).

Gametocyst and Oocyst : Freshly collected gametocysts are whitish in colour and are slightly ovoidal in shape. The partition wall between the two uniting gamonts is very distinctly observed. At 16 hours the thin ectocyst covering the gametocyst disappears and at 48 hrs it dehisces by a simple rupture releasing the oocysts. The later are released either singly or in clusters of three. A single oocyst is boat shaped and 12.0 x 4.2 μm in dimensions.

Measurements : The summary of measurements (in microns) of trophozoites and sporadins with the mean within parentheses is given below (n=20) :

Trophozoite

TL	= 195.4 - 753.5 (434.7)		WE	=	6.7 - 86.9 (40.1)
LE	= 13.3 - 58.0 (34.4)		WP	=	35.5 - 101.4 (70.8)
LP	= 44.4 - 62.2 (54.9)		WD	=	32.0 - 181.1 (106.6)
LD	= 119.9 - 637.6 (345.7)		WN	=	11.1 - 43.5 (30.0)
LEN	= 24.4 - 36.2 (30.3)				
LN	= 17.8 - 58.0 (43.0)				

Sporadin

TL	= 146.5 - 1898.2 (898.5)		WP	=	53.3 - 362.2 (156.1)
LP	= 48.8 - 188.4 (124.3)		WD	=	71.0 - 579.6 (209.6)
LD	= 97.7 - 1709.8 (774.1)		WN	=	21.7 - 86.9 (44.8)
LN	= 31.1 - 101.4 (62.3)				

LP : TL = 1 : 4.8-18.4

WP : WD = 1 : 0.6-3.9

Table 16.1 shows the R, $\bar{X}$, SD, SE and CV% of measurements of *Odonaticola playepimerita* sp. n. The same has been graphically represented in Plate III.

Table 16.1 : Showing R, X, SD, SE, CV% of measurements of *Odonaticola platyepimerita* sp. n.

	R	$\bar{X}$	SD	SE	CV%
TL	146.5 - 1898.2	828.9	445.3	99.6	53.7
LE	13.3 - 58.0	34.4	22.4	13.0	65.2
LP	44.4 - 188.4	113.9	44.8	10.0	39.3
LD	97.7 - 1709.8	709.9	409.8	91.6	57.7
LN	17.8 - 101.4	59.4	20.3	4.5	34.2
WE	6.7 - 86.9	40.1	41.8	24.1	104.2
WP	35.5 - 362.2	143.3	85.1	19.0	59.4
WD	32.0 - 579.6	194.1	135.4	30.3	69.7
WN	11.1 - 86.9	42.6	19.0	4.3	44.7

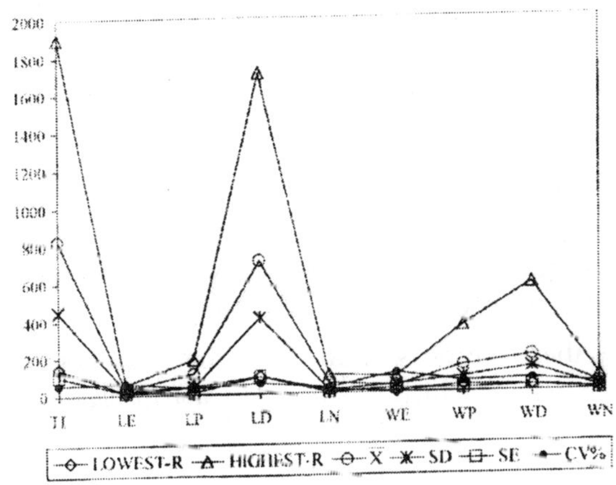

Plate III

Showing R, $\bar{X}$, SD, SE, CV% of measurements of *Odonaticola platyepimerita* sp. n.

Material

(a) **Holotype** - on slide No. DR/2 - 10, prepared from the contents of the midgut of the odonate, *Orthetryum sabina* (Drury), collected from Senpara in the district of Nadia, West Bengal, India by S. Biswas on October 11, 1995.

(b) **Paratype** - Many, on the above-numbered slide and on other slides; other particulars are the same as for the holotype material.

Affinities : In having solitary sporadins, hat-shaped epimerite with petaloid spines, long neck and smooth, boat-shape spores, the gregarine is at once included under the genus *Odonaticola* Sarkar and Haldar, 1981. *Orthetrum sabina* (Drury) has previously been reported to harbour a gregarine, *Odonaticola orthetri* by Sarkar and Haldar (1981). Both the present form and *O. orthetri* have boat-shaped oocysts, but distinct size varaitions exist between the two. Furthermore, in *O. ortheri*, there are numerous, petaloid spines in the epimerite, whereas the numbers fixed to eight in the present form. Interestingly, this particular number of spines is also present in *Odonaticola rodgii*, described as a new species by Sarkar and Haldar (1981) from *Neorothemis t. tullia* (Drury) which differs from the present form in all other characters. In summary, the following characters are listed for the gregarine under report :

(1) epimerite flattened;

(2) smaller petaloid spines;

(3) the neck is much shorter;

(4) dimensions of the various parts of the trophozoite, sporadin and gametocyst as well as the ratios of the different body parts are not in agreement.

It may, therefore, be concluded that the species of *Odonaticola* is a new species for which it is named *Odonaticola platyepimerita* sp. n.

Taxonomic Diagnosis

Host : *Orthetrum sabina* (Drury).

Locality : Senpara (Nadia) and Domkal (Murshidabad) in the state of West Bengal, India.

Site of infection : Foregut and midgut.

Incidence : 35 out of 131 (26.7%) are parasitzed during June to November; higher infection rate is noted during August to October. The higher rate of incidence during this period may be attributed to the favourable condition of rainfall, humidity and temperature as was reported by Haldar and Sengupta (1985).

Etymology : The name *platyepimeritus* has been given to emphasize the flat epimerite of the gregarine.

Odonaticola benglensis sp. n.

Trophozoite : Large numbers of trophozoites are usually encountered in preparations made from midgut. The trophzoite has the usual three segments, viz., the epimerite, protomerite and deutomerite. The epimerite consists of a long neck holding an umbrella-shaped knob at its tip. The aboral side of the knob is slightly tapering and bears nine petaloid lobes around its periphery. Each lobe is provided with a centrally located rib-like impression. Under the phase-contrast microscope, the movement of the neck was studied. And it has been noted that the neck is highly flexible as it can be bent on any side. The protomerite is dome-shaped which becomes tapering towards its anterior extremity for holding the epimerite. The deutomerite is long, tubulo-conical, wider at the middle and exhibits transverse epicyteal striations. The nucleus is variable in position and the cytoplasm is granulated.

Sporadin : In the living condition, sporadins appear as milky white bodies under the phase-contrast microscope and are of highly variable shapes and sizes. The longest sporadin recorded so far by us measures 2028.6 μm and is one of the longest of the species of *Odonaticola* described by any author so far. The protomerite is dome-shaped and its width is generally greater than length. The pellicle is highly flexible and the gregarine is capable of twisting its body to a great extent. The nucleus is ovoidal or ellipsoidal and the cytoplasm is densely granulated; transverse epicyteal striations are clearly visible.

Association : This takes place between two fully mature sporadins. The associating partners vary in their shapes, sizes as well as in the position of their nuclei. After a brief sidewise existence, the individuals surround themselves within a cyst wall secreted by them.

Gametocyst and oocyst : Freshly collected gametocysts are ovoid or spherical in shape and creamy white in colour. A cyst

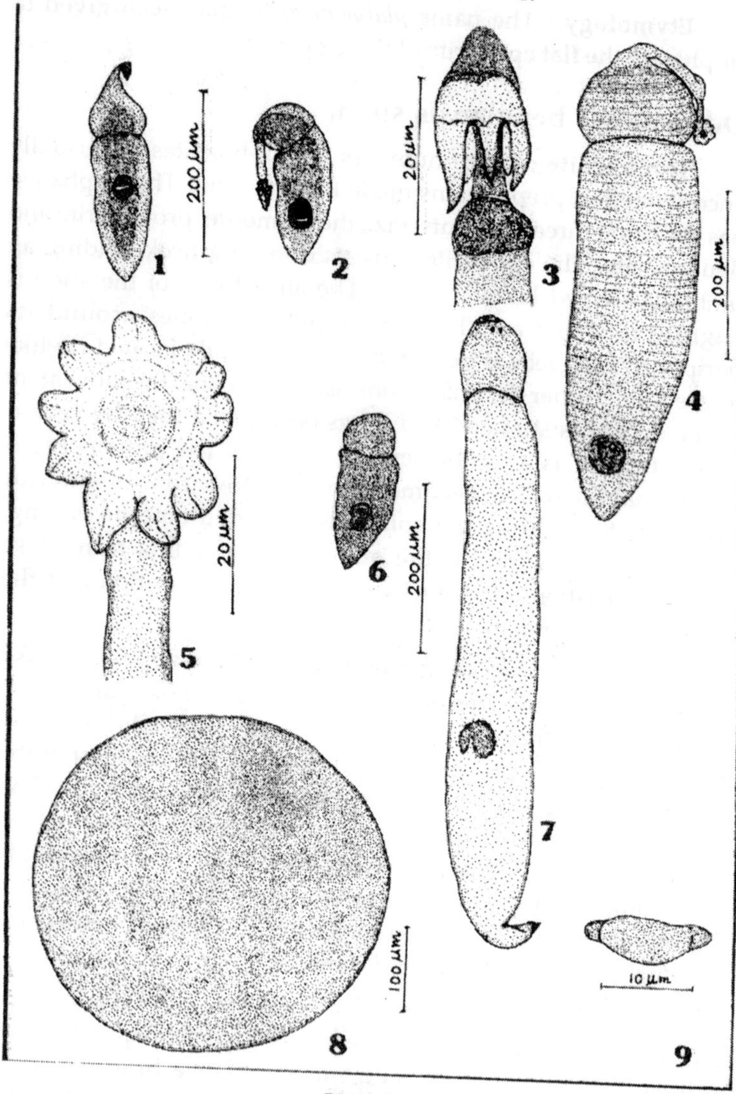

Plate IV

Figs. 16.1-16.9 : Camera lucida drawings of *Odonaticola* sp. II n. sp. (16.1) A young trophozoite; (16.2) A mature trophozoite; (16.3) Lateral view of an epimerite enlarged; (16.4) A mature trophozoite; (16.5) Frontal view of an epimerite enlarged; (16.6-16.7) Sporadins of various shapes and sizes; (16.8) A gametocyst; (16.9) A oocyst.

measures 796.9 µm x 710.0 µm in dimension. The cyst wall is thin and is devoid of any ectocyst. At 48h the cyst ruptures by simple rupture and release oocysts. Oocysts come out either singly or in clusters of three when they display a triangular appearance. These are boat-shaped and measure 11.0 µm in length and 4.2 µm in breadth.

Measurements : The summary of measurements (in microns) of trophozoites and sporadins with the mean within parentheses is given below (n=20) :

Trophozoites

TI	=	199.8 - 811.4 (398.5)	WE	=	8.9 - 21.7 (12.3)
LE	=	15.5 - 188.4 (94.1)	WP	=	53.3 - 159.4 (86.9)
LP	=	48.8 - 115.9 (74.7)	WD	=	57.7 - 173.9 (93.4)
LD	=	133.2 - 507.1 (229.7)	WN	=	17.8 - 43.5 (29.1)
LEN	=	8.9 - 159.4 (97.0)			
LN	=	26.6 - 58.0 (37.8)			

Sporadin

TL	=	213.1 - 2028.6 (1328.9)	WP	=	75.5 - 246.3 (195.3)
LP	=	66.6 - 246.3 (157.1)	WD	=	75.5 - 449.2 (283.2)
LD	=	146.5 - 1782.3 (1171.9)	WN	=	29.0 - 72.4 (53.1)
LN	=	35.5 - 115.9 (76.7)			

LP : TL = 1 : 5.4-19.9

WP : WD = 1 : 1.1-3.7

Table 16.2 shows the R, $\bar{X}$, SD, SE and CV% of measurements of *Odonaticola bengalensis* sp. n. The same has been graphically represented in Plate V.

Material

(a) **Holotype :** On slide No. DR 3/6, prepared from the mdigut contents of the odonate, *Trithemis aurora* (Burmeister), collected from Karimpur, District Nadia, West Bengal, India by S. Biswas on September 11, 1995.

(b) **Paratype :** Many; on the above-numbered slide and on other slides; other particulars are the same as for the holotype material.

Table 16.2 : Showing R, $\overline{X}$, SD, SE, CV% of measurements of *Odonaticola bengalensis* sp. n.

	R	X	SD	SE	CV%
TL	199.8 - 2028.6	1096.3	545.1	121.9	49.7
LE	15.5 - 188.4	94.1	75.6	33.8	80.3
LP	48.8 - 246.3	136.5	56.7	12.7	41.5
LD	133.2 - 1782.3	936.3	524.0	117.2	56.0
LN	26.6 - 115.9	67.0	23.9	5.3	35.7
WE	8.9 - 21.7	12.3	5.3	2.4	43.1
WP	53.3 - 246.3	168.2	67.4	15.1	40.1
WD	57.7 - 449.2	235.8	129.4	28.9	54.9
WN	17.8 - 72.4	47.1	17.0	3.8	36.1

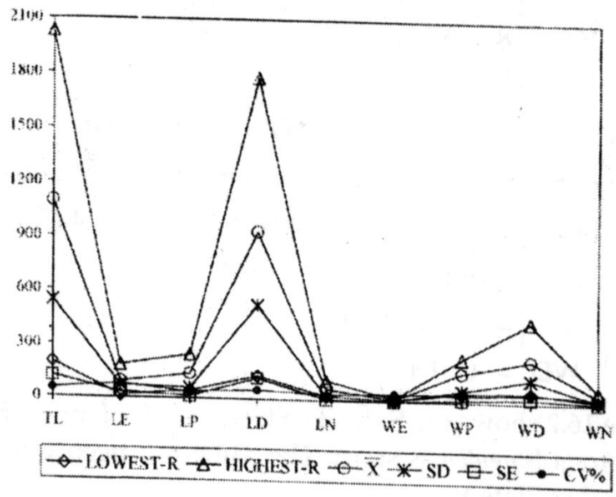

Plate V

Showing R, $\overline{X}$, SD, SE, CV% of measurements of *Odonaticola bengalensis* sp. n.

Affinities : In possessing characters like umbrella-shaped epimerite with petaloid lobes on a long neck, dehiscence of the cyst by simple rupture and boat-shaped oocysts, the species under report demands to lay under the genus *Odonaticola* Sarkar and Haldar, 1981. Kori and Amoji (1994) described a new species of gregarines, *Odonatocola haldari* from *Trithemis aurora* (Burmeister) in Karnataka, the same host species from which the gregarine has been recovered by us. The localities of collection are separated by a distance of over 1500 km differing in climatic conditions altogether. Even then, the two gregarines exhibit similarities in the structure of the spines on the epimerite and boat-shaped oocysts. But differences are many : shorter length of the trophozoite, conico-umbrella shaped epimerite with a much shorter neck and the ratios of the different body parts in the present gregarine; an ectocyst was reported outside the cyst in *O. haldari* by Kori and Amoji (1984) which could not be observed by us. We, therefore, consider the gregarine obtained from *T. aurora* in West Bengal as a new species and propose the name *Odonaticola bengalensis* sp. n. for it.

Taxonomic Diagnosis

Host : *Trithemis auroa* (Burmeister)

Locality : Kalyani and Karimpur in the district of Nadia, West Bengal, India

Site of Infection : Midgut

Incidence of infection : 89 odonates were examined out of which 31 (34.8%) were found infected during July to middle of November.

Etymology : The name *bengalensis* has been given to the present gregarine for its unique distribution in the state of West Bengal for the first time.

Odonaticola contaminata sp. n.

Trophozoite : The gregarine, as usual, possesses three segments, *viz.*, epimerite, protomerite and deutomerite. The epimerite appears as a hat-shaped conical knob at the apex of a long (107.9 μm), slender neck. The knob is furnished with 12 recurved petaloid hooks at its peripheral edge with which the trophozoite anchors to the host gut-epithelium. The protomerite is conical and dome-shaped. The deutomerite is elongated, fusifrom and uniformly granulated. Its epicyte is beset with transverse striations. The ovoid nucleus may occupy and position within the deutomerite.

Sporadin : Solitary sporadins are variously shaped. The protomerite at this stage may be conical, dome-shaped or hemisperical. In some specimens its width is greater than length.

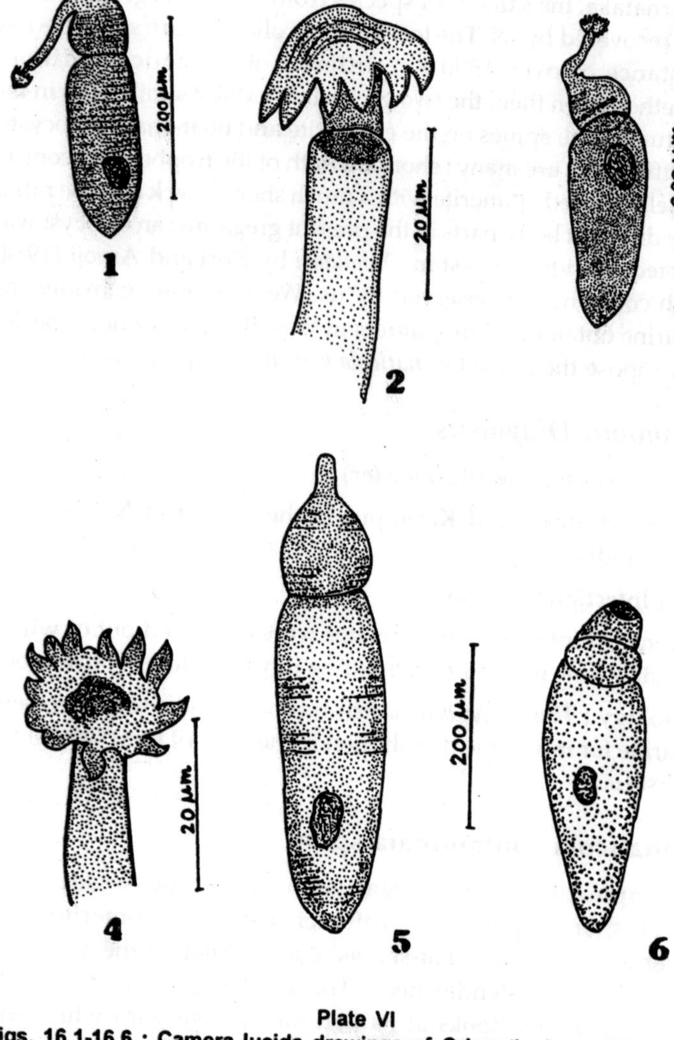

Plate VI

Figs. 16.1-16.6 : Camera lucida drawings of *Odonaticola* sp. III n. sp. (16.1) A trophozoite; (16.2) Lateral view of an epimerite enlarged; (16.3) A trophozoite; (16.4) Fronto-lateral view of an epimerite enlarged;
(16.5) An early sporadin; (16.6) A fully grown sporadin.

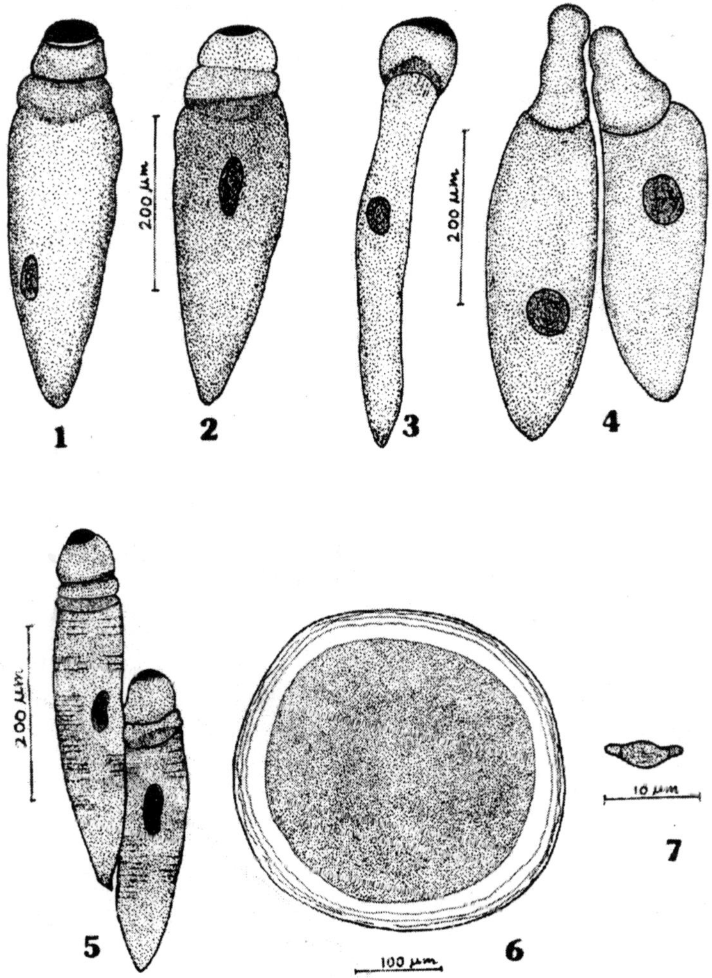

Plate VII

Figs. 16.1-16.7 : Camera lucida drawings of *Odonaticola* sp. III n. sp. (16.1-16.3) Three sporadins of various shapes and sizes; (16.4) Two sporadins coming to form a side by side association; (16.5) A side by side association; (16.6) A gametocyst of early stage; (16.7) A oocyst.

In some specimens, reminiscence of the degenerated epimerite may still be observed at the apices of the protomerite. The deutomerite is also extremely variable in shape and may be fusiform, cylindro-conical, rectangular or slender tubular. The nucleus may be ovoid, round or ellipsoid and is usually embedded within the densely granulated cytoplasm. Transverse striations are very characteristically observed throughout the deutomerite.

Association : Biassociative forms, attached sidewise, are vary comonly encountered in which the associating partners are almost identical. Only the nuclei may occupy variable positions in the deutomerite.

Gametocyst and oocyst : Milky white gametocysts ar slightly oval in appearance and measure 770.3 µm x 753.9 µm in dimensions. A thick gelatinous ectocyst surrounds the cyst which disappears shortly after the cyst is formed. Oocysts are liberated by simple rupture at 120h of development. These are normally released singly but those in clusters of three are also occasionally seen. These have a dimension of 8.0 µm x 3.5 µm.

Measurements : The summary of measurements (in microns) of trophozoites and sporadins with the mean within parenthess is given below (n=20) :

Trophozoite

TI	=	399.6 - 623.1 (495.3)	WE	=	13.3 - 21.7 (16.8)
LE	=	36.2 - 151.0 (99.7)	WP	=	75.5 - 115.9 (93.2)
LP	=	57.7 - 115.9 (87.3)	WD	=	93.2 - 173.9 (120.2)
LD	=	253.3 - 391.2 (308.2)	WN	=	26.6 - 43.5 (32.5)
LEN	=	53.3 - 139.9 (107.9)			
LN	=	35.5 - 65.2 (54.1)			

Sporadin

TL	=	362.2 - 977.8 (588.2)	WP	=	43.5 - 217.3 (116.9)
LP	=	72.4 - 144.9 (101.8)	WD	=	72.4 - 275.3 (164.1)
LD	=	275.3 - 825.9 (486.4)	WN	=	21.7 - 58.0 (34.1)
LN	=	36.2 - 86.9 (52.1)			

LP : TL = 1 : 4.0 - 8.1

WP : WD = 1 : 0.7 - 3.8

Table 16.3 shows the R, $\overline{X}$, SD, SE and CV% of measurements of *Odonaticola contaminata* sp.

Table 16.3 : Showing R, $\overline{X}$, SD, SE, CV% of measurements of *Odonaticola contaminata* sp. n.

	R	$\overline{X}$	SD	SE	CV%
TL	362.2 - 977.8	569.6	211.4	47.3	37.1
LE	36.2 - 151.0	99.7	57.2	28.6	57.4
LP	57.7 - 144.9	98.9	22.9	5.1	23.2
LD	235.3 - 825.9	450.7	206.3	46.1	45.8
LN	35.5 - 86.9	52.5	12.4	2.8	23.6
WE	13.3 - 21.7	16.8	3.7	1.9	22.0
WP	43.5 - 217.3	112.2	43.1	9.6	38.4
WD	72.4 - 275.3	155.3	65.4	14.6	42.1
WN	21.7 - 58.0	33.8	9.9	2.2	29.3

Material

(a) **Holotype :** On slide No. DR-4/6, prepared from midgut contents of *rachythemis contaminata* (Fabricius), collected at Karimpur in the district of Nadia, West Bengal, India by S. Biswas on September 16, 1995.

(b) **Paratype :** Many, on the above numbered slide as well as on other slides; other particulars are the same as for holotype material.

Affinities : The presently described gregarine has been assigned under the genus *Odonaticola* Sarkar and Haldar, 1981 for possessing characters such as hat-shaped epimerite bearing 12 petaloid spines, long neck, solitary sporadins, dehiscence of gametocyst by simple rupture of the cyst wall and boat-shaped oocyst.

The odonate, *Brachytehmis contaminata* (Fabricius), was earlier reported to harbour *Odonaticolo hexacantha* by Sarkar and Haldar in 1981. The authors characterized the species as having epimerite hat-shaped with six petaloid spines; gametocyst spherical dehiscing at 72 hours; oocysts boat-shaped with two short projections on the sides; LP : TL = 1 : 4.2 WP : WD = 1 : 1.4.

The gregarine reported hereunder has the only resemblance in parasitizing the same host insect. Otherwise, it differs from *O. hexacantha* in possessing features like 12 petaloid sines in the epimerite; gametocyst ovoidal; dehiscence period 120 hours;

oocysts without any projections; LP : TL = 1 : 4.0 - 8.1; WP : WD = 1 : 0.7-3.8; the trophozoites as well as sporadins are much large in dimensions. It is possible that two distinctly separate species parasitise the same host insect at two localities differing in their ecological niches. Considering what has been stated above, the gregarine is name as *Odonaticola contaminata n. sp.*

Taxonomic Diagnosis

Host : *Brachythemis contaminata* (Fabricius).

Locality : Karimpur in the district of Nadia and Domkal in the district of Murshidabad, West Bengal, India.

Site of Infection : Midgut

Incidence of infection : 39 out of 124 host odonates (31.4%) were infected. The hosts prevailed from April to December and the occurrence of infection was mostly found from June to middle of November.

Etymology : The name *contaminata* has been given after the host species from which gregarine has been obtained.

Discussion

In the year 1981, Sarkar and Haldar established a new genus *Odonaticola* with the following diagnostic features :

(i) epimerite hat-shaped with petaloid spines at the margin, neck long;

(ii) sporadins solitary;

(iii) gametocysts dehisce by simple rupture;

(iv) oocysts boat-shaped;

(v) development extracellular.

The genus was as named because of its incidence exclusively among the odoantes.

The combination of these characters is distinctly different from the other two genera *Menospora* Léger, 1892 and *Hoplorhynchynchus* Carus, 1863 under the subfamily Menosporinae Léger, 1892 Emend Grassé, 1953 of the family Actinocephalidae Léger, 1892 where the epimerite is a large cup at the end of a long neck, bordered with many hooks or spines and oocysts are smooth and crescentic. Thus, Sarkar and Haldar (1981) included the genus *Odonaticola* beside the two previously reported genera one of *Menospora* and the other of *Hoplorhynchus* under the subfamily Menosporinae of the family Actinocephalidae.

Sarkar and Haldar (1981) described four species under this genus, namely, *Odonaticola hexacantha* (type species), *O. longicollara*, *O. orthetri* and *O. rodgii*. During the same year, Sarkar (1981) described another species, *O. elliptica* and in addition, transferred the gregarine, *Menospora nonacantha*, described by Devdhar and Deshapande (1971) to the genus *Odonaticola*. Kori and Amoji (1983, 1984) reported three species under this genus, *viz.*, *O. crocothemis*, *O. haldari* and *O. diplacodi* obtained from odonates. In 1986 they reported another new species, *O. diplacodi* also obtained from an odonate insect. In 1991 Prema and Janardanan reported a new species *O. pantalae* which was the first report of *Odonaticola* from Kerala.

In a major revision of the genus *Odonaticola* Sarkar and Haldar, 1981, Amoji and Kori (1992) transferred four species of gregarine parasites of odonate insects, originally placed in the genus *Hoplorhynchus* Carus, 1863 and listed the following 14 species under it :

Parasite species	**Host species**
O. crocothemis Kori and Amoji, 1983	*Crocothemis servilia* (Drury)
O. diplacodi Kori and Amoji, 1986	*Diplacodes trivialis* (Ramb.)
O. elliptica Sarkar, 1981	*Crocothemis servilia* (Drury)
O. gracilis (H. Hoshide, 1953) nov. comb.	*Aciagrion hisopa* Selys
O. haldari Kori and Amoji, 1984	*Trithemis aurora* (Burmeister)
O. hexacantha Sarkar and Haldar, 1981	*Brachythemis contaminata* (Fabr.)
O. hoshidei Amoji and Kori nov. nom. [*Hoplorhynchus orthetri* (Hoshide, 1953)] [*O. orthetri* (Hoshide, 1953) nov. comb.]	*Orthetrum albistylum speciosum* (Uhler)
O. longicollara Sarkar and Haldar, 1981	*Diplacodes trivialis* (Rambur)
O. magnus (Hoshide, 1953) nov. comb. [*Hoplorhnchus magnus*]	*Crocothemis servilia* (Drury)
O. nonacantha (Devdhar and Deshapande, 1971)	*Urothemis, s. signata* (Rambur)
O. orthetri Sarkar and Haldar, 1981	*Orthetrum sabina* (Drury)
O. polyhematus (K. Hoshide, 1977) nov. comb. [*Hoplorhynchus polyhymatus*]	*Munais strigata* (Selys)
O. rodgii Sarkar and Haldar, 1981	*Neurothemis tullia* (Drury)
O. pantalae Prema and Janardanan, 1991	*Pantala flavescens*

Taking into consideration the gregarine described by Prema and Janardanan (1991), the genus *Odonaticola* now contains 14 described species. The present communication has added three more to the list.

References

Amoji, S. D. and Kori, S.S. 1992. Revision of the genus *Odonaticola* Sarkar *et. al.*, 1981 (Apicomplexa : Sporozoasida : Gregarinasina). *Acta Protozool.*, 31 : 169-171.

Devdhar, M.J. and Deshpande, S.B. 1971. On a new gregarine *Menospora nonacantha* n. sp. from the odonate *Agriocnemis* sp. (Fam. Coenagriidae) from Dharwar, India. *Ind. Zool.*, 2 : 41-45.

Haldar, D. P., Ghose, S., Sengupta, T. and Bose, Rupa. 1988. Use of Eugregarines (Protozoa : Apicomplexa) for Biological Control of Insect Pests. *Env. and Ecol.*, 6(2) : 306-310.

Kori, S.S and Amoji, S.D. 1983. On the life history of a new actinocephalid gregarine, *Odonaticola crocothemis*, sp. nov. from the midgut of an odonate insect, *Crocothemis servilia* (Drury). *Ind. Zool.*, 7(1&2) : 155-158.

Kori, S.S. and Amoji, S.D. 1984. A new actinocephalid gregarine, *Odonaticola haldari*, n. sp., from odonate insect *Trithemis aurora* (Burmeister). *Acta Protozool.*, 23 (1) : 63-66.

Kori, S.S. and Amoji, S.D. 1986. On an actinocephalid gregarine (Apicomplexa : Sporozoea) *Odnaticola diploacodi* sp. nov., from odonate insect, *Diplacodes trivialis* (Rambur). *Arch. Protistenk.*, 131 : 165-169.

Levine, N.D. 1988. *The Protozoan Phylum Apicomplexa.* 2 vols. 203 and 154 pp. CRC Press, Boca Raton, Florida.

Lipa, J.J. 1967. Studies on gregarine (Gregarinomorpha) of arthropods in Poland. *Acta Protozool.*, 5(8) : 97-179.

Prema, S. and Janardanan, K.P. 1991. Morphology and life cycles of two new species of cephaline gregarines (Apicomplexa : Cephalina) from Odonate insects in Kerala, India. *Acta Soc. Zool. Bohemoslov.*, 55 : 60-64.

Sarkar, N.K. 1981. On some actinocephalid gregarines (Apicomplexa : Eugregarinida) of odonate insects from West Bengal, India. Proc. Ind. *Acad. Sci. (Anim. Sci.)*, 90 (6) : 649-657.

Sarkar, N.K. and Haldar, D.P. 1981. Observations on a new cephaline gregarine (Protozoa : Sporozoa) *Acanthospora bengalensis* n. sp. from Odonate insect. *Arch. Protistenk.*, 124 : 278-284.

Sprague, V. 1941. Studies on *Gregarina blattaram* with particular reference to the chromosome cycle. Univ. Ill. *Biol. Monogr.*, 18(2) : 5-57.

Srivastava, J. P. and Agrawal, Neerja. 1992. Prevalence and abundance of odonates in Kanpur, India. 4th South Asian Symp. *Odonatol.*, (ABSTRACT) : 3-4.

Tillyard, R.J. 1917. *The Biology of Dragonflies*. Cambridge University Press, Cambridge.